How Revolutionary Was the Digital Revolution?

INNOVATION *and* TECHNOLOGY *in the* WORLD ECONOMY

Editors

MARTIN KENNEY
*University of California, Davis/Berkeley Roundtable
on the International Economy*

BRUCE KOGUT
Wharton School, University of Pennsylvania

Other titles in the series

How Revolutionary Was the Digital Revolution?

National Responses, Market Transitions, and Global Technology

EDITED BY JOHN ZYSMAN *and* ABRAHAM NEWMAN

A **BRIE / ETLA** *Project*

Stanford Business Books

An Imprint of Stanford University Press
Stanford, California   2006

Stanford University Press
Stanford, California

Printed in the United States of America on acid-free, archival-quality paper

Library of Congress Cataloging-in-Publication Data

How revolutionary was the digital revolution? : national responses, market transitions, and global technology / edited by John Zysman and Abraham Newman.
     p. cm. — (Innovation and technology in the world economy)
     Includes bibliographical references and index.
     ISBN 0-8047-5334-2 (cloth : alk. paper) —
     ISBN 0-8047-5335-0 (pbk. : alk. paper)
     1. Information technology—Economic aspects.   2. Technological innovations—Economic aspects.   3. High technology industries.   4. Globalization—Economic aspects.   I. Zysman, John.
II. Newman, Abraham, 1973–   III. Series.

HC79.I55H686 2006
303.48'33—dc22   200600489

Original Printing 2006

Last figure below indicates year of this printing:
15   14   13   12   11   10   09   08   07   06

Typeset by G&S Typesetters, Inc. in 10/12.5 Electra

Special discounts for bulk quantities of Stanford Business Books are available to corporations, professional associations, and other organizations. For details and discount information, contact the special sales department of Stanford University Press.
Tel: (650) 736-1783, Fax: (650) 736-1784

Contents

Contents

Figures and Tables

Acknowledgments

This book grew out of a Finnish-American research dialogue intent on understanding the development of the global digital economy. Teams from the Research Institute of the Finnish Economy (ETLA), headed by Pekka Ylä-Anttila; the University of Helsinki Institute for European Studies, headed by Olli Rehn; and the Berkeley Roundtable on the International Economy (BRIE) headed by John Zysman engaged in an ongoing discussion over the implications of information technology for economic, political, and social change. The following pages demonstrate the results of this collaboration.

This research effort would not have been possible without the support of a number of organizations and individuals that committed valuable resources to the project. On the European side, the Finnish team would like to thank Erkki Ormala and Ilkka Lakaniemi (Nokia Corporation), Otto Toivanen (HECER—Helsinki Centre of Economic Research), Olli Martikainen (University of Oulu), Tobias Kretschmer (London School of Economics), Aija Leiponen (Cornell University), Jukka Jalava (Statistics Finland), Kerem Tomak (University of Texas at Austin), and Jari Hyvärinen (Tekes—National Technology Agency of Finland) for their comments and contributions to the project.

The U.S. team has received invaluable assistance from a host of colleagues from the University of California, Berkeley, as well as other universities and private sector institutions. We would particularly like to thank François Bar (University of Southern California), Stephen Cohen (BRIE, Berkeley), Peter Cowhey (University of California, San Diego), Stuart Feldman (IBM), Martin Kenney (University of California, Davis), Jonah Levy (Berkeley), Howard Shelanski (Berkeley), and Steven Vogel (Berkeley).

We want to make a special note of our personal and intellectual indebtedness to Andy Schwartz. Andy's insights contributed to the framing of the notion of transformations as difficult political transitions from one partial equilibrium to another. As important, his energy, insight, and fundamental decency influenced the entire texture of BRIE during the years he spent with us. We miss him very much.

Financial support from the Nokia Corporation is gratefully acknowledged. Tekes has supported ETLA's involvement in the effort. Aspects of the work done for this project were supported by funds from the European Union Fifth Framework program. Additionally, the U.S. team would like to thank the NTT DoCoMo Mobile Society Research Institute for their valuable support of the research program.

This cooperative effort would not have been possible without the team that guided its development and realization. The editors would like to thank Emilie Lasseron for her commitment to the early stages and for creating the outline of and materials for the original compilation that preceded this book. We would also like to thank the staff at BRIE for all their support.

The editors are grateful to their partners at Stanford University Press, in particular the excellent work of Martha Cooley and Jared Smith.

Abbreviations

ADR	Alternative dispute resolution
AICPA	American Institute of Certified Public Accountants
AMPS	Advance Mobile Phone Service
ARIB	Association for Radio Industries and Business
ARPA	Advanced Research Projects Agency
ASP	Application service provider
ATM	Asynchronous transfer mode
B2B	Business to business
BPO	Business process outsourcing
BTO	Bandwidth trading organization
CATI	Cooperative agreements and technology indicators
CDMA	Code division multiple access
CEPT	European Conference of Posts and Telecommunications
CERN	Center for Nuclear Research
CLEC	Competitive local exchange carriers
CME	Coordinated market economies
CNPN	Cross-national production networks
CPE	Constrained partial equilibrium
DMCA	Digital Millennium Copyright Act
DOT Force	Digital Opportunity Task Force
DRUID	Danish Research Industry for Industrial Dynamics
DSL	Digital subscriber line
DSP	Digital signaling processors
EBS	Enron Broadband Services

ECJ	European Court of Justice
EEA	European Economic Area
ETLA	Research Institute of the Finnish Economy
ETSI	European Telecommunications Standards Institute
FCC	Federal Communications Commission
FDA	Food and Drug Administration
FDI	Foreign direct investment
FERC	Federal Regulation and Oversight of Energy
FIPP	Fair Information Practice Principles
FTC	Federal Trade Commission
FTTH	Fiber to the home
GATT	General Agreement on Tariffs and Trade
GPL	General public license
GSM	Global Systems for Mobile Communications
ICT	Information and communication technology
IETF	Internet Engineering Task Force
IMT	International Mobile Telecommunications
IOS	Internetwork Operating Systems
IPR	Intellectual property rights
IPV6	Internet Protocol Version 6
IRC	Internet Relay Chat
IRU	Infeasible rights of use
ISDN	Integrated services digital network
ISP	Internet service provider
IT	Information technology
ITES	Information technology–enabled services
ITU	International Telecommunication Union
LAN	Local area network
LME	Liberal market economies
M&A	Mergers and acquisitions
METI	Ministry of Economy, Technology, and Industry
MoC	Ministry of Communications
MPHPT	Ministry of Public Management, Home Affairs, Posts and Telecommunications
MPT	Ministry of Post and Telecommunications
MSA	Master services agreement
NGO	Nongovernmental organization
NMT	Nordic Mobile Telephone

NTT	Nippon Telegraph and Telephone Corporation
OECD	Organisation for Economic Co-operation and Development
OEM	Original equipment manufacturer
OMB	Office of Management and Budget
OSFS	Open source and free software
OSS	Operations and billings software systems
POI	Point of interface
POP	Points of presence
POTS	Plain old telephone services
PPP	Purchasing power parity
PTO	Public telecommunications operator
QoS	Quality of service
R&D	Research and development
RASC	Recreational Software Advisory Council
RBOC	Regional Bell Operating Companies
RFC	Request for comments
RPP	Receiving party pays
SDO	Standards development organization
SEC	Securities and Exchange Commission
SLA	Service-level agreement
SMS	Short messaging system
SONET	Synchronous optical networking
SSH	Secure shell
STPI	Software and Technology Parks of India
TAO	Telecommunication Advancement Organization
TCP/IP	Transmission control protocol/Internet protocol
TDMA	Time division multiple access
TD-SCDMA	Time division–synchronous code division multiple access
TNC	Transnational corporation
TRIPS	Trade-related aspects of intellectual property rights
TSAG	Telecommunication Sector Advisory Group
UMTS	Universal Mobile Telecommunication System
UNCTAD	United Nations Conference on Trade and Development
UNDP	United Nations Development Program
USPTO	United States Patent and Trademark Office
VAN	Value-added network
VoIP	Voice over Internet protocol
WAP	Wireless application protocol

Abbreviations

W-CDMA	Wideband code division multiple access
Wi-Fi	Wireless fidelity
WIPO	World Intellectual Property Organization
WLAN	Wireless local area connection
WTO	World Trade Organization

Contributors

Jonathan D. Aronson is a professor at the University of Southern California and the director of USC's School of International Relations. He also teaches at the Annenberg School of Communications. Professor Aronson's research focuses on international political economy, with special attention to trade negotiations, trade in services, comparative regulation, international strategic alliances, and especially international telecommunications. He also researches the impact of the globalization of telecommunications networks on international financial activities and the implications of these changes for regulation.

Naazneen Barma is a doctoral student in the Department of Political Science at the University of California, Berkeley. Her research interests lie in exploring how different types of political and institutional governance mechanisms affect political stability and economic development outcomes in poor countries. Her dissertation is focused on reconstructing state capacity in post-conflict countries.

Taylor C. Boas is a Ph.D. candidate in political science at the University of California, Berkeley, and co-author of *Open Networks, Closed Regimes: The Impact of the Internet on Authoritarian Rule* (Carnegie Endowment for International Peace, 2003).

Robert E. Cole is professor emeritus at the Haas School of Business, Organizational Behavior and Industrial Relations Group, and the co-director of the Management of Technology Program, a joint program between the Haas School of Business and the College of Engineering. He also serves as the Omron Professor of Technology Management at Doshisha University, Kyoto, Japan. His current research interests include management of technology, Japanese work organization, quality, organizational learning, knowledge management, and organizational transformation. His published papers include "From a Firm-Based to a Community-Based Model of Knowledge Creation: The Case of the Linux Kernel Development," *Organization Science* 14 (2003): 633–649, and "Case Study: Software Quality in the Development of Linux," in *Quality into the 21st Century*, ed. Tito Conti, Yoshio Kondo, and Gregory Watson (ASQ Press, 2003).

Peter F. Cowhey was named director of the Institute on Global Conflict and Cooperation on July 1, 2000. He holds a joint appointment as professor in the Department of Political Science and at the Graduate School of International Relations and Pacific Studies, University of California, San Diego. His major fields of research are international political economy, comparative foreign policy, and international relations theory.

Rafiq Dossani is a senior research scholar at the Asia-Pacific Research Center and is responsible for developing and directing the South Asia Initiative. His research interests include financial, technology, and energy sector reform in India. He is currently undertaking projects on business process outsourcing (with the support of the Sloan Foundation), innovation and entrepreneurship in information technology in India, the institutional phasing-in of power sector reform in Andhra Pradesh, and security in the Indian subcontinent.

Ari Hyytinen, Dr. Sc. (Econ.), is a research supervisor at the Research Institute of the Finnish Economy (ETLA) and a docent of financial economics at the Helsinki School of Economics. His research interests include financial intermediation, corporate finance, and industrial economics.

Brodi Kemp is a doctoral student in the government department at Harvard University. She received a J.D. from Yale Law School in 2004 and a B.A. from the University of California, Berkeley, in 2001, where she was a research affiliate for the Berkeley Roundtable on the International Economy. Her field of research is political theory, with special interest in the topics of liberalism and authority.

Martin Kenney is a professor in the Department of Human and Community Development at the University of California, Davis. His recent research has been on the history and development of the venture capital industry, the development of high-technology clusters (particularly Silicon Valley), and the offshoring of services and of R&D to India and China. He is the author of approximately one hundred articles and five books, including the edited volumes *Understanding Silicon Valley* (Stanford University Press, 2000) and *Locating Global Advantage* (Stanford University Press, 2004).

Heli Koski is affiliated with the Helsinki School of Economics as professor of technology management and policy and with the Research Institute of the Finnish Economy (ETLA). She has worked at the London School of

Economics and as a special advisor to the European Union Commissioner for Enterprise and Information Society. She has published articles in various books and journals, such as *Review of Industrial Organization, Journal of Industry, Competition and Trade, Information Economics and Policy,* and *Economics of Innovation and New Technology and Research Policy.*

Kenji Kushida is currently pursuing a Ph.D. in political science at the University of California, Berkeley, and is a graduate student researcher at the Berkeley Roundtable for the International Economy. He is currently working on research examining the role of next-generation telecommunications networks in Asia on innovation and global competition. He is also examining the effect of foreign direct investment on bureaucratic control and political power in Japan.

Aija Leiponen joined the Department of Applied Economics and Management at Cornell University in 2001. Her teaching and research focuses on the sources and effects of technological change in the economy. The overarching goal of her research program is to understand the interactions between organizational arrangements and innovation. Most recently she has examined the creation and transfer of knowledge between business service firms (such as management consulting, engineering, and R&D services) and their clients.

Olli Martikainen is the director of R&D at Necsom and a docent at Helsinki University of Technology and Lappeenranta University of Technology. His main interest areas are telecommunication software methods and tools, network architectures, performance analysis, and new industrial and economic structures in telecommunications.

Abraham Newman is an assistant professor at the Edmund Walsh School of Foreign Service at Georgetown University, a member of the BMW Center for German and European Studies, and a research associate at the Berkeley Roundtable on the International Economy. He is interested in the international politics of the digital era. His current research focuses on the international politics of privacy.

Maj Cecilie Nielsen currently studies philosophy and economics at Copenhagen Business School. She works as an organizer of professional management courses, with a particular focus on research management. She has a special interest in practical knowledge processes in knowledge-intensive companies and institutions. She is the co-author with Niels Christian Nielsen of *Verdens bedste uddannelsessystem* (Fremad Press, 1997), on the Danish educational system.

Niels Christian Nielsen is the chairman and CEO of Q. He is also a nonexecutive director of a number of companies, including Codan, Unimerco, MondayMorning, Prophet, and Development Alternatives, Inc. (DAI). He was behind the creation of the Danish government's Learning Lab. Before conceiving Q, Niels created Catenas, a global network of specialist consulting firms. Niels is also a leading thinker on knowledge, innovation, and networks. He is an adjunct professor at Copenhagen Business School and a visiting scholar at the University of California, Berkeley.

Darius Ornston is a Ph.D. student in the Department of Political Science at the University of California, Berkeley. Prior to arriving at Berkeley, Darius spent a year as a Fulbright Fellow at the University of Helsinki. He is interested in social concertation in the Western European political economy and its transformation during the 1990s.

Laura Paija is a researcher specializing in network economy, with special focus on ICT industry developments, at the Research Institute of the Finnish Economy (ETLA). In her work she has examined the evolution and network dynamics in the Finnish ICT cluster, Nokia's supplier strategies and implications for the Finnish ICT cluster, and the total employment effects of liberalization on the ICT cluster, among other questions. In 2000 she worked as the co-coordinator of the ICT Focus Group under the OECD NIS (National Innovation Systems) Program.

Christopher Palmberg is a researcher at Etlatieto and the Research Institute of the Finnish Economy (ETLA). His recent work focuses on the evolution and future challenges of the ICT sector and on the position of Finland in the new global division of R&D. Prior to this he was a senior researcher at VTT Technology Studies, where he focused on issues related to innovation processes and industrial renewal. He holds a doctoral degree from Sweden's Royal Institute of Technology, Department of Industrial Organisation and Management.

Olli Rehn is a director of research and holds a doctorate in international political economy from the University of Oxford. Currently his research and teaching interests focus on institutions and policy networks of the European Union, as well as on the political economy of the knowledge-driven economy. In 2002 he conducted an evaluation project and strategic assessment of the Finnish Communications Regulatory Agency. He served as head of the Cabinet for EU Commissioner Erkki Liikanen from 1998 to 2002 and in that capacity was a member of the commission's internal cross-departmental steering group of the eEurope Action Plan from 1999 to 2002. Prior to that he led two major research projects in 1996–97, one focusing on the political economy of the eastern enlargement of the EU, the other concentrating on coordination and policy networks of EU decision making in Finland. He was a member of the European Parliament in 1995–96 and of the Parliament of Finland from 1991 to 1995. He is currently EU Commissioner for Enlargement.

John E. Richards is director of marketing in the information worker busi- ness unit at Microsoft and has responsibility for collaboration marketing. Prior to coming to Microsoft he was a consultant with McKinsey & Company and director of the Stanford Computer Industry Project in the Department of Economics at Stanford University. His current research interests include the regulation of international services and markets, with a particular focus on wireless telecommunications and aviation services. His research has been published in books of collected studies and in such journals as *International Organization*.

Petri Rouvinen is a research director at the Research Institute of the Finnish Economy (ETLA). Dr. Rouvinen's research interests include ICT and technology in general, innovation, R&D, globalization, competitiveness, entrepreneurship, and economic policy. He has published many books and contributed to several collective volumes; his scholarly work has been published in *Economics of Innovation and New Technology, Information Economics and Policy,* and *Journal of Applied Economics,* among other journals.

Tobias Schulze-Cleven is a Research Associate of the Berkeley Roundtable on the International Economy (BRIE) and a doctoral candidate in the Department of Political Science at the University of California, Berkeley. His dissertation compares the national pathways toward more labor market flexibility in Great Britain, Denmark, and Germany. His research interests include comparative and international political economy, the politics of the knowledge society and learning economy, and the future of social democracy.

Andrew Schwartz received his B.A. in Economics from Swarthmore College and his Ph.D. in political science from the University of California, Berkeley. His publications include *The Politics of Greed: Privitization, Neo-Liberalism, and Plutocratic Capitalism in Central and Eastern Europe* (Rowman & Littlefield, forthcoming), *The Best-Laid Plan: Privatization and Neo-Liberalism in the Czech Republic* (Cambridge University Press, forthcoming), *The Tunnel at the End of the Light: Privatization, Business Networks, and Economic Transformation in Russia* (University of California Press, 1998), and *Enlarging Europe: The Industrial Foundations of a New Political Reality* (University of California Press, 1998). Before entering academics Schwartz was a NYMEX/COMEX member, an independent floor trader specializing in gold, energy products, and futures options.

Steven Weber is the director of the Institute of International Studies at the University of California, Berkeley, and a professor of political science. His areas of special interest include international political economy, political and social change in the new economy, and the political economy of globalization and European integration.

Pekka Ylä-Anttila is currently the research director of ETLA and the managing director of Etlatieto (the research and information services unit of ETLA). He has authored or co-authored some twenty books and dozens of articles in the fields of competitiveness analysis, industrial and technology policy, industrial economics, technological change, and internationalization of business.

John Zysman is a professor of political science at the University of California, Berkeley, and has been co-director of the Berkeley Roundtable on the International Economy (BRIE) since its establishment in 1982. His publications include *The Highest Stakes: The Economic Foundations of the Next Security System* (Oxford University Press, 1992), *Manufacturing Matters: The Myth of the Post-Industrial Economy* (Basic Books, 1987), and *Governments, Markets, and Growth: Finance and the Politics of Industrial Change* (Cornell University Press, 1983).

How Revolutionary Was the Digital Revolution?

INTRODUCTION

1

FRAMEWORKS FOR UNDERSTANDING THE
POLITICAL ECONOMY OF THE DIGITAL ERA

Abraham Newman and John Zysman

In the spring of 2000 one might have asked: "Are the extraordinary expansion of computing intelligence, the pervasive spread of digital networks, and the recent arrival of the commercial Internet, the edge of an historical revolution, a transformation?" The sudden interconnection of disparate networks into a single "cyber world" and broad consumer participation in those networks through vehicles such as AOL or Yahoo seemed to augur a new era. The pace at which individuals, not just firms, were being connected to the Internet in the United States was explosive. Businesses were reorganizing and extending internal activities to capture the possibilities of the new network of networks. Together this rapid diffusion of networks and consumer engagement encouraged the fantasy that these information technologies could transform the terms of competition and restructure a broad range of the economy.

By the summer of 2003, the conventional question had become different: "Was this the revolution that never happened?" Dreams evaporated with stock values, first during the dotcom collapse and then in the telecoms debacle. It seemed that the digital revolution might have more in common with tulip speculation, pure ephemera, than with the railroad expansion and transportation revolution that took place at the dawn of the industrial era.[1]

Now, in 2006, as this book goes to press, the crash itself is being reevaluated.[2] Despite the ebb and flow of technological exuberance, we believe that digital technology has dramatically altered political and economic dynamics. The primary goal of this book is to investigate the political economy of the digital era and, in turn, to better understand its implications for polities and markets.

The dominant conversation about how information technology (IT) has affected political economy focuses on technology's role in constraining the

4   choices governments face when shaping their domestic economies. As digital networks facilitate international communication and virtual markets, the transaction costs of conducting global business fall. International firms can then leverage their mobility to constrain regulatory options. Politicians who attempt to aggressively steer their national response to the digital era confront the very real threat that business will locate abroad, taking with it electorally valuable jobs (Tonnelson 2000). Additionally, the decentralized, nonhierarchical character of technology complicates any attempt to intervene. The very nature of digital networks, according to this argument, is that they route around control. Nimble capital and the particular technological architecture of the Internet, it is argued, have thus severely weakened governments' ability to shape the political and social character of the digital era (Castells 1996; Rosenau 2002).

A central objective of this book is to reconsider this discussion. It proposes to construct a framework for analysis about the international digital era, one that examines the ability of political actors to innovate and experiment in spite of, or maybe because of, the constraints posed by digital technology. IT does more than just change the costs of transportation and communication: it alters the manner in which economic value is created, changes how international production is organized, and reopens basic societal bargains struck around individual liberty and economic rights. There is no inevitable political path driven by the technology; rather, evolving technology shakes up the political order, creating the foundations for fundamental fights over the organization of markets and polities.

In order to understand this profound transformation, the book builds four distinctive arguments about the interplay of digital technology, corporate strategy, government policy, and global marketplaces. First, we argue that rather than global forces sweeping away national structures and local traditions, distinctive national tales of development create and are an integral part of the global story. The digital era is one in which an increasingly global market coexists with enduring national foundations of distinctive economic growth trajectories and corporate strategies. As IT facilitates the interaction of national political economies in international marketplaces, political experimentation and innovation rooted in specific domestic institutional environments have global consequences. Second, the effective use of information technology, we all understand, requires investment in innovative business strategies and organizations. These corporate experiments drive the innovative core of the digital era. Firms and governments face the difficult task of encouraging this risky experimentation and recognizing the educational benefits of failure for future innovation. Third, the implementation of new technologies and the adoption of new business models and strategies involve a complex market transition. The digital era forces us to understand those transitions as more than just a shift from one market equilibrium to another; rather they are indicative of a broader shift from one universe of policy and market signals to another. Alongside that transition, national and international policies, and the technological standards adopted by agreement or market place outcomes powerfully influence global competitive realities. Fourth, the choices made about digital policy shape the national polity and social community in significant ways. We therefore investigate whether, as the rules for the new digital era are being made, there is a fundamental and basic shift in the rules of society that alters the way economy and polity operate. The rest of the introduction builds on these four points and expands on the main argument of each section of the book.

PART I: NATIONAL STORIES AND GLOBAL MARKETS IN THE DIGITAL ERA

Certainly information technology expands communications capacities and reduces the costs of transaction over distance. These digital tools, and the networks that interconnect them, facilitate the communication and data exchange required for integrating operations and markets that are widely dispersed geographically into a single global marketplace. But technology does not unfold autonomously, playing out an inevitable logic. Nor does globalization lead to the ineluctable elimination of national systems of economic governance and technological development. Again and again the waves of innovation that disturb global markets emerge from enduring national structures. The national stories, the national innovations, are amplified and reverberate across global markets through information technology. Digital technology permits national political economies to more readily access and compete in international markets and thereby facilitates and demands national economic experimentation and innovation.

This view of globalization rests on a conception of the dynamics of national market systems and of their interactions (Zysman 1994). Let us sketch the outlines of this perspective. First, each economy consists of *an institutional structure* that is a function of that country's distinct political and industrial development. Many critical institutions, social arrangements, and social groups predate modern societies and market economies; others are given a modern character, often by force, in a struggle over a variety of non-market issues. These institutions and arrangements, which often shape modern markets, cannot be understood simply by a narrow analysis of economic calculus

6   (Evans, Rueschemeyer, and Skocpol 1985; Thelen and Steinmo 1992). Second, in a rather conventional step, the institutional structure of the economy, combined with its industrial structure in a more classic industrial-organization sense, *creates a distinct pattern of constraints and incentives.* This frames the interests of the actors as well as shaping and channeling their behavior.[3] The interaction of the major players generates a particular "market logic" and "policy logic" (e.g., J. Levy 1999). Third, the national *market logic* shapes the particular character of strategy, product development, and production processes in a national system. Each market economy is defined by the institutions and rules that permit it to function; or, said differently, each national system can be defined by the "institutional structure" of the economy. Because the national institutional structures are different, there are, as a consequence, many different kinds of market economies (Streeck 1991; Hollingsworth 1997; Hall and Soskice 2001).

A specific market logic (and political logic) then induces distinct patterns of corporate strategy (and government policy) and therefore encourages internal features of companies (and the government) that are distinctive to that country. There are typical strategies, routine approaches to problems, and shared-decision rules that create predictable patterns in the way governments and companies go about their business in a particular political economy. Those institutions represent specific capacities and weaknesses within each system. The French case illuminates how the institutional structure acts to generate policy routines. French political-economic institutions produced constant policy responses to a diverse set of industrial problems from the end of World War II until the mid-1980s (Zysman 1977; Hall 1986).

The consequence, and central to this chapter, is that the global dynamic and trade dynamics must in significant part be understood as an *interaction of these national market logics.* Institutionally rooted differences in corporate strategy and access to markets influence patterns of international economic competition. As national strategies play out on the global stage, they force others to adjust. These adjustments proceed as institutionally rooted adaptations to changing international and technological environments.

In sum, a national institutional structure creates the foundation for nationally specific patterns of political and industrial development. Each particular structure sets a definable pattern of incentive and constraint for the actors within the system; the interaction of the actors creates distinctive national market logic. Nationally specific patterns of government policy and corporate strategy, distinctive routines that characterize one country and not another, are the result. These national logics forge particular patterns of *interaction* between national systems. The resultant international environment may then challenge other national systems, opening up the possibility for institutional adaptation and innovation. National systems then evolve as they interact with one another in internationally competitive markets, with important consequences for national economic adjustment and the domestic political arena.

The Japanese production revolution highlights the relationship between national market dynamics and patterns of international interaction (Tyson and Zysman 1989). The market and policy logic rested on three aspects of the Japanese political economy. First, the Japanese market was relatively closed to the implantation of foreign firms; consequently, competition was restricted to Japanese firms. Second, there was rapidly expanding domestic demand; financial resources channeled to expanding sectors by government policy permitted firms to satisfy demand by building production capacity. Third, foreign technology was easily and readily borrowed. Under these conditions, the market logic encouraged Japanese firms to aggressively pursue market share as a means of maximizing profits—goals traditionally assumed to be contradictory. Formally, firms faced long-term declining cost curves (Murakami and Yammaura 1982). That meant that as firms increased volumes—ideally capturing more market share in the expanding market—costs would fall, allowing prices to drop to increase sales, thus starting the cycle over.

The pursuit of market share spilled over into international markets (Yamamura 1982). Companies in Japan competed for market share, which required them to build production capacity in anticipation of demand. Excess capacity was almost inevitably the result. Because much of the production capacity was then a fixed cost, the temptation was to sell at marginal production cost in foreign markets. As long as the domestic market was insulated and foreign markets were open for sale of excess capacity, Japanese firms had a constant incentive to build in anticipation of demand and offload the consequences of overly ambitious judgments onto foreign markets. In fact, when the domestic market became saturated, a group of firms would begin to export at the same time. The result, in the phrase translated from the Japanese debate, was a "downpouring of exports." The sudden flood of exports into the major export market—the United States—caused intense political conflict with America in a series of sectors beginning with textiles and continuing through automobiles and, later, semiconductors. Political actors in the United States were forced to confront the Japanese challenge and in some cases, such as the semiconductor industry, forged innovative political deals. The periodic international disputes over Japanese dumping were thus a function of the domestic pattern of competition in which market share was key. The Japanese example leads to

8   the issue of how one nation's policy routines and market logics influence the options of another.

As national market strategies amplified by communications technology shape the terms of international economic competition, others face the challenge of adapting. Although the choice has been cast as one to be made between national convergence around a constrained set of adjustment strategies and diverse national paths that reproduce past institutional solutions, external pressures often open up the space for political and institutional innovation.[4] These innovations naturally rest on preexisting institutional legacies but may include considerable reinterpretation and redeployment. Institutional capacity may be adapted, layered, and reconfigured to address the new international economic environment (Thelen 2003; Streeck and Thelen 2005). Old political tools find important new roles in unexpected terrain at the same time that truly new institutions may be integrated into the national system. Policy responses may unintentionally arise from the convergence of past decisions, but the character of adjustment is often strongly tied to the domestic political arena and the ability of political actors to forge a viable adaptation bargain. Where a political bargain is achieved, institutional innovations may result that effect economic development, and where it fails, national systems may be doomed to muddle along. In this volume we explore this relationship between national strategies, information technology, and international markets by examining the cases of Finland, Japan, and the emerging markets.

The Finnish story is a dramatic surprise and little understood. International markets and information technology made possible Finland's emergence as a significant economic player with a communications technology cluster and world-class communications company, Nokia. As a small country of 5 million people, it did not have the domestic economy or supply base to support a world-scale and world-class company. Its success is a national story, and not one just of market processes or corporate strategy, but of a conscious national strategy to reposition itself in the global economy by investing in information technology. At the same time, it is an international story whereby a small Nordic country emerged as a dominant player redefining the terms of global IT competition.

Finland shifted in the 1990s from being a Soviet supplier and basic forest products provider to a communications leader and a producer of sophisticated equipment. Reeling from an economic depression in the early 1990s, when real gross domestic product (GDP) fell more than 10 percent and unemployment hovered at nearly 20 percent, Finland seized an opportune moment in the global electronics industry and the process of European integration to accomplish significant structural change of its economy. Nokia and its associated cluster of firms became a major player in the world communications industry, and its wood industry modernized, becoming ever more competitive in product and equipment markets. Finland now posts annual growth rates around 5 percent, among the highest in the world.

What explains Finland's surprising success in the digital era? Importantly, the economic transformation was consciously engineered as part of an explicit bargain among social partners that included raising the levels of research and development (R&D) spending in the country to historic levels. The Finns relied on, and indeed created, a form of adaptive corporatism that allowed new directions in policy and industrial organization to forge this policy action. Finnish society then relied on domestic political institutions to attempt an economic gamble on repositioning themselves in the international economy. Corporatist political institutions were used to liberalize critical sectors and can thus be seen as a support to reform and growth, not just a rigid system of labor market privileges. The nature of production in a digital era allowed Finnish firms to rely on homegrown R&D, European standards, and free trade to find real economic success internationally.

Two chapters address these issues, one examining the economic and market story and the other the political and policy dynamics. "Finland's Emergence as a Global Information and Communications Technology Player: Lessons from the Finnish Wireless Cluster," by Ari Hyytinen, Laura Paija of ETLA, the Research Institute of the Finnish Economy, Petri Rouvinen, and Pekka Ylä-Anttila, tell the economic story. Ylä-Anttila was part of the policy-making team for the original transformation and has been a member of several working groups developing new strategies for the twenty-first century. The political story is presented by Darius Ornston and Olli Rehn in "An Old Consensus in the 'New' Economy? Institutional Adaptation, Technological Innovation, and Economic Restructuring in Finland." This chapter grows from Ornston's recent work on Finland and Ireland and Rehn's earlier work on small-country adaptation. Importantly, Rehn was responsible for economic policy in the Finnish prime minister's office and then was appointed EU Commissioner for Information Society and Enterprise and then for European Enlargement. The chapters explain how corporatist political institutions were redeployed to focus policy efforts around success in technology-centered international markets.

Although the Finnish case provides an unexpected but successful national experiment in the digital era, Japan offers an important counterpoint. The first part of the Japanese story, seemingly, has been told many times and is sketched above. In a sense, the global era began with a national story, Japanese production innovation. Under the label of lean production, Japan dominated

10   many world markets in the 1980s. The critical factor for our purposes is that this set of production innovations hinged on the internal dynamic of Japanese competition and policy (Zysman and Doherty 1996).

Then the Japanese miracle foundered. Japan was the dominant force in microelectronics during the 1980s, and its failure to become an information technology leader seems somewhat surprising. What prevented Japan from leveraging its lead in the previous round to innovating in the digital era? Just as in the case of Finland, the nature of digital technology when combined with national policy choices significantly shaped economic outcomes. The Japanese implosion over the past few years turned principally on national policy choices that were driven by internal dynamics. The financial collapse, for example, that came with the breaking of the economic bubble was a product of the particular credit-based system of finance that served well for catch-up in a closed market, but worked less well for a maturing industrial sector with an excess of savings. Note by contrast that in the 1980s the French adapted a similar system; their success in transforming their capital markets and system of monetary control permitted the close financial linkages with Germany and later the European Union.

The bursting of the bubble and the domestic deflation that ensued came at a difficult moment in international market competition, the beginning of the digital era. It was the moment of transition from the electromechanical, with significant Japanese consumer goods leadership, to a digital era of American leadership. Routers replaced switches in communications networks; Apple iPods replaced Sony Walkmen. National policy further weakened Japanese international position by bolstering those interests that were most threatened by the digital transformation.

Two chapters depict the Japanese story in the digital era. Robert E. Cole, in "Telecommunications Competition in World Markets: Understanding Japan's Decline," analyzes the causes of Japan's decline in international competitiveness in the telecommunications sector since the 1990s. Cole argues that in the face of global liberalization of the sector, technological innovations of the 1990s, and the convergence of telecommunications and information technology sectors, Japan made serious policy mistakes. First, it was unable to deregulate successfully, standing by a commitment to relationship contracting. The government supported established firms and products in an environment characterized by disruptive technologies. Second, newcomers were critical in this phase of the industry's evolution. Cisco's router command interface had become a de facto industry standard by the time Japanese equipment manufacturers could shift to the production of routers. Cisco's powerful first-mover advantage forced Japanese firms to give up global markets. Third, the Japanese took non-strategic approaches to setting standards. Japanese man-ufacturers had dominated global markets in wireless telecommunications in the 1980s, but Nippon Telegraph and Telephone's (NTT's) proprietary domestic personal digital cellular (PDC) standard, deployed in the 1990s, isolated the Japanese market. The PDC standard not only dissuaded foreign players from entering, but also made it difficult for Japanese equipment producers to export abroad.

Kenji Kushida, in a complementary chapter, "Japan's Telecommunications Regime Shift: Understanding Japan's Potential Resurgence," sets out to ex-plain likely conditions under which rapid domestic developments from the late 1990s might reverse this decline. He asks what might lead to a sudden in-crease in the presence of Japanese firms in global IT markets. He considers how the Japanese "regime" in telecommunications is changing in a way that might make a competitive rebound possible. The concept of regime includes the primary industry and government actors involved, how these actors inter-act with one another, and sources of standard setting. In this new regime, the Ministry of Post and Telecommunications (MPT) is no longer the sole pro-mulgator of telecommunications industrial policies; NTT has lost its techno-logical dominance and the power to set standards, thanks to the advent of the Internet; and electronics firms can no longer count on NTT's procurement budgets for networking equipment or start-ups.

Kushida argues that the decline in international competitiveness in the 1990s was a result of discontinuous technology entering a malconfigured telecommunications regime. The regime shift currently underway is the re-sulting adjustment process — a politically mediated process in which state and industry actors iteratively create new constraints and opportunities. In terms of a possible Japanese resurgence, domestic developments associated with this regime shift are causing Japanese manufacturers to refocus on taking back in-ternational market share.

Which unexpected national stories might appear on the global stage to de-fine the trajectory of international IT markets? Japan's arrival was startling and affected the terms and rules of global competition; its loss of leadership was unexpected. Finland's rise was certainly unforeseen, but it followed trajecto-ries established in Europe and elsewhere. What surprises lie ahead?

Naazneen Barma addresses this question in "The Emerging Economies in the Digital Era: Marketplaces, Market Players, and Market Makers." Much empirical and policy attention has been devoted to IT access and usage di-mensions of the global digital divide. This chapter directs analytic focus to the other side of the coin, that of global digital innovation. It thereby challenges the conventional wisdom that developing countries are merely marketplaces

12 for digital products innovated in the industrialized world. Empirical examination of recent developments in IT production and use in several newly industrializing economies—in particular, China, India, and the East Asian newly industrialized countries (NICs)—reveals the different roles that emerging economies can and do play in global IT innovation. Probing the concept of innovation demonstrates that it often comes through invention, yet it can also emerge from learning by doing and from learning by using. Understanding the trajectories of these different ecologies of innovation and the stakeholders associated with them are central to an analysis of the role emerging economies play in global digital competition. In particular, it appears that successful innovations in emerging economies are often user focused. Furthermore, the state plays a more visible role in digital innovation in emerging economies, be it through heavier regulation, private-public cooperation, or outright government IT policies. These more incremental ecologies of innovation may not pose a direct challenge to dominant digital producers; they do, however, have the potential to alter the structure of future global digital markets.

PART II: THE EXPERIMENTS

The technologies that underpin the digital revolution provide new ways of organizing, storing, analyzing, and transferring information. The catch is that what needs to be done is not always evident. What matters for productivity and growth is the capacity to imagine how the underlying digital technology can be used, to envision the tools, to design their sophisticated application. The imagination and the applications evolve as an array of experiments—experiments not only in technology or tools, but in the organizations that employ the tools and the business models for establishing new ways of creating value. Some of those experiments will succeed; some will fail. All provide rich material from which to learn about innovation in a digital era.

We consider three such categories of experiments in this book. The first exemplifies the failure of imagination; old models were applied to new situations. Enron is treated as a failure to understand the collaborative possibilities of the new networks. The second category of experiments considers the reorganization of production in the digital era: offshoring/outsourcing, on the one hand, and development of open-source software, on the other. These represent experiments that rely on the opportunities inherent in digital technology. The third category includes experiments with the management of knowledge—experiments that question the fundamental nature of knowledge. Without explicitly addressing the issue, the chapters all suggest that national contexts are producing unique streams of experiments that may generate distinctive strategies for firms with different national origins.

The first category of experiments deals with business strategy. Andrew Schwartz tells the story in "Missed Opportunity: Enron's Disastrous Refusal to Build a Collaborative Market." The classic Enron story focuses on the easy answer, fraud; Enron as a Ponzi scheme designed to enrich scoundrels. But Schwartz argues that beneath the off-balance-sheet transactions and partnerships that have drawn such intense scrutiny, Enron's efforts to reduce complex products into tradable commodities represented one of the most promising ideas of the past twenty-five years. Enron's failure, the argument goes, was due in part to a business strategy that missed the collaborative opportunities represented by the new network marketplaces. Enron saw competitors as ruthless and uncompromising, a mentality that rejected the very real possibility that rivals could, working together, create new markets with tremendous profit opportunities. Enron's brilliant vision of the New Economy, contends Schwartz, did not go far enough; it required a new business model that emphasized cooperation among competitors.

Work reorganization constitutes a second set of experiments. The introduction and application of networks that permitted easier communication and exchange of data, even in the years before the Internet, followed a clear pattern. François Bar and Michael Borrus pointed out that first existing processes were automated; then applications in new networks were launched; and finally work processes were reorganized (Bar and Borrus 1997). Hence we need to consider both the reorganization of existing work processes and the creation of fundamentally new processes of production.

In "The Relocation of Service Provision to Developing Nations: The Case of India," Rafiq Dossani and Martin Kenney examine the reorganization of service work. It is really a story about the conditions that contribute to offshoring work in a service era. Apart from the capacity to store and transmit information, a variety of factors facilitate outsourcing from a company and offshoring the work away from a core location. Increasingly, production — whether production of objects, software, or service — is converted into standard constituent elements. Those constituent elements can then be addressed as modules with standard interfaces. One consequence is that work can be segmented organizationally and geographically.

Dossani and Kenney's chapter highlights the notion that the reorganization afoot is only partly about cost; more fundamentally, it is about imagining and implementing new approaches to the organization of production. The story parallels the earlier reorganization of manufacturing but raises potentially a more fundamental challenge. A few years ago, it seemed that the United States was supposedly moving from an industrial economy to a postindustrial or service economy. A decline in manufacturing employment was offset, so the argument went, by the growth in the service sector. Many argued that we

14 were living through a digital revolution, and thus, as a national economy, we could safely exit manufacturing through a secure economic afterlife developing software and providing services — a whole array of activities that do not involve making physical goods.

Now, as American corporations move some white-collar and technology jobs to regions with lower-cost yet highly educated workforces, there is a fear that the scale of the offshoring will be great enough to affect the fate and future of higher-wage skilled workers in the United States. The real debate needs to be about jobs tomorrow and the capacity to adjust and innovate in global markets. It hinges on whether the further diffusion of IT abroad will continue to enhance economic productivity at home and whether that increased productivity at home will in turn generate new high-wage jobs at home. Whether or not job creation offsets job losses, as the optimists would propose, the adjustment will be substantial.

In "From Linux to Lipitor: Pharma and the Coming Reconfiguration of Intellectual Property," Steven Weber considers the possibilities of radically new production systems. The outsourcing/offshoring debate, whether about services or manufacturing, inherently considers the reorganization and relocation, and then adjustment, of existing production structures. Open source as a principle of organization hinges on distinct approaches to mobilization and coordination of work, not a vague voluntarism but replicable rules of participation and gain. But the principles and rules on which it rests are new. For example, it rests on foundations that turn notions of property from ones of control over the use of an object into control over the processes of distribution. The collaborative work arrangements it points to are about production of software and made possible by digital networks. One critical question is whether the open-source strategy of production that emerged in software in fact is applicable in other domains, such as the pharmaceuticals sector. If it is, what implications would it have for the future productivity of the industry?

The third set of experiments concerns knowledge, the very fundamentals of an information or digital society. Information in digital form can be formalized, stored, searched, transmitted, and used to control the operations of physical processes. The complex relationships whereby engines operate or planes fly can be stated as algorithms represented in digital form. But how do we know, in an avalanche of facts and stated relationships, which ones we care about? How do we manage the knowledge we have? That ultimately forces two questions: What is the nature of knowledge? And how will knowledge contribute to the creation of value in companies and the economy?

Knowledge, particularly theoretical knowledge, has been recognized as an essential element of the contemporary economy. Critically, though, it is the expression of information, data, and knowledge in digital form that is truly distinctive, permitting the application of digital tools. Digital technology represents a set of tools for thought, tools that manipulate, organize, transmit, and store information in digital form. In so doing they extend the range of what can be represented in formal data. We both gather an awesome amount of data and formalize the know-how of communities of practice. In one sense the flood of data made possible by these tools can drown the recipient, but oddly, the same tools for thought make easier the creation of meaningful information and the generation of knowledge from that flood of data.

Niels Christian Nielsen and Maj Cecilie Nielsen, in "Spoken-about Knowledge: Why It Takes Much More than Knowledge Management to Manage Knowledge," confront the question of what knowledge is in order to address the question of how it should be managed in companies. They note that both the conventional categories of knowledge that we use, stored as formally stated sets of facts and relationships and as know-how embodied in communities of practice, are philosophically incomplete or misleading conceptions of what knowledge is. Only a recognition that knowledge is embedded in often fundamentally metaphoric frameworks will allow us to confront the fact that knowledge takes on value in the constant interplay of formal and embodied knowledge. This conversation recreates and recasts the frameworks and metaphors.

The growing importance of knowledge and its management are reflected externally in strategic R&D relationships and internally in quite innovative approaches to company organization. Given the diffuse and specialized character of much of the knowledge required for any product and market, the way companies relate to each other is critical. Christopher Palmberg and Olli Martikainen, in "Pooling Knowledge: Trends and Characteristics of R&D Alliances in the ICT Sector," demonstrate the diverse means by which this external knowledge management may occur. Especially vital to information technology sectors, these relationships vary cross-nationally and within sectors. These experiments, then, reflect in part the environmental setting within which firms are embedded.

Internally, the company organizations required for efficient manufacturing may not be the same as those required for effective exploitation of knowledge. It appears a distinctive form of organization is emerging in the digital era, the learning organization. It may be distinct from the traditional categories of craft, Taylorist, or lean production. The work of Lorenz and Valeyre, as explicated by Tobias Schulze-Cleven, suggests that a distinctive organizational form is emerging in northern Europe, principally in the Nordic countries. Social welfare policies that may have been complications in an era of mass production

16   provide the social basis for the learning organization needed in the knowledge
economy. As Zysman argues in the next chapter, and the cases here support,
the route to adaptive flexibility can come through a strategy of both reinforcing social protections and eliminating them. The question is not necessarily
the level of protection, but rather the form those protections take.

PART III: MARKET TRANSITIONS

Corporate strategies are chosen given a web of policies that set market signals
and define market opportunities.[5] Moving from one set of market signals to
another is not simply a matter of whether a new equilibrium can be defined
and can produce better outcomes, but whether one can navigate successfully
the transition to the desired end point. In short, corporate experiments implementing new digital technologies are entangled in political debates about the
reorganization of markets and marketplace rules. Some of these debates are
about applications of specific rules: what may be done with bits of personal information that can be gathered, stored, referenced. Other debates are about
the networks themselves and about the rules for use and competition that
structure those networks. These debates delimit the politically viable menu of
possible market transitions. Some market endpoints become impossible because the transition to that endpoint does not exist. An analysis concerned
with the political economy of the digital era, then, must not only examine the
characteristics of static marketplaces but also the transitions that form the
bases of such markets.

Conventionally, in the West such transitions often began with a process of
deregulation, a term that was rather loosely applied to a grab bag of very different objectives and stories. The term implied a call for a reduction in the extent of government intervention and rule making in marketplaces, and for the
substitution of competition for government's direct regulatory or administrative hand on what were perceived as natural monopolies. In the United States
the classic story of deregulation was the breakup of the regulated monopoly
AT&T. Indeed, as Steve Vogel has argued, deregulation really meant reregulation, a change in the character and purposes of rules — changes that often reinforced the authority of bureaucrats — rather than a reduction in total government intervention (S. Vogel 1996). In Western Europe and elsewhere,
public administrations performing supposedly public services have been converted into private firms operating in newly formed marketplaces. This was a
process of de-administration, privatization, and market creation.

A diverse set of political and economic factors drives the formation and
re-formation of markets. There are at least two sets of debates: a political one

concerned with who wins and who loses from the process of reregulation, and a technical economic one focused on defining market rules that will optimize welfare. The classical argument about regulation begins with a concern for how government rules overcome market failure. The notion of market failure is one in which agents pursuing private interest in a market will not produce an optimal outcome. A natural example is R&D, where public investment can offset a private tendency to underinvest. In response, of course, others argued that state regulation of the economy was not a market corrective but rather reflected government failures and the use of government rules by specific interests to capture private rents. And more recently, it has been argued that self-managed marketplaces and private competition serve the public interest better than regulation.

Can these stories and arguments be unified in a political economy of market transition? To do so, we should ask why some of these market transitions succeed, while others that seem quite similar on the surface fail. A marketplace can be understood as a constrained or partial equilibrium, a market outcome resting on the specific set of rules and institutional arrangements that at a given moment define constraints and induce a logic of market competition. Each marketplace, then, is constructed by a contingent set of rules, rules that are usually embedded in a set of institutionalized organizations such as firms, banks, and governments (North 1990). Marketplace transition or creation — be it deregulation, reregulation, de-administration and privatization, or creation — involves a shift from one constrained equilibrium to another. The creation and transformation of markets will always be a story about establishing rules and the marketplace structure. These rule sets and institutional arrangements rest on explicit, and sometimes implicit, political bargains.[6]

The transitions are not automatic. What is interesting is the route: how we go from one point to the other. Let us define a successful transition as arriving at a new stable constrained partial equilibrium (CPE), a market clearing within constraints, without dramatic and continuing political disruption. The French financial reforms in the 1980s reached a stable end point; we no longer hear about the reforms or their consequences. The Japanese reforms in the same period have not reached a stable endpoint; they have created economic difficulty and continuing political struggle. There can be several CPEs, of course, each with different implications for distribution and innovation. But to represent an endpoint all CPEs must have prices that allow the demand for goods to clear.

How do we understand the politics that allow that transition, that shift in the rules and institutions of the marketplace, to happen? Such a political process can be addressed from two vantages, rents that motivate politics and

18   sensitive prices that dislocate politics. The first view focuses on the rents that motivate market actors to seek changes in the rules. The rules of a marketplace can be redefined to create new opportunities for existing players or possibilities for new entrants. Stability requires that the rents, or possible rents, be largely extracted or no longer politically available. This is a sequence, with moments of stability and periods of aggressive dislocation. It is not just that rent seekers create new rules, but rather that each rule change creates new rent seekers, new market players who are likely political activists in the rule game. As long as the rents are there or significant new rents and new political contestants are created, then the bargains will not be stable, and the rules will fluctuate and evolve. As the rules evolve, market dynamics, prices, and supply will also fluctuate. Arriving at a market clearing equilibrium is not obvious.

A second vantage would examine the power of politically sensitive prices. If energy prices jump too high or too abruptly, many consumers and producers are profoundly dislocated, without the capacity or time to adapt. If rents move upward abruptly, then many residents are forced to move or alter expenditure patterns significantly. Such price dislocations define affected groups that may mobilize politically. They will likely mobilize to mute or reshape markets.[7] If the market process of moving from one equilibrium to another involves movements in those prices that dislocate society, then significant disruptive politics are the result. We often think about how massive layoffs or firm closures generate demands for government intervention. Energy costs, whether electricity or oil, can only move so far so fast without disrupting the community. Changing the regulation in the California energy markets produced both shortages and price increases. Whether those swings in supply were expressions of a shortage of generating capacity or market manipulations that were not possible in the regulated utility market, the result was an intense political struggle.

The route from one CPE to another is evidently an interplay between markets and politics. Imagine two sets of prices: those prices that must move significantly for market adjustment, and those prices that are politically very sensitive. If there is extensive overlap and significant movement between radical price moves and prices that are politically sensitive, then there is a real risk of political controversy over the new rules. If there is no overlap between market-significant and politically sensitive prices, then the losers can be ignored or perhaps compensated, depending on the style of the political system. As overlap grows, the political problem of managing the discomfited increases. Again, however, this is not likely to be a single step, but a series of moves and adjustments.

One lesson from these two vantages is that the market transition is not over until the politics settle down; otherwise the rules are likely to change again to mollify the dislocated and to capture rents. Or, put differently, each marketplace rule structure is a political bargain or imposed outcome that by either intention or result settles questions of who gets what. And conversely, each new CPE constitutes a new set of political bargains; hence the politics of a marketplace transition is an outcome, not just a cause.

There are significant implications for policy makers debating rule changes in markets and for academic analysts. Policy makers who simply compare static outcomes without considering the route by which markets must move from one equilibrium to the next can be profoundly misleading. It is critical to analyze the transition as well as the possible outcomes; the transition to some outcomes may simply not exist, and the price of others may simply be too high.[8] For academic analysts, this vantage emphasizes that distributional consequences are central to understanding developmental trajectories, and it points to the political strategies one must assess in understanding outcomes.

We consider here the market transitions associated with the emerging cellular networks, with the stories told twice, once from an American perspective and then from the European vantage. In "The Peculiar Evolution of 3G Wireless Networks: Institutional Logic, Politics, and Property Rights," Peter Cowhey, Jonathan Aronson, and John Richards argue that wireless telecommunication developments demonstrate the difficulties associated with such moves. Predictable roadblocks will slow the transition from one set of powerful embedded interests to new property rights and marketplace rules. The new property rights will benefit a reformulated set of stakeholders and political entrepreneurs and create their own market distortions. As this unfolds, the evolution of institutional mechanisms for organizing markets may lag behind the changes in property rights, producing marketplace discontinuities. The basic political economy of the 3G wireless marketplace described here differs in fundamental ways from the stories on 3G in the business press. The set of contributions from the Research Institute of the Finnish Economy opens a similar set of issues. As information technology continually disrupts sector equilibrium, market rules — often in the form of standards — influence corporate responses. Both in the diffusion of second-generation wireless, described by Heli Koski in "Factors for Success in Mobile Telephony: Why Diffusion in the United States and Europe Differs," and in the innovation strategies held by firms focusing on third-generation technology, depicted by Aija Leiponen in "National Styles in the Setting of Global Standards: The Relationship between Firms' Standardization Strategies and National Origin,"

20   government intervention has shaped patterns of consumer uptake and industry cooperation. Market transitions, then, are integrally linked to the institutional settings in which they are embedded.

PART IV: SOCIAL TRANSFORMATIONS

The unfolding digital revolution and the creation of a digital infrastructure are not just about national economic strategy, corporate experimentation, or market transitions. Even as information technology creates new market possibilities, it has the potential to redistribute power, force the reexamination of basic political bargains, and threaten influential political interests. The very opportunities inherent in the digital era raise concerns that could threaten to disrupt the economic potential of the digital era. Governments, businesses, and citizens are left to struggle with devising policy responses that resolve these disputes, and the resultant actions inevitably shape the character of the national information society. A critical feature of the digital era, then, is managing the sociopolitical repercussions of technological change.

Decisions about the rules of information — be they intellectual property, privacy, or market transactions — are at once decisions about broader social values, whether they are intended to be or not. Polities and communities are at their very root based on flows of information. And shifts in those information flows affect the character of human interactions and, in turn, political organizations. If information that was once available only in local groups is suddenly available through a network in the isolation of the home, interaction patterns change. Seemingly clear and agreed-upon technical purposes, such as how to provide a system of addresses for e-mail or to provide credit or payment, almost inevitably involve more fundamental issues such as privacy or freedom of speech.

The effect of information rules does not play out in a single inevitable logic. The architecture of an information society (or the political and technological rules that govern the use of digital tools) may very considerably. Just as decisions about how to construct a freeway or public transportation system influence the character of a city's labor or housing market, variations in information rules may produce radically different digital societies (Lessig 1999). Whether or not an individual has the ability to log on to a computer anonymously has significant implications for use patterns. Societies and governments, then, confront a series of intense debates over how to structure political and economic rights in a digital environment.

This section begins by disposing of the old debate that the digitally networked world would be a Wild West, immune to government control. Taylor C.

Boas, in "Weaving the Authoritarian Web: The Control of Internet Use in Non-democratic Regimes," uses the case of free speech in authoritarian regimes to demonstrate the level of control possible in a digital architecture. Rather than purely limiting technology diffusion, authoritarian regimes have constructed mechanisms to embed surveillance into their national version of the digital era. Far from being forced from power by innovation, these regimes have been able to mobilize Internet technology to expand their political reach.

Three issues — intellectual property, free speech, and privacy — show the interconnection between business strategy and broader public values and how the resolution of these issues influences the texture of our political and social communities. Brodi Kemp, in "Copyright's Digital Reformation," shows how incumbent interests were able to use legislation to reassert their interest in the digital world. Far from ushering in a peer-to-peer, user-based intellectual property system, privileging an individual's right to share purchased goods, the entertainment industry successfully lobbied to bolster their rights of control over digital media. With the recent prosecution of individuals by the music industry for sharing files, it seems clear that the balance of power over intellectual property has (at least for the near term) shifted even further to content providers and away from consumers — a rather ironic result from a technology that was heralded as bolstering individual power over corporate interests. Contributing to a broader debate, Kemp highlights how international agreements have been deployed in the digital era to bolster domestic economic interests.

We try to sum up the issues and cases by asking, do the political rules of state and polity shift in significant ways with the digital revolution? To drive the discussion in heuristic exaggeration, Newman and Zysman ask, in "Transforming Politics in the Digital Era," whether the digital era marks a second great transformation (K. Polanyi 1944). The Industrial Revolution was not a technological story, but the outcome of a basic transformation in the organization of economy and society. The authors examine how digital technologies shift the fundamental logic of politics in this new era and attempt to describe the variety of information societies that are emerging.

Information technology has unleashed a set of national, corporate, and societal experiments. The effects of these experiments range from the terms of global competition to the character of everyday political life. And the success and failure of these efforts have been in large part influenced by domestic and international politics and their interaction. As the digital era enters its second decade, it is vital to assess its progress as well as its future — to learn from these initial moves. We believe that the contributions in this volume make a significant contribution to this effort.

NOTES

1. For the political version of this argument see Margolis and Resnick 2000.

2. Brad de Long and Konstantin Magin, "The Last Bubble Was Brief, but It Was Still Irrational," *Financial Times*, April 19, 2005, 19.

3. Alexis de Tocqueville makes the classic argument. See Bendix 1964.

4. For a summary of the convergence and diversity debate, see Berger and Dore 1996.

5. Andrew Schwartz and John Zysman were developing this argument together at the time Dr. Schwartz's illness became more serious. Zysman wants to acknowledge the depth of Andrew Schwartz's insight, his consistent and persistent analytic drive, and his support. In addition to the argument presented in his contribution to this volume, Schwartz's view on market transition can be found in his book *The Politics of Greed: Privatization, Neo-liberalism, and Plutocratic Capitalism in Central and Eastern Europe* (Lanham: Rowman & Littlefield, forthcoming).

6. For the importance of political bargains to the rules of the market see Fligstein 1996.

7. This follows the logic of the double movement elaborated by Polanyi 1944.

8. Howard Shelanski's comments on Cowhey, Aronson, and Richards's paper at the conference "The Digital Era: National Responses, Market Transitions, and Global Technology," Berkeley, CA, October 22, 2004, makes this point.

2

CREATING VALUE IN A DIGITAL ERA

How Do Wealthy Nations Stay Wealthy?

John Zysman

To clarify the character of the digital era, this chapter asks how wealthy nations stay wealthy amidst radical changes in competitive markets. To address this question, it considers how the firms that make up these economies stay competitive as the mechanisms of value creation, the engines of productivity and growth, evolve. Twin drivers, the global and the digital, constantly shift the sources of market advantage, forcing companies and countries to adapt. A firm's internal functions suddenly become products to be bought in the market, products that generated premium prices suddenly become commodities, and the sources of differentiation for products and production processes evolve. It is not just that there is an increased pace of change, but that the market environment is inherently less predictable. Traditional tools of strategy and policy analysis will not suffice. Conscious experimentation will be central to both corporate and national adaptation. Companies will have to look at their initiatives as experiments, attempts to find a way through a maze of quite fundamental uncertainty. Each company effort, and each effort of a competitor, must be culled and systematically assessed for lessons. Governments must consider what an "experimental economy" will require and how an environment can be created for individual firms and networks or clusters of firms to experiment effectively. This chapter first situates the present digital era in historical perspective. A brief review of the material, some of it familiar, clarifies the character of issues being confronted today. It then considers how the global and the digital change the problem of value creation in the marketplace.

EVOLVING MODELS OF PRODUCTION AND COMPETITION:
THE DIGITAL ERA IN HISTORICAL PERSPECTIVE

The influence of the digital revolution is visible in the productive economy, through the evolution in how we make and distribute goods and services (Zysman 2004b). Let us examine production and value creation in historical perspective. Each historical phase involved different business problems, a different role of the "abroad" in the dynamics of the national economy, and a different emphasis on the state's role in the economy.

American Dominance: Fordism and Mass Manufacture

Mass manufacture, epitomized by Henry Ford and the Model T, was the first twentieth-century production revolution, though its roots were formed in the nineteenth century. In this system, large-scale manufacture implied rigidity. Fixed costs in the production line and design were high; consequently, changes in products or reductions in volume were difficult and expensive. This rigidity created not just technical but also political problems.

Mass manufacture is broadly understood to mean the high-volume output of standard products made with interchangeable parts connected using machines dedicated to particular tasks and manned by semiskilled labor (Hirst and Zeitlin 1997; Jones, Roos, and Womack 1991). Traditionally noted features of this basic definition include:

- the separation of conception from execution: managers design systems that are operated by workers in rigidly defined roles that match the worker to the machine's function;
- the "push" of products through these systems and into the market; and
- large-scale integrated corporations, whose size and market dominance reflect mass manufacture's economies of scale.

There was a political consequence. Drops in demand were difficult to absorb for companies built on Fordist models, which made the national economy rigid as well. An initial downturn in demand could snowball into sharper economic downturns. Booms and busts implied worker dislocations, and the counterpart of the corporate business cycle management task in the national economic policy became a political debate about how to use a public policy to cushion not only the economic but also the political dislocations that would come from mass unemployment. Demand management policies, associated with John Maynard Keynes, were born. Fordism, an American innovation, was—and I use the past tense intentionally—mass production with Keynesian demand management. With its emphasis on internal demand and domestic demand management, Fordism might have been called a strategy for "capitalism in one country."

Challenges from Lean Production and Flexible Specialization

Challenges to American manufacturing came from two different directions. The more important challenge was the interconnected set of Japanese production innovations, loosely called *flexible volume production* or *lean production* (Cohen and Zysman 1987; Jaikumar 1988; Coriat 1990; Jones, Roos, and Womack 1991). The stunning world market success of Japanese companies in consumer durable industries requiring complex assembly of a large number of component parts, principally mechanical and electromechanical goods, set the American, and secondarily European, industrial establishment on its heels. The distinctive features, and advantages, of the Japanese lean production system, a logical outcome of the dynamics of Japanese domestic competition during the rapid-growth years, were firmly in place by the time of the first oil shock in the early 1970s.[1] The Japanese lean production system seemed to provide flexibility of output in existing lines as well as the introduction of new products, permitting rapid market response. High quality came hand in hand with lower cost.

The Fordist story highlights national strategies for demand management; the Japanese story of lean production and developmentalism highlights the interaction among the markets and producers of the advanced countries in international competition. The Japanese developmental state actively promoted domestic development with closed markets at home while free-riding on the international system using exports as a domestic balance. The strategy required the combination of an open international system with controlled competition behind managed trade borders in Japan. Indeed, protected domestic markets with intense but controlled competition were decisive in the emergence of the innovative and distinctive system of lean flexible volume production.[2]

The second challenge to the classical American mass-production model had little to do with the volume-production strategies emerging in Japan. Different accounts of its development variously label this collection of innovations *diversified quality production* and *flexible specialization* (Piore and Sabel 1990; Streeck 1991; Boyer and Hollingsworth 1997; Boyer and Saillard 2002). The "Third Italy" and the Germany of Baden-Wittenberg were the first prominently displayed examples of an approach in which craft production, or at least the principles of craft production, survived and prospered in the late

26 twentieth century. The particular political economy of the two countries gave rise to distinctive patterns of company and community strategies (Berger and Piore 1980; Sabel 1982; Herrigel, Kern, and Sabel 1989; Hirst and Zeitlin 1997). Firms in these countries often competed in global markets on the basis of quality, not price; they used production methods involving short runs of products that had higher value in the marketplace because of distinctive performance or quality features. Competitive position rested on skills and flexibility, not low wages. Communities or groups of small companies arose, organized in what are perceived as twentieth-century versions of industrial districts. These communities are able, in at least some markets and circumstances, to adapt, invest, and prosper in the radical uncertainties and discontinuities of global market competition more effectively than larger, more rigidly organized companies. "These districts escape ruinous price competition with low-wage mass producers," Sabel explains, "by using flexible machinery and skilled workers to make semi-custom goods that command an affordable premium in the market" (Sabel 1994). The emphases in these discussions are the *horizontal connections*, the connections within the community or region of peers, as distinct from the *vertical or hierarchical connections* of the dominant Japanese companies. The flexible specialization model hinges on local institutions that permit the continuous combination and recombination of local activities.

These two innovative challenges to American production dominance each embedded a distinct role for policy and the state: lean production hinging on an arbitrage between closed domestic markets and the open international system; flexible specialization as originally formulated depended on local institutions which allowed quality craft production.

The Transition to a Digital Era and the American Comeback: Wintelism and Cross National Production Networks

The first chapter of the digital era can be best characterized by two elements: Wintelism and cross-national production networks (CNPNs) (Borrus and Zysman 1997a). Let us define each. Wintelism, as a code word, points to the transition between an electromechanical and a digital era.[3] It reflects the sudden importance of the constituent elements of the product in the final market competition (the Windows operating system and Intel processors are examples, whence the name) and the consequent strategic shift in competition away from final assembly and vertical control of markets by final assemblers. CNPN is a complementary concept pointing toward a corresponding change in production systems. CNPN was the label first applied to the disintegration of the industry's value chain into constituent functions that can be contracted out to independent producers, wherever those companies are located in the global economy. CNPNs permit and result from an increasingly fine division of labor. The networks permit firms to weave together the constituent elements of the value chain into new, competitively effective production systems while facilitating diverse points of innovation. This was the first production era in which we could really speak of a global economy. It was one in which competition and the critical final markets were in the advanced countries and production was organized by firms from these same advanced countries but spread across borders, principally through Asia. CNPNs were part of a process that turned large segments of complex manufacturing into a commodity available in the market.

Wintelism, as a strategic stance and production system, emerged as a response by American producers to the Japanese production challenge. Twenty years ago the story was that American electronics firms were being dominated in international markets. As the semiconductor industry joined consumer electronics and automobiles as sectors under intense competitive pressure in the late 1980s, it seemed that the fabric of advanced electronics was unraveled. That is, the array of equipment suppliers to the semiconductor industry was eroding, making it more difficult for American semiconductor producers to hold market position. With the weakening position of the semiconductor makers, many feared that final-product producers would not have access to the most innovative chip designs needed in their final products.

Then, suddenly, American producers rebounded. They did not reverse the loss of production advantage in electromechanical products; rather, a new sort of consumer electronics product emerged, defining a new segment of the industry. The then "new" consumer electronics, as Michael Borrus argued at the time, were networked, digital, and chip based (1997). They involved products from personal computers to mobile devices. The nature of production changed dramatically from the complex mechanical or electromechanical assembly to electronic chip production, board stuffing, and putting the boards into boxes. The sources of product functionality moved to chip-based systems given functionality by software. The core engineering skills shifted from mechanical to electronic.[4]

Wintelism involved new terms of competition and, linked to them, a new model of production. Consider the personal computer. What part of the value chain confers the most added value and leverage in the market? It is not the producer of the final product, the metal box we call the PC, even if, like Gateway or Hewlett Packard, the box carries the company logo. Much of the

28 added value is in the components or subsystems: the chip, the screen, and the operating system. This has several implications.

1. Each point in the value chain can involve significant competition among independent producers of the constituent elements of the system (e.g., components, subsystems). Control over the evolution of technology and final markets in many market segments could be exercised by the component or module companies, not just by the final assemblers. The pace at which technology evolved was increasingly dictated by Microsoft or Intel, and not by the assemblers of computers in the personnel computer segment; similarly, Cisco, the newcomer and independent equipment provider, drove the emergence of Internet technology.

2. Competition in the Wintelist era is often a struggle over setting and evolving de facto product market standards, with market power over those standards lodged anywhere in the value chain, including product architectures, components, and software. Components and subsystems are built to generally agreed standards that emerge in the marketplace. Thus part of their value lies in the standards, in partially open but owned standards that create de facto intellectual property (IP)–based monopolies or dominant positions.

3. These fundamentals of Wintelism have evolved in such a way that the constituent elements of the product become modules. Even if distinctive intellectual property remains in the modules, production becomes modularized as the knowledge about the elements and components and how they interconnect becomes codifiable, that is, formally stated and expressed in code, and then diffused.

4. As a result, products can be easily outsourced because they are increasingly built as modular systems in which many components and subsystems are clearly defined. Modularization further encourages a vertical disintegration of production. Outsourcing, a tactical response usually aimed at cost savings through a decision to procure a particular component or service outside the organization, evolved into cross-national production networks (CNPNs) that could produce the entire system or final product.

5. The core engineering skills moved to chip-based systems given functionality by software. The range of production skills needed to produce an optical film camera is much greater than that needed to produce a digital camera, whether in a cell phone or not.

The Wintelist era of the 1980s and 1990s, the moment of the American comeback in electronics, turned, politically, on domestic — initially American — deregulations and international deals that created an ever more open international trade system. At home in the United States, domestic deregulation and

competition policy in a variety of sectors—especially telecommunications
and computers—contributed to significant market competition among, and
shift in market leverage toward, component makers. Those initially domestic
American competitions in software and microelectronics as well as in
telecommunications reshaped the electronics industries worldwide. Ever
more extensive and dispersed networks of investment, trade, and production
were the first step in an evolution of complex production networks and supply
chain management. The emerging production structure and trade structure
contributed to and perhaps even drove the expansion of something loosely
called globalism.

COMPETING IN A GLOBAL AND DIGITAL ERA

The distinctive features of the current era, the global and the digital, are
changing the mechanisms for creating value. Let us consider each in turn.

Globalization with Borders

The classic version of the globalization story begins with reduced costs for
transport and communication that lower "transaction costs." Lowered costs of
doing business over distance, it is argued, create incentives for companies to
expand trade and drive financial interconnection. Government choices are often constrained by the evident consequences of policy decisions, such as
macro-policy and efforts to manage exchange rates, and by lobbying of mobile
capital.

From an alternate vantage, globalization is a story of national innovations
played out on a larger stage (Borrus and Zysman 1997a). A sequence of new
competitors, new and often unexpected loci of innovation and production,
bring new processes, new products, and new business models to the international marketplace. The dramatic marketplace developments have usually
been cooking inside national systems of innovation and competition, largely
unobserved by the outside. Consequently, they are startling when they burst
onto the global marketplace. This gives the global era the feel of a seemingly
increasing pace of unexpected competitive challenges.

Tools for Thought as the Foundation of a Digital Era

This digital era is best characterized by a new set of distinctive tools, tools for
thought: "Information technology builds the most all-purpose tools ever, tools
for thought. . . . These tools for thought amplify brainpower by manipulating,
organizing, transmitting, and storing information in the way the technologies

of the Industrial Revolution amplified muscle power" (Cohen, DeLong, Weber, and Zysman 2001; Cohen, Delong, and Zysman 2000).

The tool set rests on a conception of information as something that can be expressed in binary form and manipulated (Weiner 1954, 1965; Shannon 1993). It is made up of the hardware, consisting of equipment that executes the processing instructions; the software, consisting of written programs including procedures and rules, which guides how the hardware processes information; data networks that interlink the processing nodes; and the network of networks that together create a digital community and society.

The digital tools constitute a leading sector that has reshaped the economy as a whole. Demand for the products and services made possible by the new digital technology have been part of growth and transformation in the advanced economies in the latter part of the twentieth century.[5] IT is not unique. Demand for the goods in a leading sector grows faster than the economy; the surge initiated by the leading sector involves not only new technologies embedded in leading sector products, but also new infrastructures for making and using the technologies. Producing innovative goods creates chains of linked and interlinked activities. The production chains are evident; for example, steel for trains and cars, petrocarbons (coal and petroleum products) to drive them, and coal to make the steel.[6] Many would argue that the significance of information technology for the contemporary economy is greater than that of earlier leading sectors in their era. Resolving that argument about scale does not matter to this essay. What does matter is that the IT tools can affect economic activity in which information sensing, organizing, processing, or communication is important — in short, virtually every single economic activity.[7]

The IT revolution is transformative, changing the character of products, processes, marketplaces, and competition throughout the economy. The capabilities of processing and distributing digital data multiply the scale and speed with which thought and information can be applied. Because the expression and manipulation of information is now possible in a common digital electronic form, a range of previously separate information and communication sectors merge or become entangled with each other ever more intimately. For example, print, broadcast, and communications suddenly become integrated, with the possibilities of search and storage of information thrown in. Some argue that movable type contributed to the social revolution of the Renaissance, with the obvious question of whether the social consequences of these radical information technologies will be on a similar historical scale. As important, the knowledge component of much industrial activity can now be formalized, codified, and embedded in equipment. Industrial processes once defined loosely as know-how can more readily be expressed and implemented in digital code. Examples would include auto braking, which could be understood abstractly but acted on only imprecisely by human intervention or through analog control solutions.[8] Embedding functionality in digital controls rather than in electromechanical form makes it easier to vary the functionality of many goods, to create a variety of functionally distinct versions from one electromechanical foundation that retains scale. Information technology has moved inside of machines, controlling their functionality, but also moved out into the communications networks, altering not only how and at what price we talk, but how we share, store, and use information.

The logic of cost as well as functionality changes. The cost of creating digital information, producing the information in the first place, remains a fixed and often high cost, while the cost of reproducing and transmitting content in digital form drops toward zero. The consequences of often nonexistent replication costs are amplified by the very nature of information goods. How do I price and value what you know and want to sell me without seeing it? But if I see it, and thus possess it, how can you still sell it to me? And if I can reproduce and distribute that knowledge widely at low cost, what happens to your market? New business models have to be invented and older models and the forms of distribution and IP defended through contracts and courts.

As important, the application of information within machines makes the trade-off between IT and other forms of capital possible. Use more information technology and you need less fuel or simpler machines. However, before real productivity or strategic gains can be achieved, change has to be forced into business strategies, in, for example, basic parameters of how factories operate.

All of this tells us that tools for thought, or information technologies, alter the product development, production, and competition throughout the economy, but not how companies might take advantage of these changes, nor how governments might support IT development and diffusion to capture gain for their communities.

One might list the mechanisms through which the digital tools affect business strategy, noting in turn network effects, the changing character of content products when functionally identical copies can be made, copied, and distributed at marginal cost, as well as the capacity to identify and create multiple product versions. But in this approach, listing the tactical and strategic consequences of the IT tools rather quickly reaches limits without contributing distinctive insight. Information tools and information goods have a unique logic, "information rules," to use the clever phrasing and insightful arguments of Shapiro and Varian (1999). But when does that logic apply? Certainly an "information rules" logic applies in the competition over browsers, such as in the Windows/Netscape competition. It may apply in the case of search engines such as Google, where Varian is an advisor. But which elements of information goods, or digital tools or network economics, apply in the case of the automobile industry? And how do we decide which issues matter in a particular setting? If we can't deduce the answers from first principles, we need an alternative strategy to understand value creation in a digital era.

CREATING VALUE: PRODUCTS, COMMODITIES, AND DIFFERENTIATED ASSETS

To understand the influence of the global and the digital on market dynamics, business interests, and hence policy and politics, let us begin with the basic notion of creating value. Created market value, oversimplified, is price minus cost.[9] (Let us set aside for the moment all the necessary qualifications about externalities and politically set rules.) If we are to locate the influence of digital tools, there are two obvious questions about value creation. First, how do digital tools and information products change the task of generating something for which people will pay a premium? In other words, how does a company avoid having its products become commodities? How does the company create unique or differentiated goods so that a premium price can be charged? There is an array of means: create distinctive products, be early to market, or own a standard defining what a product must look like. Second, how do these tools affect the cost of providing a product or service to customers; if you cannot charge a premium, can one generate distinctive margins by being a low-cost producer? The argument here is that the points of competitive leverage, of strategic advantage, are now constantly shifting and moving.

To address these questions we need to define explicitly three notions with which we are generally familiar: product, commodity, and differentiated asset.

- A *product*, whether an object or a service, is that which can be bought and sold in the market.
- A *commodity* is a good or service that is exchanged in competitive markets with little advantage to any particular buyer or seller. A product becomes a commodity when it is generally available from a number of suppliers on common terms in the market.
- A *differentiated asset* creates the basis for premium price, distinctive sales advantage, or cost advantage in production or distribution.

There is a constant reshuffling among products, commodities, and differentiated assets. As reshuffling occurs, business models must change as well. Globalization accelerates the reshuffling, and digital tools often are the means of accomplishing the reshuffle. Globalization represents new competitors who may transform a premium good into a commodity with low-cost production or

generate advantage by adding value to what seemed to be a commodity good, 33
as when the Japanese made quality a "free" good. Digital tools change the
levers of advantage and value creation. Consider finance, wherein the application of sophisticated mathematical tools to the creation of financial products and online transactions replace the ties to our local banker, transforming
distinctive advantages into commodities and creating a new basis for premium
products. Consider versioning, wherein a properly programmed microprocessor distinguishes the functionality of a high-cost product from a low-cost
product though they are similar except for the programming. The continuing
reshuffle includes the transformation of internal company functions into
products available on the market. There is a constant question of whether
the function is a commodity that should be sourced in the market or a strategic asset that must be developed in house or in carefully nurtured supply
relationships.

R&D and production provide examples, not only of this constant reshuffling, but of internal company operations that became first products, and then,
sometimes, commodities.

R&D

R&D, traditionally an internal function differentiating a company's products
from its rivals, can now be sourced outside the company. The presumption has
been that product development, and the R&D to support that development,
are at their core strategic assets, the foundation of innovation and a powerful
antidote to commodification. But R&D, and thus innovation itself, has taken
on aspects of a product, something that can be purchased in the marketplace.
Even as innovation and continuous product or production improvement become more critical, major corporations are shrinking their core research departments. Simply put, companies can buy much of what they had previously
developed internally.

There are a variety of sources from which to buy R&D. First, in the United
States, universities have become a source of technology and joint technology
development. Many of the engineering schools are rooted in science-based
engineering, solving engineering problems by working with fundamental
principles. The Bayh-Dole Act pushed universities into developing "mar-
ketable" technologies using federal funding. An array of mechanisms, from
licensing through facilitating spin-offs to institutions for joint development,
has been established at the major technology universities to facilitate ties to
industry. Second, of course there are start-ups that spin out the development
of particular elements of products or services. Many projects are best developed outside the traditional hierarchy of a major company. Firms from Intel

through Nokia to IBM establish mechanisms, including their own investment companies, support start-ups as an approach to technology development and an alternative to internal development. Third, companies set up joint product development projects with other companies, basically combining technology strengths. Fourth, major companies establish technology development outposts both to monitor developments and to tap into distinctive pools of talent and technology around the world. Fifth, a wide range of countries is entering the development game, investing in R&D both in public labs and in support of industrial labs. Hence the number of points of purchase for "technology" and "development" has grown.

Major firms become, at least in part, technology integrators, not just technology developers. Firms cannot be at the cutting edge in all the technology developments that affect them and must therefore look outside. Often disruptive technologies, which are capable of supporting newcomer entry into the market, are difficult for established companies to develop in house (Christensen 1997). Firms have to decide and continuously reassess which elements of development are effectively high-end commodities, which technologies are strategic assets best acquired (procured on an exclusive basis or developed in house), and how to move to capture those distinctive technological assets.

Production in a Digital Era

Production in a digital era, for companies and countries alike, can similarly be either a strategic asset or a vulnerable commodity. Over the past decades, production has increasingly become a commodity, a product bought in competitive markets. Manufacturing firms went offshore for reasons of cost or to have access to local markets but discovered abroad a widely distributed capacity for technical and management innovation. Outsourcing led to CNPNs and eventually skills of supply chain management, each step making the next phase of outsourcing, that is, commodifying production, easier. It may be easier for services to move offshore today than it was for manufacturers to do so twenty years ago. The required tool set, consisting of computers, software, and communications, is available in the market and easily transported. These are largely general-purpose tools that can be adapted to particular service tasks. How far, we may ask, will this geographic dispersion go? Can all activities be placed just anywhere? Is there any geographic stickiness to production?

Not all of production is a commodity, however. In fact, production skills are often a strategic asset that creates distinctive advantage. Firms must consider under what circumstances the lack of in-house world-class production skills will represent a strategic vulnerability. Note that because of the ability to segment supply chains, the questions would need to be asked not only about control of the whole process of producing a good or service, but about each individual element of the process. For the nation, or perhaps the region, the question becomes, "What can be done to make this country (or region) an attractive location for world-class manufacturing and an attractive place for companies to use production to create strategic advantage?"

In a world in which services as well as manufacturing are being outsourced, we need to clarify some definitions. Although manufacturing implies manipulating things and materials, its definitions in my online dictionary more generally concerns "the organized action of making goods and services for sale" and putting a product together from components and parts.[10] Certainly a software product, Quicken, qualifies as manufacturing by this definition, as do the creation of the Yahoo website and the assembly of the software tools that allow that website to function. But the word manufacturing implies smoke and factories. Why not just talk of production as the general case and manufacturing as the specific case of physical production? In that case, production — the know-how, skills, and mastery of the tools required — is absolutely central to the products in the digital sector. But when we do speak of production in this way, the corporate strategy questions remain. The traditional questions — what should be produced or built in house? what can be outsourced? what skills are required to produce the digital product? does outsourcing influence the quality? — do not disappear. They are just posed in a new context.[11]

There are at least three circumstances in which in-house control of production, or elements of production, can be a strategic advantage: first, when the in-house control of production provides advantage in cost, timing of goods to market, quality, or distribution that cannot be obtained by outsourced production; second, when knowledge about existing production processes is required to develop next-generation product entry, whether design of the products themselves or design of the processes to produce them (or put differently, if in-house production mastery may be required for rapid product innovation); and third, when critical intellectual property about the products themselves is so tightly woven into the production process that commodity outsourcing is tantamount to transferring product knowledge to competitors.

Evidently, these same issues pose themselves differently in each market or industrial context, and those contexts evolve. Let us consider emerging sectors, which are based on new processes and new materials. An emerging sector such as nanotechnology or biotechnology is all about how you make a thing. Product knowledge and process knowledge are intertwined. In these sectors the question of production, product innovation, value creation, and market control remain entangled. More generally, the strategic place of production in these emerging industries is evident if we ask, who will dominate the new sectors? Will those who generate or even own, in the form of intellectual property rights, the original science-based engineering on which the nanotechnology or biotechnology rests be able to create new and innovative firms that become the significant players in the market? Or will established players in pharmaceuticals and materials absorb the science and science-based engineering knowledge and techniques by purchasing firms that have spun out from a university, or alternately by parallel internal development by employees hired from those same universities? There is an ongoing, critical interaction among: the emerging science-based engineering principles, the reconceived production tasks, and the interplay with lead users that permits product definition and debugging of early production. Arguably, learning is more critical in the early phases of the technology cycle, and outsourcing may hinder the learning process. When they outsource production, firms may lose the learning that comes from the interplay of development and production.

As noted before, the rapid entry of diverse new competitors into global markets contributes to the process of commodifying production and the transformation of innovation or R&D into a product that can be purchased in the market. The new entrants into markets and the ever-evolving competitive position of others — globalization — represent new opportunities, challenges, and threats that come from unexpected directions. Initially, the notion of globalization came with the entry of Asian (really Japanese) producers as fierce competitors in the established European and American markets. Third-tier Asian producers — Korea, Taiwan, Hong Kong, and Singapore — entered global markets as part of supply chains for Western producers before establishing their own positions. Now India, China, and the countries from the former Soviet bloc are all finding their positions in world markets. The new entrants represent both new markets and new competitors who create not only new sources of production and R&D, but often new product, production, and management strategies.

DIFFERENTIATED ASSETS AND CORPORATE EXPERIMENTATION

How, then, can firms escape from the world of commodities, from new competitors from new places nipping at their heels? What can companies do in an era of hypercompetition, when everything threatens to become a commodity? It is easy to say that they must create differentiated products for which the customers will pay premiums, and differentiated processes that can create distinctive advantages. However, the answers will not be arrived at in such a straightforward way. A traditional analytic approach to strategy will only be a starting point in the process of corporate adaptation. Companies will have to look at their initiatives as experiments, attempts to find their way through a maze of uncertainty. They will need to learn how to evaluate their own experiments and interpret experiments of others. Doing so, of course, creates dilemmas. Effective execution is what distinguishes a good idea from a real success, and effective execution is all the harder if an initiative is seen as tentative, a feeler. So the management of committed "experiments" will be a real and required skill.

The Classical Approaches to Differentiation

Branding and design are classical, and increasingly important, strategies for differentiation that need to be acknowledged. They are quite evidently mechanisms for segmenting the market in an era of potential commodities. In a digital era, of course, many electronic products are constructed from very similar modules, thereby achieving very similar functionality; hence design and branding become critical.

Branding, the creation of an identity for a product or set of products, serves as a critical instrument to differentiate branded products from a pool of commodities. For example, amongst an array of similar products that tend toward commodity, the question of whom you trust matters. Hyundai's efforts to establish the once low-end Korean cars as high quality and General Motors Saturn's efforts to establish a no-trickery sales identity are examples of efforts to create trust through branding. Online, the issue of trust is even more important. Here the possible anonymity of the market participants, the difficulty of imagining recourse to a virtual participant, makes trust essential. This is the problem E-bay has so cleverly addressed. As important, there is an ever-greater array of culture products, fashion products, identity products — choose your label — that give expression to a customer's sense of self. And, of course, it is not simply the object, but how the object is perceived by others that matters to that projection of an individual's identity. The "brand" identity in part states the "presentation of self" that the client chooses. For example, Gap Inc. owns Banana Republic, Gap, and Old Navy; the differences in the clothing offered by the three stores are in the quality of the material, the price, and the brand-name identity. Similarly, design takes on ever-greater importance in differentiating products that might otherwise be fundamentally commodities. The Danes for decades have been selling Bauhaus, the source of Danish modern product style. An extreme example of value created by design is the Danish company Bang and Olufson, which sells high-end commodity technology sold at an extraordinarily high price as a lifestyle good. The "brand" identity is based on its exceptional electromechanical characteristics and pure design.

Experiments and Digital Tools

The tools for thought that underpin the digital revolution provide new ways of organizing, storing, analyzing, and transferring information. Investments in training, reorganization, and strategic reorientation are required. The critical question is what to do with those underlying digital capacities and how to use their potential (Brynjolfsson and Hitt 2004; Dorgan and Dowdy 2004). In other words, to capture competitive advantage or to generate productivity gains, you cannot just buy the tools and store them in a closet.

Some of the digital approaches to creating value and to differentiating products have become very well known. First, and now widely understood, are digital approaches to segmenting the market and then attacking specific segments with functionally varied, and for the most part distinctively branded, products. A fundamental feature of the digital era is that analytic tools of database management permit the consumer community to be segmented into subcomponents, each with distinct needs and wishes. At an extreme, individuals and their particular needs can be targeted. Early on, the insurance industry moved from using computers exclusively for back-office operations to using them to create customized products for particular consumers (Baran 1986). Thus, collecting detailed information about customers as groups or individuals in a variety of forms — credit cards and grocery store purchases are obvious examples — is a critical matter. The result, of course, is a policy struggle about what information can be gathered, shared, and combined. The wishes of companies and governments to assemble information from diverse sources into consumer profiles or threat assessments is set against individual rights to privacy and community needs for the integrity of the individual. Once the market segments are defined, digital tools help firms create functional variety in products. Standard products can be given diverse functionality. The coffee maker that automatically turns on at a particular time in the morning depends on simple digital functionality. The difference between many higher-speed, higher-priced printers and their slower, lower-priced brethren is in the software that tells the printer how to operate (Shapiro and Varian 1999). Firms have new ways to identify who will pay how much for what and then create products or give functionality to commodity products that people are willing to pay for.

Secondly, digitally rooted online sales and marketing and supply chain management alter the links between a firm and its customers as well as its suppliers. The Dell story tells how innovative uses of the Internet that tie customers from sales through production can create dramatic advantage (Kenney and Mayer 2002; Fields 2003). And as development and production processes are woven together to speed up the time to market and improve design choices, the lines between production, design, and development become even more blurred.

The Need for Experimentation

There is a catch. It is not always evident what needs to be done, which strategies and organizations are required to create value or generate productivity. What matters for productivity increases and growth is the capacity to imagine how the underlying digital technology can be used.

The imagination and the applications evolve as an array of experiments — experiments not only in technology or tools, but also in the organizations that employ the tools and the business models to establish new ways of creating value. Many of those experiments will fail; some will succeed. Analytically, we cannot just add up anecdotes of success and failure. So how should we proceed to make sense of the transition to a digital era? We proceed here by considering three categories of experiments: work organization, the use of knowledge, and business strategy.

Reinventing Production in the Digital Era

In the continuing reshuffle of the levers of advantage, reorganization and reinvention of production represents a first category of experimentation. The introduction and application of networks that permitted easier communication and exchange of data, even in the years before the Internet, followed a clear three-step pattern. François Bar and Michael Borrus pointed out that first existing processes were automated; secondly, from the initial but automated base, experiments in the use of the new networks were launched; finally, work processes were reorganized (Bar and Borrus 1997). Critical in their story is the question of where, and by whom, experimentation and learning takes place. The same processes are evident now. Consider production and the drive to outsource work in the service sector. The capacity to store and transmit information digitally means companies can segment and distribute work geographically and organizationally. And in the current round in the United States of outsourcing service functions offshore, lower wages have been the primary driver. Martin Kenney and Rafiq Dossani have argued that in the case of India, although lower costs drove the initial move offshore —which largely meant reproducing existing activity at lower cost, as it did in the early days of offshoring manufacturing — many companies found that possibilities for higher quality emerged abroad (Dossani and Kenney 2004). Yet management capacity of the contract producer to manage outsourced offshore projects is as critical a variable as cost in explaining the location of tasks. When an Indian company such as Wipro opens outsourced production activities in the United States, it is clear that management skill, experience with outsourcing, and experimentation with automation of existing processes, rather than the cost of labor alone, underlie the move. The conclusion must be that the service sector reorganization currently afoot is only partly about cost, but more fundamentally about imagining and implementing new approaches to the organization of production. Sometimes for the buyer of outsourcing services, outsourcing is an excuse to avoid tough internal choices about product strategy or internal organization. Sometimes, as in finance, outsourcing obscures the possibility of delivering distinctive services. Sometimes, as in software development, outsourcing creates risks of losing intellectual property or propagating competitors. Hence the issues of who experiments and learns, and of what should be done in house and what should be outsourced, all reemerge with each step.

But of course there are also radically new production systems, such as the lean production systems that emerged in the 1980s and perhaps open-source software in the digital era. Open source as a principle of organization hinges on distinct approaches to mobilization and coordination of work, not a vague voluntarism but replicable rules of participation and gain. But the principles and rules on which it rests are new. For example, it rests on foundations that turn notions of property from ideas of control of the use of an object or of an objectified body of code or knowledge into ideas of control of the processes of distribution. The collaborative work arrangements to which open source points are about the production of software and made possible by the digital networks (Weber 2004).

Managing Knowledge

Knowledge, particularly theoretical knowledge, has been recognized as an essential element of the contemporary economy. Critically, though, it is the expression of information, data, and knowledge in digital form that is truly distinct, permitting the application of digital tools, the suite of tools for thought. In a digital form information can be formalized, stored, searched, transmitted, and used to control the operations of physical processes (Cohen, DeLong, and Zysman 2000). We can put the entire contents of the Library of

Congress onto a single digital memory stick and transmit it in a flash. The complex relationships on which engines operate or planes fly can be stated as algorithms represented in digital form. In one sense the flood of data made possible by these tools can drown the recipient, but oddly, the same "tools for thought" make easier the creation of meaningful information and the generation of knowledge from that flood of data. But codified knowledge, whether stored digitally or embedded in equipment, is only a piece of knowledge that cannot stand alone. For example, how do we know, in an avalanche of facts and stated relationships, which ones we care about? Analytically, there are limits to both the value of piling up and searching documented knowledge and to formalizing the tacit knowledge embedded in individuals and communities of practice. Experiments with knowledge management in this information-rich era force open the very fundamental question of what knowledge is. As Nielsen and Nielsen (2004) have argued:

> Knowledge unfolds in the iterative processes between tacit and codified forms, and optimizing knowledge in organizations is essentially an issue of optimizing these iterative processes. Put in a more grandiose way: Only a recognition that knowledge is embedded in often fundamentally metaphoric frameworks, will allow us to confront the question that knowledge takes on value in the constant interplay of those who cart around both formal and embodied knowledge, in the constant conversation that recreates and recasts the frameworks and metaphors, in the perpetual resorting of knowledge in context that reveals potential relationships and reforms the contexts itself.

There is an organizational implication of this consideration of the nature of knowledge. Internally, the company organizations required for most efficient manufacturing may not be the same as those required for effective exploitation of knowledge. In the 1980s the Japanese innovations of flexible volume production using lean, just-in-time techniques created distinctive production advantage and rocked market competition. Is a similar revolution afoot now? Lorenz and Valeyre (2004) point to the traditional craft organization, Taylorist organization, lean production systems, and an emerging distinctive learning organization. That distinctive organizational form is emerging in northern Europe, principally the Nordic countries. We can only speculate as to why, pointing to experiments in work organization in an era of mass manufacturing that may be paying off in a knowledge era.

Experiments in Business Strategy

The tactical experiments—branding, design, versioning, production reorganization, and knowledge management—have to find expression in new business models, the underlying strategies for creating and capturing value. Those new business models must reflect the shifting location of leverage in creating value. But that is not easy. Many of the most spectacular failures of the bubble era were simply business strategy experiments gone awry, failures of conception and execution. Recall that the dotcom investment wave hinged on the notion that the network tools would "disintermediate" traditional distributors, that brick-and-mortar relationships would be replaced by electronic links, or that wholesale intermediaries would be eliminated by electronic markets. Often the fantasy was that new entrants, new companies, using these digital tools could displace established companies. There are some evident successes; the travel industry, from travel agents through airlines, is being reformed by online operations. But Borders and Barnes and Noble in their brick-and-mortar form are more of a threat to the local bookseller than Amazon is. Indeed, venture capitalists behind Amazon report that the original investment was an experiment in the consequences of Net-based retail marketing by new entrants: disintermediation. The conclusion they drew early on from Amazon was that there were sharp limits to the retail possibilities the tools provided. However, concluding that online companies and markets would transform commerce, the venture capitalists community made a whole array of largely unsuccessful bets. Similarly, the telecom collapse hinged on faulty or false notions of how data networks would be used. A most evident false notion was the asserted belief in the staggering and continuing expansion in the use of bandwidth to carry entertainment content. The image was often that the consumer net would become a sophisticated vehicle for centrally distributed content. However, the error is evident in the history of the American post office. The post office in the United States was established to distribute newspapers, but the killer application that supported the system was letters — peer-to-peer communication, to use today's vocabulary. Communication, not just voice but messaging and video meetings, and peer-to-peer exchanges are likely to be the killer applications.

By contrast, consider IBM's two fundamental shifts. IBM's first fundamental shift was from a product company that wrapped its products in high-value service support to a service company that sells solutions that embed its products. As IBM migrated from electromechanical to digital information processing, it established itself as the dominant player in the market. Consequently, its per-unit development costs were radically lower than its competitors', making its margins substantial. That allowed service to be bundled into costs, offering a sense of certainty and reliability to its customers. Its market share allowed it to keep its core software, operating systems and the like, closed and privileged. That model of competition was no longer viable as the era of the mainframe

and even of the minicomputer passed. Networks emerged to support business 43
services composed of multiple networks and varied suppliers. IBM began to
offer service solutions.

More generally, the IBM story points to the blurring distinction between
service and product in the digital era. The distinction between service and
product has never been very clear. Once, national accounts categories ob-
scured the relative importance of services and production in an evolving economy (Cohen and Zysman 1987). A window washer at Nokia or GM is a manufacturing employee; if Ace Window Washers contracts to outsource the
washing of Nokia's and GM windows, the same employees are counted in the
service sector. Now the blurred line between product and service becomes a
matter of strategic importance. Consider accounting. Accounting is a personal service provided by accountants utilizing tools from the original double-
entry bookkeeping system to computers. But if you create a digital accounting
program and put it on a CD, put it in a box, call it Quicken, and allow its unlimited use by the purchaser, then you have a product. If you put the program
on the Web for access with support for use on a fee basis, then you likely offer
a service as an ASP, or application service provider. Next, consider pharmaceuticals. If NextGenPharma sells a drug to be dispensed by a doctor or hos-
pital or sold in a pharmacy, it is producing a product. With gene mapping and
molecular analysis, we are moving toward the possibility of a service model of
therapies adapted to particular physiologies. If NextGenPharma really is a
database company with a store of detailed molecular-level drug information
and genome functionality, it could sell an online service to customize drugs
or therapy.

IBM's second fundamental shift was to support open-source software rather
than proprietary software and the development of frameworks and tools to
implement solutions within that framework. Microsoft and Unix provided
common platforms through which competitors could integrate their offerings,
limiting IBM's leverage. Selling solutions in a multi-vendor environment
suggested that a move away from closed proprietary systems might as well be
to one of hyper-openness in which a capacity to define solutions, provide an
integrated offering, and embed some distinctive proprietary modules would
be decisive in keeping customers tied to IBM.

Assume business strategies for capturing the evolving advantages of the digital era are experiments or bets whose success is uncertain, not investments
whose returns are predictable. Then the question is, of course, why some companies make better bets, or more effectively conduct the process of experimentation that must carry them into the future. Possibilities must be seen as
just that: hypotheses about the future that are to be continuously evaluated.

44 Certainly the dotcom-era bubble reflected greedy projections of assumptions that were rarely reassessed. In fear that the "moment" would pass by, images that were projections of possibilities were taken as solid facts. Each era, one must note, has its own uncertainties and its own risks. Entrepreneurs in each epoch confront those risks and transform the possibilities into profits and growth. What is distinctive about our era is the pervasive and continuous uncertainty, both technical, across an area of technologies, infrastructures, sectors, and products, and competitive, as new competitors arrive and unexpectedly dislocate market competition or established competitors reach out for the strategy that will overturn the character of industry competition.

TOWARD THE EXPERIMENTAL CORPORATION
AND THE EXPERIMENTAL ECONOMY

Let us consider in turn the implications for the corporation as a driver of the recombination of resources, for economic policy, and for the political dynamics on which economic policy rests.

The Experimental Corporation

The American venture capital firm is the quintessential experimental company. It makes a set of investments, creating a risk portfolio. It anticipates that some will simply fail, a few will do adequately, and a handful will be dramatic successes. Apart from the original analysis to determine the investments, much of the venture capitalist's task is how to judge progress, decide when to close an experiment, when to morph it into a related experiment, and how to capture value from the successful one. And although some venture-capitalist firms in some periods have done brilliantly, the venture phenomenon contributed to the faddish pursuit of the myth of disintermediation; recall the notion that clicks would replace bricks throughout the economy. Does this model, this manner of approaching investment, have general relevance?

Let us state the corporate problem. Companies trying to create value are constantly searching for the levers of advantage. The difficulty is that the optimum spots, differentiated assets of various sorts, are always moving about in a rapid and unpredictable fashion. We have noted that company internal functions become products, products become commodities, and the sources of differentiation for products and process are constantly evolving. The "global," as a set of national stories played out on a larger stage, is a constant source of new competitors, products, and processes. Since these innovations are often bred privately at home in diverse national settings before surging out onto world markets, there is a constant sense of surprise and of accelerating change; certainly, the more players there are, the more often there are radical changes. It is evident that what works today may not work tomorrow, and continuing radical change makes it is difficult to plan effectively.

Corporations must take an experimental stance toward their planning. That does not mean launching a whole new series of expensive half-baked "experiments." It is a matter of how to go about thinking about strategic planning, evaluating options and ongoing efforts as well as generating options in the first place. Certainly the traditional strategy efforts, devising a strategy after careful logical assessment and then purposefully implementing that strategy, are necessary, but they may be insufficient. Planning is often a matter of making the strongest business case possible for a particular frame of action.

By contrast, treating a strategy as an experiment changes the logic. The task then is not to prove a case to be true, which can often be self-confirming as one finds evidence for one's favorite point of view. The experimentalist is an organized skeptic who asks, "How would I know if I am wrong?" From a corporate standpoint we might rephrase this to ask not "How do I know if I am wrong?" but rather "What will convince me I am on the right path before I commit more resources to the experiment?"[12] Certainly, part of that effort is to challenge assumptions in the first place. But more is involved. As a project proceeds, one must look for the early warning signs, the indicators, that the original assumptions were wrong, or that the project will not unfold as envisioned. Or one must continuously look for evidence that might suggest the critical parameters on which the case rested have changed. Perhaps we must have continuously emergent, rather than planned, strategy. Company strategic choices must be considered experiments in the face of quite fundamental uncertainty, not as bets and gambles.[13] In an ever-evolving competition, each effort of oneself and one's competitors must be culled and systematically assessed for evidence to test the hypotheses on which a strategy rests. Generating innovative strategies and evaluating evidence on unfolding projects is not simply a matter of narrowly calculating returns on investments from defined premises. Rather, it is also a matter of creatively and imaginatively recombining what we know with what we imagine. That is critical both to generating new innovative approaches and to recognizing when a given approach may have gone radically wrong. There must be a constant interplay of frameworks and metaphors, as a fundamental unit of knowledge, to identify new possibilities (Nielsen and Nielsen, this volume). Lester and Piore (2004) write similarly about analysis and interpretation. Only interpretation and conversation amongst those with different interpretations allow the possible to be sorted out

46 from amidst the evident. Conversation within companies, amongst those with different frames and references, is the vehicle that crosses conceptual as well as organizational boundaries on the road to innovative projects.

In this chapter we need only to establish the principle that the notion of an experimental corporation is a practical reality. Firms can go beyond assuring the conversations, the walking-around knowledge required to systematically sustain innovation (Campbell and Faulkner 2003). The crucial point here is that if they are to deal with the profound uncertainties and unpredictable risks of the global digital era, firms must adopt an experimental stance.

State Action in the Experimental Economy

What does it take to create an experimental economy, an environment wherein firms alone and in networks or clusters experiment effectively and capture gains from that experimentation? First, talented, trained, educated, skilled people and centers of technology development and diffusion are simply necessary starting points. Promoting their combination in centers of creative imagination is evidently critical (Florida 2002). Assure an educated population and substantial development of science and technology or lose position in global markets. There is little controversy about the necessity, but often sharp disagreement about the method. Second, the infrastructure of the economy must be similarly assured. This is certainly a matter of the physical infrastructure of broadband lines, roads, and bridges that permit product to be generated and sold. But it is also a matter of the institutional infrastructure of the marketplace; there have to be marketplace rules that permit resources to be innovatively deployed and rewarded for successful implementation in the face of risk and imagination. Again, the necessity is not controversial, but the concrete mechanisms become the debate. Third, the lessons of diversified quality production and flexible specialization are that custom production and rapid turnaround, discussed above, suggest that tight geographical and organizational links between production and development are required. But are the same geographic linkages needed in a digital era, when inexpensive data transfers and communications render irrelevant part of the personal contact that geographic proximity permitted? What role can government play to make those connections work effectively in its jurisdiction?

The old and always controversial question is whether there are roles for the government in an innovation-centered experimental economy other than creating resources of people and technology and assuring the proper rules for experimentation and competition. The story of the build-out of the Internet in particular and digital infrastructure in general will provide abundant evidence for whatever ideological predilection you may have. In the United States, the

creation of the Internet was simultaneously the product of purposive intervention, government action by the Defense Department's Advanced Research
Projects Agency, and aggressive deregulation and reregulation. DARPA (the
original acronym was ARPA, Advanced Research Projects Agency) funded the
creation of the underlying conception and protocols of the Internet (Hafner
and Lyon 1998). But it was the aggressive introduction of competition into a private utility playing a public role, AT&T, under the label of deregulation of the
telephone system that unleashed user-led, consumer-based innovation. That
opened the way to user-generated networks and facilitated the radical and rapid
spread of Internet technology (Hafner and Lyon 1998). The European story
would likewise highlight these twin roles. Simplified, one part of the story is
deregulation of the telecommunications system, led by the European Commission. The Commission created national coalitions for Europe-wide rules
that would compel the transformation of state administrations responsible for
post and telegraph into regulated companies in at least a partly competitive
market.[14] The other side of the story is an array of directed state actions intended
to develop and diffuse digital technology. The development of the foundations
of the World Wide Web at CERN, the European Organization for Nuclear
Research, was dramatic. Choose your ideological bias, segment the story selectively, and there is evidence galore for either state intervention or market
competition.

So the relevant question remains: under what circumstances can government policy directed at particular objectives — be they technological, infrastructure, or industrial — promote a round of innovation and growth, and when
does the effort distort and misdirect trajectories of growth? Or, more precisely,
which mechanisms and policies are effective, and under what circumstances?

The Politics of Experimentation (or, The Politics of an Experimental Society)

It is the politics of growth that may be the most difficult. In textbooks, economic growth is the painless accumulation of compound gains from productivity increases and increased deployment of productive resources (savings).
All those productivity increases involve imaginative deployments, redeploy-
ments, and reorganizations. Certainly there need to be rewards as incentives
to risk and innovation, and a social capacity to make those adaptations; however, the adaptations represent not only new firms and new practices, but a shift
of resources out of some sectors into others, a movement of many peasants off
the land as they move to small towns and cities, factories closing, layoffs, and
displacement. The easy assumptions of painless or not-too-painful movement
of resources producing collective gain in the form of growth hide the reality

48   that there are real losses and real losers along the way. And the losers rarely volunteer for the role. Economic development always requires resolving a particular simultaneous equation. The technical equation is that how goods and services are produced and distributed must evolve flexibly. No investment and productivity gain, no growth. The political equation is that the allocation of gains and losses must be stably resolved or the fights over distribution will interfere with the technical processes. Economic development, to put the notion of growth back into the context of real communities, is a difficult, politically troubled, and sometimes even bloody process.

In response to the dislocations of the market, each advanced economy has created some system of social protection. It is not just a matter of demands from the Left. The surprise to some is that that great Prussian conservative Otto von Bismarck created an early system of social protection to limit the capacity of labor to organize politically. Sometimes it has been the Left that has demanded and insisted on state protection against the market. Sometimes the mechanisms of delivery have been in employment security provided by the firm.

The often-expressed concern is that social protection interferes with market adaptation, that growth slows without the flexibility to adapt. The notion is often that social protection mutes market signals, slowing or preventing adaptation. The counterpart fear is precisely that a rapidly adapting flexible economy must increase the number of losers or the costs that the losers bear, that imperative of experimentation, competitiveness, or adjustment will be claimed to justify reducing social protections. In fact we do not need to make a choice between establishing the flexibility needed to adapt to the evolving economy and sustaining the social protection against the vagaries of the market that makes the growth worthwhile?

The mechanisms of social protection, I would propose, can be the foundations of market flexibility. Of course, those who are displaced may be frightened and resist; but accepting the necessities of the broader economic adjustment is always easier if one can see the possibility of one's own place in that future. Apart from the obvious — investments in education, training, and technology — that we mention above, we need to reconcile social protection and flexibility. Making social protection a foundation for flexibility requires that all sides reconsider old debates. We need to separate out and consider the several dimensions of social protection.

A social protection system has at least four different dimensions:

- who is protected;
- the level and form of protection, which is not just a monetary amount but a matter of whether particular jobs or positions are supported;

- the mechanism of delivery—whether there is administered aid, for example, or cash grants; and
- the influence on the operations of adjustment in the economy.

The same level of protection for the same groups of people can be delivered in very different ways with very different consequences. And the obvious is not always the most important. The most politically difficult controversies are often about social identity. Often what is in dispute is not just economic well-being—the level of support—but the social place of particular groups and jobs in the economy that turns on the character and form of protection.

Let us glance for a moment at the influence of social protection on the operations of labor markets and financial markets. It is conventional to assume that labor market flexibility means stripping job protection and that social protection means rigidity. Britain and the United States are the model of that argument; they are taken to have extensive labor market flexibility and lower social protections. They constitute one model of how to achieve labor market flexibility. Germany, France, and Japan would be considered examples of social protection interfering with labor market operations. Consider Japan. Social protection is often embedded in private employment structures. One consequence is that firm failure is quite expensive socially, often leading to continued bank financing to prop up troubled companies. Cumulatively, that has contributed to the financial troubles and rigidity of the Japanese economy over time. Flexibility requires unwinding the company-finance- social protection nexus (J. Levy, Miura, and Park 2006). Consider France. Apart from the formal system of government-financed social protections, the French economy abounds with an array of "acquired rights," situations that embed privileges, from taxi licenses through café licenses to protection of job locations. Social protection is embedded in the defense of particular social and employment arrangements (Cahuc and Kamartz 2004). Now consider Denmark. The Nordic tradition of social protection as a part of citizenship rights prevails (Esping-Anderson 1990). The broad social foundation of social protections contributed to a political deal that makes firing easy and labor market flexibility simple. And easy firing means easy hiring.[15] Clearly there is more than one road to achieving economic flexibility.

The conclusion is simply that social protection and labor market flexibility are not alternatives. The task is to reconsider and reconfigure the packages of social protections so they support experimentation and adjustment. Conservatives must consider that a truly secure community may in fact be the base of a flexible economy. The Left must recognize that social protections can be reconfigured without reducing actual protection. My conclusion here is that an experimental economy will itself require imaginative policy and politics.

*

Over the last century, there has been a series of production eras, each with a corresponding logic. And just as the success of the Fordist system, for example, required Keynesian policy buffers to offset systemic political and production rigidity, the digital era poses a new set of political and production challenges. In sum, the endemic uncertainties and risks of the global digital era require corporations that are prepared to experiment, an experimental economy that can sustain and facilitate that experimentation, and a politics of growth that makes flexibility and adaptation socially acceptable and politically possible.

NOTES

1. Japan's automobile and electronics firms burst onto world markets in the 1970s and consolidated into powerful conglomerates in the 1980s. The innovators were the core auto and electronics firms, which, in a hierarchical manner, dominated tiers of suppliers and subsystem assemblers; the production innovation was the orchestration and reorganization of the assembly and component development process. The core Japanese assembly companies of the lean variety have been less vertically integrated than their American counterparts. Rather, they have been at the center of vertical *keiretsus* — loosely speaking, Japanese conglomerates conventionally understood to be headed by a major bank or consisting of companies with a common supply chain linking wholesalers and retailers, that have tightly linked the supplier companies to their clients (Tyson and Zysman 1989).

2. The argument is simple. The relationships of production and development in these production systems are, at best, delicate. Just-in-time delivery, subcontractor cost and quality responsibility, and joint component development push considerable risk onto the subcontractor in the case of demand fluctuations. The high growth rates — combined with the need to re-equip Japan in the postwar years — created the basis of the continuous expansion. But domestic growth did fluctuate, and the rivalries for market share led consistently to overinvestment, or excess capacity, in the Japanese market. The story about Japan told by Yamamura (1982), Tyson and Zysman (1989), and by Tate (1995) in the case of the auto industry, shows that the excess capacity was dumped off onto export markets. Seen differently, these exports permitted a steady and smooth expansion without which the production innovations outlined here would not have emerged.

3. By *vertical control* I mean both vertical integration from inputs through assembly to distribution, as in the case of American auto producers, and the "virtual" integration of Asian enterprise groups, as when Japanese producers of consumer durables effectively dominate market relations with semi-independent suppliers through the *keiretsu* group structure. See Aoki 1988; Gerlach 1992; Aoki and Dore 1994.

4. More or less at that same moment, products that were thought to spin off from technology investment in military goods into civilian products seemed less significant. Instead of talking about spin-off technologies, technologies that had their birth in the defense sector and were spun off to commercial applications, talk turned to spin-on technologies. Leading-edge civilian technologies contained more advanced

technologies and components than their military counterparts. Technologies began to spin off from the civilian sector to military applications. See Sandholtz et al. 1992; Stowsky 2003.

5. According to the Department of Commerce Bureau of Economic Analysis, in 1998 U.S. trade in IT was $314 billion. The total volume of American trade — imports and exports — in information technology is now doubling in less than seven years. DeLong and Summers (2002) offer the following statistics: in 1950 there were 2,000 computers in the United States. By 2002, there were 300 million computers. That is a 4-billion-fold increase in raw automated computation power, an average annual rate of growth of 56 percent.

6. There are traditional lists of leading sectors, or clusters of technological innovations, over the past two centuries. They include in some format: (1) the industrial revolution and the Arkwright mill; (2) the age of steam and railways; (3) the era of steel, electricity, and heavy engineering; (4) the automobile era of mass production; and now (5) the era of information and telecommunications. See, for example, Perez 2003.

7. The use and application of transformative technologies alters the array of activities in the economy as a whole. The diffusion of those transformative technologies is undoubtedly the critical step. It is not just the fortunes made as the leading sector expands, but the industrial development transformative technologies engender. Notably, as DeLong has argued, in the nineteenth century the several railroad bubbles brought down the price of transport and in the process, by extending the geographic size of markets, generated such innovations as mail-order retailing. Thus, ironically, the dotcom and telecommunications collapse over the last few years may, in historical perspective, prove to have accelerated use and diffusion. The collapse of major telecom carriers as a result of overbuilding of telecom networks has brought a precipitant drop in the price of network use.

8. Other examples would be hip surgery or semiconductor ovens that require temperature controls to within 1 degree C at roughly 2,000 degrees.

9. Thanks to Stuart Feldman of IBM for his presentation at the "Innovation Alliance: Succeeding in an Evolving Global Economy" conference, Berkeley Roundtable on the International Economy, Berkeley, August 27, 2004.

10. "Manufacture: To make or process (a raw material) into a finished product, especially by means of a large-scale industrial operation. To make or process (a product), especially with the use of industrial machines. To create, produce, or turn out in a mechanical manner. To concoct or invent; fabricate. To make or process goods, especially in large quantities and by means of industrial machines" (*The American Heritage Dictionary of the English Language*, 4th ed. [Houghton Mifflin, 2000]).

11. The critical question, once we acknowledge that software production is a form of manufacturing, is, what are the most effective ways of organizing software production? For this discussion, the list begins with the conventional questions of whether to outsource and of where, geographically, to locate software development. The story becomes interesting when we ask whether we should choose conventional hierarchical production structures typified by Microsoft or new alternatives such as the commercialization of Linux products developed in an open-source model.

12. Thanks to John Stopford for help with this point and this section.

13. Note that this argument is consistent with and now draws on the framing argument of Gunnar Eliasson. It was a considerable relief when Pekka Ylä-Anttila pointed out this chapter made an argument similar in language and concept to that Eliasson had innovated years earlier.

14. This comment is based on research interviews conducted by the author and Peter Cowhey.

15. The situation is well summarized in Thomas Fuller's fascinating article "Denmark: A European Welfare State par Excellence": "It comes as a surprise to many outsiders that Denmark is actually one of the easiest places to get fired in Europe. A construction worker here can be fired with as little as three days' notice. Salaried employees get a longer notice period — three to six months, depending on seniority — but do not expect severance pay, which generally does not exist. . . . 'Protection against dismissal has never been a major issue,' said Einar Edelberg, deputy permanent secretary in the Danish Ministry of Employment. 'It's easy to fire — and accordingly, it's easy to hire. . . .' 'The Danish system creates a flexible labor market,' the Danish Confederation of Trade Unions said in an official document. 'Danish companies are more willing to hire new employees in times of economic revival than their European competitors, who have trouble letting off workers when the economy goes downhill again.'" Note that the source of this last comment is the country's largest labor union confederation, a sign of the consensus surrounding the easy-to-fire policy. Changing jobs has become part of Danish work culture. About one-quarter of the Danish work force switches employers every year, a churning labor market that constantly creates new openings. The bottom line for Denmark is an unemployment rate that, at 5.3 percent, is well below the 8.9 percent average for the European Union and that of the Continent's economic heavyweights, France (9.5 percent) and Germany (9.9 percent). See Thomas Fuller, "The Workplace: Firing's Easy in Denmark; So Is Hiring," *International Herald Tribune*, December 15, 2004.

I.   NATIONAL STORIES AND GLOBAL MARKETS
      IN THE DIGITAL ERA

3

FINLAND'S EMERGENCE AS A GLOBAL
INFORMATION AND COMMUNICATIONS
TECHNOLOGY PLAYER

Lessons from the Finnish Wireless Cluster

Ari Hyytinen, Laura Paija, Petri Rouvinen,
and Pekka Ylä-Anttila

In 1990, Finland was hit by the most severe economic crisis in any country
in the Organisation for Economic Co-operation and Development (OECD)
since World War II. Real GDP dropped by over 10 percent in just three years,
and unemployment had risen to nearly 20 percent by 1994 (see, e.g., Kiander
and Vartia 1996; Honkapohja and Koskela 1999). By the end of the 1990s, however, Finland had emerged as a global information and communications technology (ICT) player. Its corporate icon, Nokia, was Europe's most valuable
company, and technology enthusiasts expected Linux, an open-source operating system with Finnish roots, to replace Windows on virtual desktops. With
its GDP increasing at an annual rate of approximately 5 percent, Finland was
also one of the fastest-growing countries in the world.

Finland's remarkable recovery is in considerable part attributable to developments in the ICT sector. The Finnish ICT sector accounts for approximately 10 percent of Finland's GDP, and Nokia's contribution alone to growth
averaged more than half a percentage point. How did this smallish country become a success story in ICT and in wireless technologies in particular? What
were the drivers of the economy's emergence as a global ICT player? Does the
Finnish experience hold lessons for other countries?

This chapter argues and documents that there was no systematic plan to restructure the Finnish economy or to build a globally competitive ICT sector.
Rather, an array of processes and policy measures were simultaneously at
work. The country itself provided particularly fertile framework conditions,
not least because of the high level of education of its labor force. It had also
accumulated a great deal of ICT-related expertise by the early 1990s. Owing
to unfavorable macroeconomic shocks, it had numerous underemployed

56 resources, and digitalization presented a technological opportunity. Furthermore, the country had early exposure to two successive generations of winning
technological standards and faced increasing competition and pressure for liberalization on its way to EU membership. On top of this, it had a company
with the vision and a strategy to make it happen.

To track how this transformation came about, we first introduce the foundations of Finland's economy in the next section. The emergence of the Finnish ICT cluster is then described. Next we consider explanations for this
emergence and single out some important drivers. We then highlight contemporary challenges facing the Finnish wireless cluster. The final section
offers some concluding remarks.

FOUNDATIONS FOR AN INNOVATION-DRIVEN ECONOMY

Finland did not emerge as a global ICT player in a vacuum. Several important preconditions made the transformation of the early 1990s possible.

A Small but Open Nordic Welfare State

Entering a phase of innovation-driven development presumes the interplay of
several factors. Preconditions including high social cohesion, a consistent and
predictable policy environment, and a sound basic infrastructure are often
emphasized. Most of these preconditions were in place before Finland's emergence as a global ICT player. Finland's relatively stable and transparent economic and social institutions, like those of other Nordic countries, have existed for centuries. It can be appropriately characterized as an open but small
Nordic welfare state: an egalitarian country with relatively even income distribution and small class distinctions.

During the twentieth century, Finnish GDP per capita grew at an annual
rate of close to 3 percent, among the fastest in Europe. Although the starting
point was relatively low compared to countries in the vanguard of the Industrial Revolution in the late 1800s, many of the basic preconditions for growth
were already present. Institutions such as well-functioning education, transportation, and banking systems were not only important in the take-off phase,
but also later when the economy moved from factor- to investment- and, later,
innovation-driven stages of industrial development. After completing the liberalization of both internal and external trade during the end of the 1870s, the
trailhead for industrial growth and new business activity had been found.

Finland's most important — and virtually its only — endowment of natural
resources, forests, proved decisive in the take-off phase — first through timber

and later pulp and paper. From the late 1950s to the late 1970s, the Finnish forest industry carried out massive investments and gradually transformed itself into an international technology leader with the most modern and efficient production capacity in the world (Raumolin 1992). By the late 1980s, the forest sector had developed into a globally competitive industrial cluster that continues to provide high-value-added paper grades, forestry technologies, and consulting services (Hernesniemi, Lammi, and Ylä-Anttila 1996; Ojainmaa 1994; Rouvinen and Ylä-Anttila 1999).

The small home market encouraged Finnish firms to specialize and seek foreign markets from the outset. It is therefore not surprising that exports have long been integral to the growth of the Finnish economy and that the export sector is relatively large. Today, the share of exports in GDP is approximately 40 percent.

The Structural Transformation of the Early 1990s

Between 1990 and 1993, Finland experienced an unprecedented economic crisis, spurred by a number of coincidental and structural factors. These included a downturn in the nationally vital forest-related industries, disruption of the country's sizable eastern trade owing to the collapse of the Soviet Union, a speculative bubble in the domestic securities and real estate markets fueled by uncontrolled credit expansion and favorable terms of trade, and mismanaged financial liberalization, which eventually led to a credit crunch and excessive private sector indebtedness (Kiander and Vartia 1996). In addition to these immediate reasons for the recession, rigidities in economic and political systems and neo-corporatist structures contributed to an economic dead end.

This deep recession initiated a structural transformation. Widespread corporate restructuring and bankruptcies modified the economy's industrial structure (Figure 3.1), and the banking crisis that coincided with the economic recession triggered a fundamental reorganization of the Finnish banking system (see, e.g., Hyytinen, Kuosa, and Takalo 2003).

The change in Finnish industrial structure and exports during the 1990s was unique both nationally and internationally. In less than a decade, electronics became the most important single branch in production and exports. The Finnish industrial structure shifted from raw material-, capital-, energy-, and scale-intensive to knowledge-intensive production.

The recession also led to a clear shift in policy thinking. Greater emphasis was placed on long-term microeconomic as opposed to short-term macroeconomic policies in an acknowledgment that the foundations of sustained

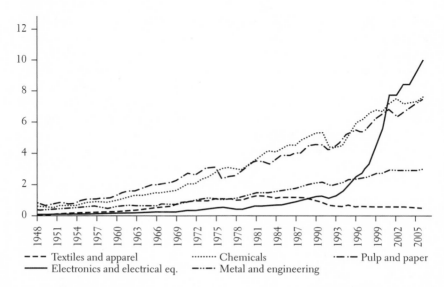

FIGURE 3.1   Finnish manufacturing production volume by industry (€ billions in 2000 prices)

Sources: ETLA database, Hjerppe et al. (1976), National industrial statistics by Statistics Finland

national competitiveness are largely created at the micro level—in firms, financial institutions, and policy agencies. The setting for industrial policy design was characterized by intense and informal communication between government, industry, academia, and labor market actors.[1] Its objective was to build and enhance businesses' operating and framework conditions and not to provide heavy subsidies or other direct government support. To advance this goal, the government assumed the role of a facilitator and coordinator. A number of forums to promote interaction between different interest groups had already been founded by the 1980s, including the prominent Science and Technology Policy Council, chaired by the prime minister. Industrial associations, acting as influential intermediaries between industry and the public sector, also played a salient role in the Finnish policy arena.

The recession of the early 1990s was a watershed between the investment- and innovation-driven stages of national development. A prime example of this change was how public R&D funding expanded in the midst of the recession: it rose at a time when virtually all other public expenditures were cut. The intensity of the country's R&D activity grew rapidly as the business sector increased expenditures on innovative activity. Today, Finland's relative R&D intensity, the share of the gross domestic research and development

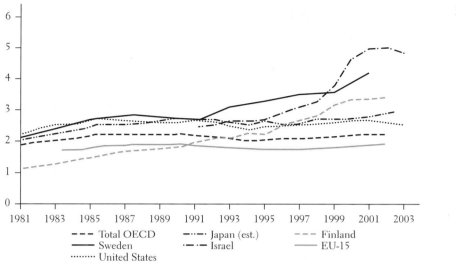

FIGURE 3.2   Ratio of gross domestic expenditure on R&D to GDP
Sources: OECD Main Science and Technology Indicators Vol 2004 release 02

expenditure (GERD) of GDP, is among the highest in the world (3.4 percent in 2003); see Figure 3.2.

The European integration process further fueled the policy shift initiated by the economic crisis of the early 1990s. Finland joined the European Union in 1995 and, unlike other Scandinavian countries, adopted the euro from the outset. For Finland, integration meant that the scope for national macroeconomic policies was considerably reduced.

Knowledge-Driven Growth

In addition to investment in R&D, a key factor in Finland's change to an innovation-driven economy was a strong commitment to education (Figure 3.3). Owing to increased investments in the education system, by the late 1980s younger generations of Finns were among the most educated in the world. Education that would enhance technological change was prioritized in the policies of the 1960s and 1970s. Among the OECD countries, the Finnish educational system lags behind only the Korean and German systems in terms of its relative emphasis on the natural sciences and engineering.

During the most intensive ICT-driven growth period the supply of skilled labor ran short of the industry's requirements. The government reacted by

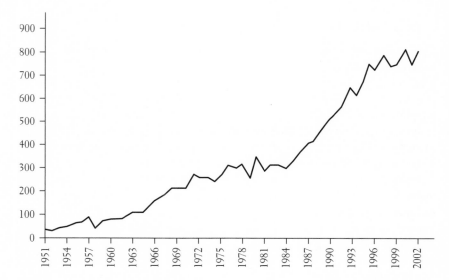

FIGURE 3.3   Postgraduate degrees in natural sciences and engineering in Finland
Source: Ministry of Education, KOTA OnLine (www.csc.fi/kota/)

increasing the number of openings in institutions of higher education. Between 1993 and 1998 the total intake in universities nearly doubled, and in polytechnics it nearly tripled. In early 1998, the government adopted a program aimed at further increasing ICT education between 1998 and 2002.

Educational resources allocated to science universities did not match the sizable growth in enrollment, owing to the retrenchment in public spending, yet the tertiary level of the education system suffered relatively little from the government's belt-tightening. Its share of total expenditure on education exceeded the OECD average even during the deepest recession (Virtanen 2002). During the same period, the budgeting system for higher education became more demanding. It was shifted toward performance measures, and research funding was based on competition for both public appropriations and external (i.e., private) funds.

THE EMERGENCE OF THE FINNISH ICT CLUSTER

In ICT, laggards rarely catch up with, let alone leapfrog over, the leaders. Originally ICT-specialized countries tend to become more so. Finland is an exception to this rule. During the 1990s, it went from being one of the least ICT-specialized industrialized countries to the single most specialized.[2] In this section, we describe some important facets of this transformation.

Origins of the Telecommunications Sector

The origins of the Finnish ICT sector and its particular features can be traced back to the Telephony Decree of the Finnish Senate in 1886, which distributed numerous private operator licenses in order to circumvent Russian telegraph regulations. Upon gaining independence in 1917, an additional public telephony operator (PTO) and regulator was established to operate the telegraph and military telephone network left behind by the Russians.[3] In the 1930s there were hundreds of private telecommunications operators in Finland. Even today, there are some forty significant operators.

From the outset, Finnish telecommunications equipment markets were open to foreign suppliers. Unlike in countries with an equipment manufacturing monopoly, such as France, Germany, and Sweden, there was no public interest in protecting domestic supply in Finland. Though they gradually caught up with foreign suppliers in technical know-how, domestic equipment manufacturing could not yet meet local demand. Independent operators were free to choose among different suppliers and thereby put small local manufacturers under competitive pressure.

Private operators' interest in state-of-the-art technology was fueled by the threat of being taken over by the PTO regulator for underperformance. The tension between the public and private camps, originating in the early days of telephony, was hence an important source of industry dynamics. This long tradition of a dual market structure provided a basis for balanced competition.

Anticipating worldwide deregulation and the liberalization efforts of the EU in particular, the development of the Finnish market culminated in the early deregulation of the telecom market at the turn of the 1990s. When deregulation began, Finland already had relatively competitive and diverse telecommunications operators and equipment markets. The process was therefore relatively smooth; extensive government intervention to induce equitable competition between the incumbent(s) and the entrants was not needed.

Formation of the ICT Cluster

Finnish telecommunications equipment manufacturing was initiated around 1920 in three separate organizations that were gradually merged under the management of one company, Nokia, by 1987. The original companies had somewhat different focuses at the outset: Salora concentrated on the resale of radio and television sets, Suomen Kaapelitehdas focused on cables and electricity production, and the radio laboratory under the Ministry of Defense, later named Televa, started with military radio systems.

62   In 1963, a call by the Finnish army for tenders for a battlefield radio spurred companies to give physical expression to their accumulated expertise. Ultimately the army did not have the resources to purchase the system, but the prototypes served as the forerunners of commercial portable phones. Other state agencies with demanding communications requirements such as the telecommunications administrator, the railways, and the coast guard also had a major influence on companies' product development efforts.[4]

The Auto Radio Puhelin (ARP [Car Radio Phone]) network was introduced in 1971 as the country's first mobile telephone network with nationwide service. It provided good geographical coverage but was not technologically sophisticated. In the mid-1970s, the service had some ten thousand subscribers, and Finnish radiophone manufacturing gained a major market share across the Nordic countries. Although ARP did not turn mobile communications into a major business, it provided experience and customer interfaces for companies such as Nokia, Salora, and Televa. It also indicated that there was commercial potential in mobile services.

In network systems, development was intense but yielded little sales revenue. Other electronics applications, such as TV sets, computers, and industrial process control systems, dominated commercial electronics until the late 1980s. The adoption of semiconductor techniques in the 1960s served as the basis for electronics development. Coupled with pioneering product development in digital transmission and digital signal processing, it produced know-how that proved pivotal for later success in digital telecommunications.

The development of the analog Nordisk Mobil Telefon (NMT [Nordic Mobile Telephone]) standard in the 1970s was a highly valuable outcome of the traditional cooperation of Nordic telecommunications administrators and industry. It aimed at creating a Nordic market for mobile telephony and inducing competition. The standard development project was open to third-country suppliers as well. Openness promoted competition in network equipment and handsets. Advanced features such as roaming were included, and, fortunately, the diffusion promoting "caller pays" practice was also adopted.

In the early 1980s, the Nordic countries formed the largest mobile communication market worldwide in terms of the number of subscribers (Table 3.1). Mobira, a joint venture of Nokia and Salora, supplied the first NMT portable phones. In contrast, Finnish companies were neither ready nor willing to supply network technology at the starting phase of the NMT project. Eventually, under pressure from the PTO, which wanted to curb the market power of the Swedish firm Ericsson and equipment prices in general, Mobira and later Telenokia started to manufacture network equipment (Palmberg 2002).

TABLE 3.1   Market shares in NMT
handsets in 1985 (total 83,525 units)

Mobira (Finland)	25.7%
Ericsson (Sweden)	16.9%
Panasonic (Japan)	8.9%
Storno (U.S. until 1977 Denmark)	7.1%
Dancall (Denmark)	6.5%
Mitsubishi (Japan)	6.1%
NEC (Japan)	6.0%
Siemens (Germany)	5.6%
Motorola (U.S.)	5.6%
Simonsen (Norway)	2.3%

Source: Nokia Mobile Phones (as cited in Häikiö 2001c, 134; countries of origin added by the authors).

In 1988, the telecommunication authorities of the European Community published the Groupe Spécial Mobile (GSM [Digital Global System for Mobile Communication]) standard. The technological challenges of GSM related to the digitalization of radio transmissions and the exponential increase in the complexity of the signaling and control software. These were fields in which Nokia had accumulated competencies with its customers in the advanced banking sector, know-how that served as a ticket of admission to the standard development project (Palmberg and Martikainen 2005).

At the same time, Nokia reorganized its telecom divisions to cater to the envisioned GSM-based growth in cellular systems and to contribute to the goal of inaugurating GSM Service in Europe in 1991. The tight deadline was met in Finland with the world's first GSM call in June 1991, even though the pan-European inauguration of the service was delayed because of technical problems. Nokia and Ericsson were among the first to adopt GSM, which eventually became almost universally accepted (see Koski, this volume).

On the operator side, the PTO had a monopoly over the NMT service. Owing to the lucrative nature of the mobile market, the private camp applied for a second license without success. In 1988, it decided, on the basis of a regulatory loophole, to construct a mobile network without a national license. The private venture chose the newly developed GSM standard, which was not yet in commercial use anywhere in the world. In 1990, a license was finally granted to a newly established company, Radiolinja, after an intense political debate on the viability of parallel networks in a small country. The digital mobile service was commercialized the following year by Nokia, which thus

64 made its global GSM premiere with Radiolinja's network. The PTO followed suit in a partnership with Ericsson.

The mobile market entrant started to erode mobile service pricing, and soon Finnish mobile services were the least expensive in the world. Together with more affordable portable phones gradually replacing common car phones, mobile communications were adopted by the masses.

Although the foundations of domestic equipment manufacturing were laid in the 1920s, foreign manufacturers dominated the market until the 1980s. During the 1970s and 1980s, Finland advanced rapidly in digital and mobile technologies. Nokia participated in these developments and became a central force in the consolidation of the industry since the 1970s. By the late 1980s, a fair part of the Finnish telecommunications equipment industry had merged into Nokia.

Nokia's Transformation

The merger of Nokia (originally a wood pulp mill), Suomen Kaapelitehdas (Finnish Cable Works), and Suomen Gummitehdas (Finnish Rubber Works) in 1966–1967 may be seen as the birth of the current Nokia Corporation (see Paija 2001). Although the forest-based company lent the name, the cable company (Suomen Kaapelitehdas) provided the core knowledge base for the new entity. Nokia pursued a conglomerate strategy early on and continued to do so in the 1980s, when it made several sizable acquisitions in consumer electronics (i.e., televisions such as the Swedish Luxor in 1984 and the German Standard Elektrik Lorenz in 1987), information systems (e.g., the Swedish Ericsson Information Systems in 1988), and other fields not directly related to telecommunications. It is telling that in 1986, Nokia had ten divisions, forty-five business units, and 180 lines of business, and that at about the same time, Nokia was the biggest manufacturer of personal computers and color TV sets in the Nordic countries and was among the top 10 in Europe. Thus, the birth of the current Nokia Corporation and the conglomerate strategy it followed explains, at least in part, Nokia's role in merging the various businesses of Salora, Suomen Kaapelitehdas, and the radio laboratory of the Ministry of Defense by 1987.

The conglomerate strategy, however, was not a success story. Together with managerial and ownership problems and the early 1990s recession, it led to a deep crisis. The company almost went bankrupt in the early 1990s, primarily because of its overly ambitious and costly acquisition and internationalization strategy. After 1992, the company changed course (see Figure 3.4), and activities outside mobile communications were divested under the leadership of a new CEO, Jorma Ollila. The process was completed by the late 1990s.

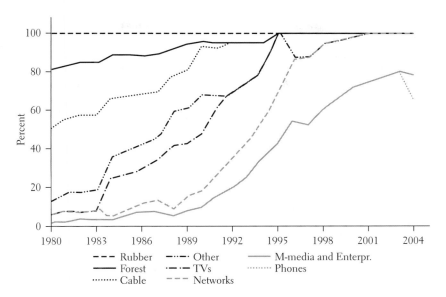

FIGURE 3.4   Nokia's sales by industry

M-media & Enterpr. refers to the combined share of two new business units — Multimedia and Enterprise Solutions — formed at the beginning of 2004

Sources: Derived by the authors from an earlier version of Paija (2001), with additional information from Häikiö (2001) and Nokia's annual reports

With the exception of UK-based Technophone, Europe's second-largest mobile phone manufacturer at the time, Nokia has not made major foreign acquisitions in communications. In fact, it retreated from its acquisition strategy almost completely after the early 1990s. But alliances were important from early on. In handsets, Nokia established joint ventures with the American firm Tandy and did private labeling with Tandy-owned Radio Shack, AT&T, and others.[5] On the network side, it initially partnered with Alcatel (France) and AEG (Germany) to provide GSM solutions.

With about 30 percent of the market, Nokia is the leader in mobile handsets and also one of the dominant players in mobile network infrastructure equipment. It has been riding the wave of exploding global mobile telecommunication markets, fueled by worldwide deregulation in telecommunications. Thanks to its narrowly defined and globally oriented strategy, it has been able to meet the market challenge somewhat better than its closest competitors. Furthermore, management has been able to build an innovation-driven culture and supportive organizational structure, flexibly exploiting both internal and

66   external networking — yet retaining, in contrast to its main competitors, most of its manufacturing in house.

In the 1990s, Nokia's challenge was to manage rapid organic growth, which was clearly aided by its agility and lack of bureaucracy. Although the company has Finnish roots and its executive board is populated mainly by Finns, Nokia's orientation has been distinctively global. While Nokia has faced its share of challenges, what seems to set Nokia apart from other giant corporations is its ability to react quickly and improvise in moments of crisis.

Today, Nokia accounts for some two-thirds of the total turnover of the domestic and foreign ICT companies operating in Finland. Nokia is responsible for one-fifth of Finland's total exports and around 3–4 percent of its GDP, making it the biggest company in the country. Its role is even more important in strategically important activities such as R&D and the internationalization of business. Nokia accounts for 45 percent of total business sector R&D and a third of total national R&D. Hence, as a performer of R&D, Nokia is bigger than the entire Finnish university sector. More than 60 percent of Nokia's R&D (3.8 billion euro in 2003) is conducted in Finland. Nokia employs about twenty thousand people in Finland, of whom more than half are in R&D.

Smaller Players in the Finnish ICT Cluster

Although Nokia's central role cannot be overemphasized, it is not the sole driver of the emergence of the Finnish ICT cluster. The Finnish ICT sector — extending from digital content provision to network infrastructure equipment manufacturing — is made up of six thousand firms (Paija and Rouvinen 2003), including three hundred first-tier subcontractors of Nokia (Ali-Yrkkö and Hermans 2004). Some of these smaller Finnish ICT companies are gaining ground in global markets, and ICT has indeed become the country's third industrial pillar, thus diversifying Finland away from the traditional metal and engineering as well as forest-based sectors. In the 1990s, ICT's share of GDP rose from 4 to 10 percent. Also, firms such as Ericsson, Fujitsu, IBM, Hewlett-Packard, and Siemens have established R&D units in Finland, which has been interpreted as a sign of the viability of the Finnish ICT cluster.

Although there is more to the Finnish ICT cluster than one highly successful company, there have been no major commercial breakthroughs in broader ICT market segments. Given the substance and sophistication of Nokia's computer-related activities,[6] coupled with the fact that Finnish universities have been the hatcheries of several pathbreaking Internet-based inventions, the weak commercial success of the Finnish ICT sector outside wireless communications is somewhat disappointing. It has also been argued that ethical and social rather than commercial motives have characterized some of the best-known Finnish innovations in the Internet domain.[7]

DRIVERS OF THE TRANSFORMATION

There is no well-defined and unambiguous set of factors that has driven the emergence of the Finnish ICT cluster. Identifying such drivers is especially problematic in this case, for many processes were embedded in global developments. Nonetheless, some interesting partial correlations can be identified.

Creative Destruction

An economy's historical developments and macroeconomic environment provide the general conditions for its economic development and microeconomic restructuring. For Finland, the recession of the early 1990s provided a decisive break from the past. This event fostered a pragmatic and straightforward culture in both politics and business. Faced with the deep recession, the Finns simply could not afford inflexibility or bureaucracy. Thanks to its stable political environment and social cohesion, political institutions remained functional, and thus the necessary policy adjustments were made even during the crisis (Ornston and Rehn, this volume).

Vast unemployment in the 1990s gave the emerging ICT cluster a large recruitment pool necessary for its expansion. The public educational system also responded to the needs of the ICT cluster. Finally, the collapse of eastern trade relaxed resource constraints within firms, which could then be targeted to the development of ICT (including GSM).

Liberalization and Reduced Financial Constraints

The liberalization of global markets for goods, services, capital, and technology, initiated by developments in the United Kingdom and the United States in the late 1970s, led to a global economic boom in the mid-1980s. Together with Europe-wide liberalization, globalization gave Finnish companies access to new markets and exposed them to increasing international competition.

Lack of capital and inefficient capital investments have been the Achilles' heel of the Finnish economy. It was therefore instrumental to the revival of the economy and the emergence of new clusters that the liberalization of capital markets relaxed capital constraints and enhanced the efficiency of the allocation of capital. After deregulation and participation in the European

68   Economic Area (EEA) and EU, Finnish companies, especially larger ones, gained direct access to foreign investors. There was a huge influx of capital to Finland in the mid-1990s, and for a couple of years the Helsinki stock exchange was the most internationalized in the world as measured by market value owned by foreigners.

The Finnish financial system moved away from a bank-centered system, when the stock market grew and intermediated debt finance contracted in the wake of the banking crisis of the early 1990s. Changes in creditor and shareholder protection also contributed to this structural transformation of the Finnish financial system (Hyytinen, Kuosa, and Takalo 2003). This development was even more visible in Finland than in other Nordic countries where similar structural changes were also taking place.

Contrary to the Israeli case, smaller Finnish companies have not used NASDAQ or other foreign stock markets as an important source of capital. They have nevertheless benefited from the rapidly increasing availability of venture-capital financing since the early 1990s. The enhanced availability of venture capital has been especially important for small Finnish ICT firms, which need to maintain conservative leverage ratios (Hyytinen and Pajarinen 2005).

Technological Opportunity

Digitalization was a major technological breakthrough in voice and data storage, processing, and transmission. It was important for Finland, because it provided an opportunity for new players with no experience or vested interests in computing or communication. Finland had sufficient expertise in digital technologies at large and in telecommunications in particular, both of which were absolutely vital for the big GSM breakthrough.

Critical Standards

Telecommunications standardization in the Nordic and European contexts may be the single biggest explanatory factor behind Finnish ICT success. Finland was an early adopter of NMT and then GSM, both of which eventually proved to be winning technologies in their eras.

Early on, NMT provided critical mass and relatively high penetration rates, which led to rapid recovery of development costs and accumulation of hands-on knowledge and economies of scale. Firms also accumulated other production and consumption-based network benefits. Upon the transition to digital technologies, Nokia bet heavily on GSM as the second-generation (2G)

standard, which eventually commanded three-quarters of the worldwide user
base. Nokia managed to capitalize on its early lead in both GSM networks and handsets.

The fact that mobile telecommunications standards were agreed upon beforehand rather than being completely or partly determined by market forces clearly aided entrants and market creation. The settlement on these and subsequent standards was in part based on demonstrations where the benefits of a given technological solution could be shown. Nokia was quite successful in these open competitions and thus able to contribute to the formation of these standards.

Advanced Users and Sophisticated Local Demand

Unlike many of their international competitors, Finnish ICT firms had ample experience operating in a competitive environment with diverse customer needs. Besides having a history of telecommunications competition that dated back over a hundred years, Finland was also some three years ahead of other industrialized countries in taking the final steps toward completely deregulating communications markets.

In mobile telecommunications in particular, deregulation brought about eager second-tier operators and service providers that wanted to deploy new, technically advanced networks rapidly. Former monopoly operators were forced to respond by upgrading their networks. This actual and potential competition lowered prices and fueled demand, which in turn led to further investments. Thus, the industry was in a virtuous demand cycle in the 1990s.

Scandinavians seem to be accustomed and therefore quite willing to test new technologies. In the early years of mobile telecommunications, new generations of phones always caused quite a stir and in a sense forced many users to shop for an upgrade. Fortunately, customer needs in these markets preceded those elsewhere, thus giving something of a first-mover advantage to Scandinavian firms. ICT penetration rates have been high, and Finland and Sweden are, or at least were, leaders in certain types of ICT usage, such as online banking and mobile payments. Thus, the early Scandinavian market was a rather happy marriage of technological competence in both production and use.

In handsets, Nokia was among the first to offer curvy "pocket-fitting" designs with integrated antennas, screens with sufficient contrast and size for comfortable reading, end-user customization such as exchangeable covers, and downloadable ring tones and logos, as well as entertainment such as off- and online games. Though all these seem obvious now, it took a surprisingly long time for them to become part of the standard setup.

70   More recently, the strong Finnish forest cluster with roots in traditional factor-driven industries has found interfaces with the knowledge-driven ICT cluster, as evidenced by the integration of ICT into pulp and paper-making processes and maintenance services. Furthermore, global consolidation in pulp and paper, as well as in other traditional industries, has spawned new ICT markets in integrating geographically dispersed activities.

Cooperation and Visionary Management

Competition brings about efficient and lean organization. Somewhat paradoxically, cooperation has been equally important for Finnish ICT success. Indeed, international comparisons (European Union [EU] 2000; OECD 1999c) suggest that intense interorganizational cooperation is an essential feature of the Finnish innovation system (Palmberg and Martikainen, this volume).

As shown above, a diverse set—but not all—of Finnish communications expertise was eventually merged into Nokia. Nokia took two important steps that turned out to be far-reaching. The first one was that Nokia was (despite its roots) able to give up its forest-related activities and realized quite early that the Soviet trade was best treated as a cash cow to finance developments elsewhere. The second important step was the decision to focus on mobile communications, and not on something else.

Although Finns are often accused of being too engineer oriented, Nokia has been less so than its closest competitors, Ericsson and the U.S. firm Motorola. This may be due to its historically broader customer interface in both the operator and end-user side, Nokia's early lead in the handset market, and initial industry developments. Nokia started to emphasize design and branding before its competitors did because it anticipated that the mobile phone was going to become a mass-market consumer product.[8] It seems that from early on Ericsson envisioned itself as a system company, while Nokia identified itself as a handset company even as the network side commanded a large share of turnover. As compared to Ericsson and Motorola, which had long traditions in the field, Nokia was clearly the challenger and thus had to be modest.

The Role of Technology Policy

The institutionalization and strengthening of science and technology policies began in the early 1960s. Important changes that contributed to the knowledge-driven growth and expansion of the ICT sector took place throughout the following decades. The main target of these policies was to strengthen the science and technology base of industry (Lemola 2002).

In the beginning of the 1980s, technology policy became increasingly target oriented and systematic. The National Technology Agency (Tekes) was established in 1982 to coordinate public R&D support and related efforts, such as national technology programs. Tekes and its programs became important instruments for implementing policies emphasizing technology transfer and the commercialization of research results, especially as it applied to information technology. In fact, two extensive information technology programs had been initiated even before Tekes was established.

Toward the end of the 1980s, a more systemic view on policymaking was adopted, and in the early 1990s, the deep recession fostered the relative importance of microeconomic policies. In the 1990s, the Science and Technology Policy Council, a high-level body advising the cabinet and the president on science and technology matters, introduced the national innovation system as a basic framework for policymaking. Innovation was seen as having a systemic character, contrary to the traditional linear innovation model. This enhanced cooperation between various policy agencies and improved possibilities for making use of emerging complex ICT. The systemic view also emphasized the role of education in adopting, diffusing, and utilizing new technologies (Georghiou et al. 2003).

CHALLENGES

Recent industry turbulence, the rise of various technological uncertainties, and patterns of changing demand in the global digital economy present clear challenges to the future of the Finnish ICT cluster.

Industry Turbulence

There was overinvestment in virtually all ICT-related activities in the late 1990s. In part, these were driven by one-time events such as deregulation and liberalization in major markets, the Y2K computer glitch, the introduction of the euro, and the commercialization of the Internet — not to mention the new economy bubble. With the benefit of hindsight it is easy to say that the market participants should have anticipated some leveling off in demand. But few could have anticipated the near collapse of the ICT market after the new economy bubble.

The bear market that followed the bubble took a heavy toll on ICT companies in Finland and abroad. However, as of early 2005, the market prospects are again good, and selective new entry is going on.

Technological Uncertainty

One of the key challenges for the Finnish ICT cluster is the ongoing convergence of voice and data communications, information systems, and consumer electronics, as well as the digital content being tailored for these various channels and devices. Mobile Internet, perhaps more appropriately labeled "whatever, wherever, and however desired," will introduce a new playing field with diverse players. Indeed, participants in the respective industries are already competing in both handsets and networks, and this tendency will only strengthen as Internet protocols (IP) increasingly form the basis for all electronic communication. Over time the focus on equipment weakens as it becomes more diffused and shifts to applications and content.

Finland has two major weaknesses in the all-IP future. First, it has little clout outside mobile telecommunication equipment and thus cannot leverage significant market power in other domains as the industry is transformed. Second, the all-IP world is not likely to favor the integrated and closed architectures and business models of the telecommunications world. The first problem can be addressed by acquiring a broader set of competencies and forming alliances with the leaders of the respective industries. The second problem can only be addressed by actually competing in an ever more open and fragmented operating environment.

Another well-known challenge is related to how the uncertainty over next-generation networks is resolved. In the mid-1980s, the International Telecommunications Union (ITU) assumed an active role in the introduction of the next-generation (3G) standards. Although the ITU pushed for one worldwide standard, eventually three became accepted in International Mobile Telecommunications (IMT-2000) guidelines: W-CDMA (better known as UMTS [Universal Mobile Telecommunication System]), CDMA2000 (promoted in particular by the American company Qualcomm), and the Chinese firm TD-SCDMA. Originally ITU's decision was considered a win for the Nokia-Ericsson camp promoting UMTS, but the early market signals were favorable to CDMA2000. However, not all promises conveyed by these early signals were kept; as of 2005, it seems that W-CDMA has gained a lucrative market position.

Europe attempted to maintain its lead in mobile telecommunications by pushing for rapid deployment of UMTS. By making a public commitment to one technology over another, the auctions violated technology neutrality, which is often considered one of the golden rules in technology policy making. In many European countries radio spectrums for 3G operations were auctioned for over €100 billion. It soon became clear that deployment and diffusion would be slower, network building costs higher, and expected revenue per user lower than the licensees had anticipated. Though the radio spectrum auctions generated large immediate returns to some governments, these midterm effects were unanticipated. The operators' indebtedness due to auctions, combined with the bearish financial market after the new economy bubble, held back the deployment of 3G for some time. The unanticipated midterm effects of the auctions on 3G rollout sparked requests for public actions to "reverse the damage." However, as of early 2005, those requests have trailed off, and 3G is being deployed on a commercial scale throughout Europe and elsewhere. Early signals of the demand for the new wireless services enabled by 3G are not discouraging.

The main benefit of the first-generation digital (post-analog) system was improved voice quality. The key promise of 3G is improved data communication. So far voice has been the key driver of mobile communication, although data is gaining ground. Upon bidding for a spectrum, the operators seem to have assumed a rapid and large shift from voice to data. This shift is indeed taking place but, from the European point of view, somewhat unexpectedly. Whereas Europeans seem to have assumed that the mobile Internet would be an extension of mobile telecommunications, the American route of extending wire-line data communications architectures to wireless local area networks (WLANs, also known as Wi-Fi or 802.11x, where x refers to the incarnation) seems to have an early market lead.

Arguably, a combination of WLAN and an intermediate generation (2.5G, e.g., general packet radio service [GPRS]) mobile telecommunications system having the "always on" feature could be used to reach the goals of 3G. WLAN nevertheless has a number of unsolved problems, such as control for login and access rights, payment, and coverage, that have already been solved in 3G. It is too early to say how the market will unfold, but most likely 3G and WLAN will coexist with in-between roaming. Only one thing is certain: how the market will unfold will crucially depend on consumers and business representatives and on how they want the wireless culture to evolve. The early experiences with wireless application protocol (WAP) provide support for this view and suggest, moreover, that pricing can matter a great deal as well (Gao, Hyytinen, and Toivanen 2005). If the technology does not meet customers' expectations, the demand is likely to be (price) elastic.

Lack of Lead Users?

Consumers' and businesses' tastes and their willingness to pay will ultimately determine who wins in the marketplace. A considerable challenge for the Finnish wireless cluster is therefore a lack or the wrong kind of lead users in the domestic market. Demanding lead users can provide important and immediate feedback and even drive innovation to a significant extent. Both our analysis and those of Berggren and Laestadius (2003) and Richards (2004) emphasize the importance of local demand for innovation and the role of local lead users as an explanation for Finland's and Sweden's emergence as global ICT players. With their aging populations, however, Nordic countries and Europe can probably serve as only a partial test bed during an era when the key growth markets are in Asia and China in particular.

As lead users, Finnish and European corporations are not as aggressive as would be optimal for the Finnish ICT cluster. Translating technical preconditions into improved performance seems to have been challenging for the European corporate sector. Nor is the picture in Finland very bright: several studies indicate a gap—albeit not exceptional in the international context (Dutta, Lanvin, and Paua 2003)—between the Finns' technical capabilities and effective use of ICT.[9] Although Finland is also an advanced user of ICT in many respects, it nevertheless seems that the future's lead corporate users will be elsewhere. This is alarming from the perspective of the Finnish wireless cluster and its smallest players in particular.

CONCLUSIONS

The performance of the Finnish economy in the 1990s was remarkable. It looked as though the country had found a unique way to combine high social security, dynamism, and growth. The rapid turnaround of the Finnish economy would not have been possible without the rise of the ICT cluster, which was facilitated by the convergence of a number of factors. Although Finland did well, its experiences are not unique in the Nordic context: other Nordic countries, particularly Sweden, were able to build competitive ICT sectors during the 1990s (Berggren and Laestadius 2003; Richards 2004).

There was no systematic plan to restructure the Finnish economy or to build a globally competitive ICT sector. Rather, a number of private initiatives, processes, and policy measures were working simultaneously. The country possessed fertile framework conditions and had accumulated a great deal of ICT-related expertise. Owing to unfavorable macroeconomic shocks, its resources were underemployed and something new was desperately needed. Digitalization presented a technological opportunity. Furthermore, the country had early exposure to two successive generations of winning standards. Finnish firms had already laboratory-tested competition when deregulation created a wide-open world market. On top of this, there was a company that had the vision and a strategy to make it happen. These factors, combined with quite a few lucky breaks, put Finland out in front of the pack.

Though in hindsight the Finnish public policies of the 1990s were successful, the Finnish miracle can only be partially explained by the policies pursued. Many of the necessary policy changes had already been made in the 1980s, with some having come much earlier. For example, the high-road strategy of innovation and technology was initiated in the 1970s and 1980s. In the 1980s, Finnish technology policy began to give high priority to ICT. These policies were continued in the following decade, when a new set of policy measures was initiated as a part of Finland's path to EEA and later to EU. As a result of these policy changes, a major shift in priorities took place: the focus changed from short-term macroeconomic to long-term microeconomic policies. Though we cannot document what businesses thought about the various policy initiatives nor show that they got the framework conditions they desired, we believe that these changes contributed to the Finnish economic success story of the 1990s.

What can be learned from the Finnish experience? It seems that a deep crisis often precedes considerable and lasting shifts in economic and social structures. People seldom have a desire to move rapidly into the unknown, but a crisis may bring about a willingness to accept the inevitable. The Finnish response to the most recent crisis was to open up the economy, modernize social structures, strengthen public R&D finance, and shift policies from direct business involvement to building framework conditions for private business. Moreover, it appears that Finland was able to leverage its small size: thanks to the homogeneity of its population and close interaction (networking) among policy makers and industry, Finland was able to adjust to new technologies and initiate a transformation from the challenging situation it faced in the early 1990s. This is a kind of small-country paradox: one might expect that larger countries would grow faster and achieve higher levels of income than smaller ones due to economies of scale in the production and use of knowledge.

Overall, the case of Finland is a good example of the interaction of several intertwined growth-generating factors. The Finnish model probably cannot be replicated as such. One thing, however, is sure: the long-term strategic perspective of education and innovation policies has been integral to Finland's emergence as a wireless player. These policies were relatively consistent over the long term and were not dictated by short-term cyclical or political considerations. Finland's favorable position when the wireless opportunity arose had nothing to do with luck.

There are nevertheless major challenges on the horizon. First, the Finnish ICT cluster remains highly specialized in mobile communications even though its scope has broadened in recent years. Although smaller Finnish companies have made efforts to decrease their dependency on Nokia (their

76   key customer), their fortunes are still tied to it. Second, the Finnish economy faces a rapidly aging population. This creates at least two specific challenges. First, there are increasing needs for labor market flexibility. The working population will inevitably start to decline in only a couple of years. This will weaken one of the economy's most important competitive advantages as the growth of the highly educated labor force slows. Moreover, Finland's (and Europe's) aging population can provide only a limited number of lead users (and thus serve as only a partial test laboratory) for new ICT goods that will be aimed at the growing Asian and Chinese markets and their relatively young consumers.

NOTES

We would like to thank the editors, Jonah Levy, and participants at the various BRIE-ETLA project meetings for helpful comments. This research is part of the wireless communication research program of ETLA, the Research Institute of the Finnish Economy, and BRIE, the Berkeley Roundtable on the International Economy at the University of California, Berkeley. Financial support from Nokia and Tekes is gratefully acknowledged. This work was completed while Rouvinen was a Jean Monnet Fellow at the European University Institute in Florence, Italy. He gratefully acknowledges the hospitality of the Robert Schuman Centre for Advanced Studies as well as the financial support of the Academy of Finland and the Yrjö Jahnsson Foundation.

1. Finland ranked second in frequency of collaborative arrangements between business and the public sector, with more than 40 percent of firms participating (Organisation for Economic Co-operation and Development [OECD] 1999c).

2. Finland ranked number 1 in 2000 in terms of ICT employment as a share of business sector total and ICT-related R&D relative to GDP. Finnish ICT value added as a share of business sector was second to Ireland, whose ICT production is based largely on foreign outsourcing (OECD 2002b).

3. Coincidentally, in the very same year, Eric Tigerstedt, a Finnish inventor who was well ahead of his time, attempted to patent a "pocket-size folding telephone with a very thin carbon microphone."

4. Important additional demand in terms of both learning and sales revenue came from Soviet authorities, who appreciated not only Finnish technical innovativeness and flexibility, but also Finland's political neutrality in the postwar era.

5. According to a Nokia director, Kari-Pekka Wilska, the Tandy cooperation considerably enhanced the company's customer orientation. In the leading Finnish daily newspaper, *Helsingin Sanomat*, he noted: "We had a Finnish engineer's mindset. As a major distributor of consumer products, Tandy's view was totally different. . . . We learned that even though the product can command a high price in the marketplace, it does not have to be expensive to produce" (April 7, 2002; authors' translation).

6. Nokia electronics, established in 1960, resold computers, provided computing services, and also manufactured some of its own electronic devices. Sales were modest,

but the 1960s may be seen as an era of competence building in digital technologies. The real breakthrough and expansion came in the 1970s. In 1972, Nokia signed a contract to deliver a large computer system for the Loviisa nuclear plant. In 1973, Nokia decided to start its own computer manufacturing after a major order from a local bank (Kansallis Banking Group). In order to capitalize on accumulated computer expertise and to leverage its phone cable business, Nokia became involved in fixed-line digital telecommunications by acquiring a license for a central telephone exchange from CIT-Alcatel in 1976. Its own (in part developed at Televa) digital exchange, the now legendary DX200, was introduced in 1982. It was based on a standard Intel microprocessor and was thus easily programmable and upgradable. With its distributed processing power, all-digital silicon architecture, and industry-standard components and programming language, it went against prevailing beliefs about telecommunications. DX200 was amazingly profitable in fixed networks and later formed the foundations of Nokia's wireless network systems. With a total of two thousand person-years of R&D effort spanning more than ten years (Keijo Olkkola, as cited in Häikiö 2001a, 275), it may be the biggest single R&D project in Finnish history (see also Palmberg 2002).

7. The best-known example of noncommercial pursuits is the open-source software Linux, which is now challenging the predominance of Microsoft. The revolutionary operating system, which today includes contributions of thousands of programmers, was initiated by a twenty-two-year-old student, Linus Torvalds, in 1991. The next year, Linux was combined with U.S.-based open software, GNU, to produce a complete free operating system. Another widespread Internet application, the real-time chat environment (Internet relay chat [IRC]), also originated in a Finnish university classroom. Furthermore, the first Internet information browser with a graphical user interface was created by a group of Finnish students in 1992, a year before Mosaic was released by the National Center for Supercomputing Applications at the University of Illinois. The students, however, shelved the software for lack of business interest. Another invention in the category of "missed Internet business opportunities" include the router connecting local and general networks, created by Nokia's engineers in 1983 before Cisco developed it but neglected at the expense of modem development. Some of the student-initiated inventions have grown into viable businesses, particularly in the domain of secure network solutions (e.g., SSH Communications Security, F-Secure), but even the most successful internationally operating companies fall into the class of small and medium-sized enterprises.

8. The name Nokia became the centerpiece of the company's branding strategy in 1991. Relatively early, "lifestyle consumption," as opposed to technological excellence, became the focal point in branding. A decade later, Nokia had become the strongest brand in the mobile market and one of the ten most valuable brands in the world.

9. Maliranta and Rouvinen (2003) use firm-level data to study the effects of ICT usage. The average effect in Finland is almost exactly the same as the mean estimate calculated across dozens of similar international studies. There is, however, huge variation across firms.

4

AN OLD CONSENSUS IN THE "NEW" ECONOMY?

*Institutional Adaptation, Technological
Innovation, and Economic
Restructuring in Finland*

Darius Ornston and Olli Rehn

The Finnish experience of the 1990s presents the scholar of political economy with a remarkable economic transformation and an intriguing political puzzle. Stated most provocatively, a diminutive postwar paper producer and high-end Soviet supplier emerged at the end of the millennium as a global telecommunications leader. In the early 1980s, over a quarter of Finnish exports went to the Soviet Union (Economist Intelligence Unit 2000, 21). Bilateral trade contributed to industrialization during the early postwar period, but it was nursing internationally uncompetitive sectors such as textiles, televisions, and transportation equipment by the 1980s (Pehkonen and Kangasharju 2001, 218). In 1992, this trade collapsed to 2.8 percent of exports, exacerbating a financial crisis and contributing to the deepest recession in the nation's history (Economist Intelligence Unit 1995, 28). By the time trade with Russia had recovered to a rapidly rising 7.5 percent of exports in 2003, its character had changed radically and irrevocably. Telecommunications had replaced textiles and transportation equipment as Finland's leading manufactured export. Finland was no longer known as a paper producer or a high-end Soviet supplier, but rather as a leading innovator and "wireless giant" (Rouvinen and Ylä-Anttila 2003).

Bilateral trade with the Soviet Union represents one facet of a more complex economic transformation that unfolded over the course of the 1990s. Industrial restructuring was pervasive, part of a national "innovation-based" strategy that witnessed rapidly growing R&D expenditures and per capita high-tech patent applications that ranked among the highest in the European Union. As described by Ari Hyytinen, Laura Paija, Petri Rouvinen, and Pekka Ylä-Anttila (this volume), many factors, including sound macroeconomic management, long-term investment in education, common telecommunications standards, and a favorable regulatory environment, contributed to these developments. This chapter focuses on the political bargains that underpinned radical economic restructuring and their role in facilitating change within an ostensibly stable and incremental organized economy.

At first glance, the Finnish experience would appear to support one of two common understandings about economic restructuring and institutional change. For some, Finnish restructuring represents successful liberalization, or a movement "from cartels to competition" based on privatization and deregulation (Steinbock 1998). Others cite Finland as an example of "neo-corporatist continuity," whereby Finland relied on traditional wage-setting and policymaking instruments to facilitate economic adjustment (Wilensky 2002; Katzenstein 2003). Careful analysis of two liberal reforms (financial and trade liberalization) and two forms of neo-corporatist continuity (fiscal policy and pay determination) reveals that neither narrative fully describes the process of institutional experimentation and adaptation that transpired during the 1990s. Rather, long-standing patterns of tripartite concertation were adapted to perform radically different functions within fundamentally different fora. This shift in concertation, most vividly evident in Finnish "technology policy," emerged from an explicitly political process of institutional renegotiation that simultaneously facilitated and constrained adjustment. This chapter generalizes findings with reference to similar developments in Denmark and Ireland, but begins by situating the Finnish experience within a broader literature on contemporary political economy.

STORYTELLING IN A GLOBAL ECONOMY

Contemporary literature on comparative political economy is characterized by a sharp disjuncture between liberal and organized market economies. While organized market economies come in a wide variety of shapes and sizes, this chapter focuses on concertation, or collaborative forms of policy making and production between organized actors and state agencies that deviate from free market competition and price signals (Lehmbruch 1979). Historically, concertation has occurred within neo-corporatist channels, characterized by highly centralized, encompassing, monopolistic producer groups such as employer associations, universal banks, trade unions, and state agencies (Schmitter 1979). Though it is closely associated with neo-corporatism, concertation refers to a cooperative policy-making process, and neo-corporatism refers to the specific constellation or institutions or policies within which it occurs (Baccaro 2003).

Both manifestations of the organized economy are closely linked to social stability, incremental innovation, and gradual adjustment. Even the most articulate efforts to link corporatism and change betray significant continuity. Thus, Peter Katzenstein's insightful analysis of Austria described redistributive institutions that facilitated a gradual upgrading of preexisting sectors rather than radical shifts to new industries (Katzenstein 1984). Economic discontinuity and radical innovation, by contrast, are commonly located within the purview of the liberal market economy. Liberal financial markets facilitate the rapid reallocation of capital to emerging enterprises, while decentralized wage setting fluidly and nimbly accommodates changing economic conditions. By virtue of its emphasis on competition between individual firms, the liberal market economy is said to possess a "comparative institutional advantage" in fast and flexible adjustment, even if its unstable character generates long-run inefficiencies (Hall and Soskice 2001).

Descriptions of institutional continuity and change in contemporary capitalism reflect this dichotomy between liberal change on the one hand and neo-corporatist continuity on the other. Consider the literature regarding national responses to technological change and economic globalization. Many scholars have described a wave of liberal reforms designed to increase flexibility and encourage innovation, citing evidence of privatization, product market deregulation, financial liberalization, and/or the decentralization of collective bargaining to support their claims (Iversen 1996; Pontusson and Swenson 1996; Rehn 1996). Others have emphasized institutional continuity by pointing to specific neo-corporatist features such as financial concertation, the limited decentralization of collective bargaining, and an enduring state role in important parts of the economy (Wallerstein and Golden 1997; Wilensky 2002; Katzenstein 2003).

Though compelling, this dichotomous characterization of liberal change and neo-corporatist continuity is problematic in two respects. First, it neglects the way in which concertation can contribute to ostensibly liberal objectives such as radical innovation and extensive economic restructuring. Although it is commonly linked to stability and incremental adjustment, concertation varies significantly over time and space. It has been alternately associated with wage restraint, welfare state expansion, fiscal retrenchment, labor market reform, technological diffusion, skill accumulation, and a number of other "redundant capacities" on the supply side of the economy (see Calmfors and Driffill 1988; Cameron 1984; Ebbinghaus and Hassel 1999; Rhodes 2001; Ziegler 1997; Hall and Soskice 2001; and Streeck 1991, respectively). In Finland, it will be demonstrated that a distinctive pattern of concertation, oriented around innovations in technology policy, contributed to a dramatic repositioning of the Finnish economy during the 1990s.

This contribution was predicated on institutional renegotiation, whereby 81
concertation was adapted to pursue radically different ends within fundamentally different fora. Indeed, the liberalization of preexisting neo-corporatist institutions in Finland had the paradoxical effect of triggering intensive consultation and concertation between firms, employer associations, trade unions, and state agencies during the 1990s. The aforementioned dichotomy between liberalization and neo-corporatist continuity obscures these dynamic and fundamentally political efforts to leverage inherited institutional capacities in responding to new economic challenges. That said, institutional adaptation is not an automatic or functionalist response to crisis. The political bargains that underpinned radical restructuring in Finland constrained adjustment in other, problematic ways, giving rise to an ongoing process of "perpetual renegotiation" (Ornston 2004). In fact, a similar process of perpetual renegotiation could be said to have characterized Finnish economic development for most of the postwar period.

A CHRONOLOGY OF CHANGE: THE INSTITUTIONAL AND ECONOMIC CONTEXT

Before evaluating rival narratives based on liberal change and neo-corporatist continuity, it is first necessary to establish a historical context for understanding recent developments in Finnish political economy. This section introduces institutional legacies accumulated over the course of the postwar period before describing the challenges that emerged during the early 1990s and the remarkable degree of economic restructuring that ensued.

State Intervention and Tripartite
Concertation in Postwar Finland

Finland entered the 1990s as a highly organized and densely institutionalized economy bearing the imprint of several successive developmental strategies. With the economy in a condition of pronounced backwardness at the beginning of the postwar period, the Finnish state assumed a leading position in encouraging industrialization. This early postwar industrialization program, precipitated in part by heavy war indemnities imposed by the Soviet Union, was framed as an exercise in nation building. The government at the time deliberately decided not to decrease taxes to prewar levels, while Finland's statutory pension system generated additional capital for investment. Public savings, 30 percent of aggregate savings in the early postwar period, was then ploughed back into firms in the form of subsidized credit (Vartiainen 1999, 228–229). Postwar investment in Finland was 25.2 percent of GDP in the early 1950s, 7.9 percent above the average for the Organisation for Economic Co-operation and Development (OECD) Europe, and second only to Norway (Boltho 1982, 33). The state assumed an even more direct role in economic development through public companies such as Outokumpu (mining), Kermia (chemicals), and Neste (oil refining, petrochemicals, and plastics). State-owned industry increased in salience over the course of the postwar period, generating 23 percent of manufacturing value added, 27 percent of manufacturing investment, and 30 percent of manufacturing exports by 1993 (Rehn 1996, 282).

Private business was deeply engaged in this national project through dense interfirm networks and close connections to the state. Immediately after the war, the state constructed joint committees of civil servants, private industrialists, and bankers to organize war indemnities to the Soviet Union (Vartiainen 1998). In subsequent years, firms were linked to the state through industrial development agencies such as the Ministry of Trade and Industry. Firms were also connected within expansive universal banks and encompassing employer associations. Finland thus began to conform to the German stakeholding model of corporate governance and finance. Until the 1990s, Finnish industry was dominated by four main banks, each with its own industrial "family." Simultaneously, Finnish firms also developed a dense network of employer associations, closely tied to the state and trade unions through their counseling, lobbying, and collective bargaining activities. These associations originated from interwar export cartels but merged into a unified Confederation of Finnish Industries (Teollisuuden Keskoliitto [TKL]) in 1974 following developments in the Finnish trade union movement (Rehn 1996, 282).

Labor played a marginal role in early industrialization, partly as a legacy of the civil war of 1918 and partly because of the resultant division of the trade union movement into social democratic and communist camps. Growing trade union clout as well as the desire to create a (social) democratic bulwark against the communists inspired periodic but opportunistic recourse to national collective bargaining and income policies following strategic devaluations during the 1950s and 1960s. Centralized collective bargaining became a permanent feature of Finland's institutional landscape in 1967 after the election of a social democratic coalition government and the creation of the more encompassing Confederation of Finnish Trade Unions (Suomen Ammattiliittojen Keskusjärjestö [SAK]). Collective bargaining increased in prominence in subsequent decades and was closely linked to rapidly expanding social protection and labor market regulation. That said, Finnish neocorporatism never transcended its "fair-weather" bias and was most success-

fully deployed only after recourse to a forced devaluation (Rehn 1996).
Notwithstanding its inconsistent character, Finland could be said to have possessed all the dense organizational and institutional structures of a neo-
corporatist economy by the end of the 1980s. Centralized employer associations
and trade unions played a key role in pay determination and policy formulation, while universal banks linked firms together within dense, long-term financial networks. Meanwhile, the state continued to play an unusually prominent
role in the economy by building up interest associations such as the SAK, supporting collective bargaining with policy and tax concessions and allocating
preferential credit and industrial subsidies to targeted firms and sectors.

The Finnish Recession and Economic
Transformation of the 1990s

Institutional arrangements inherited from four decades of postwar development resonated with Finland's comparative advantage. The Finnish economy
enjoyed a comfortable, if unremarkable, place internationally. Pulp, paper, and
forestry industries accounted for over a third of Finnish exports while bilateral
trade with the Soviet Union provided a secure market for manufactured goods
such as textiles, televisions, and transportation equipment. A devastating economic recession in the early 1990s, however, shattered Finland's stable position.
The immediate cause of the recession was poorly managed financial deregulation and liberalization, but the crisis was aggravated by the collapse of the Soviet Union and a 70 percent decline in trade with Russia. By 1993, output had
fallen 14 percent and unemployment approached 17 percent (Honkapohja and
Koskela 1999). The recession was described as a "collective nightmare," casting
doubt on traditional economic structures and strategies as concerns about
"economic competitiveness" rose in political salience (Kiander 2002).

By the late 1990s, however, Finland was being quietly heralded as a success
story. From 1995 through 1999, annual GDP growth averaged 4.4 percent,
peaking at 6.3 percent in 1997. Growth had declined to 2.0 percent by 2003,
but it was still fourth highest in the Eurozone and well above the regional
average of 0.5 percent. More significantly for the purposes of this chapter,
Finland had negotiated a remarkable structural transformation toward high-
technology production. High-technology production increased as a share of
exports from 7 percent in 1990 to 23 percent by 2000, surpassing even the
dominant pulp and paper industry in production volume.

The success of the ICT cluster reflects a more fundamental national transformation from "investment-driven" to "innovation-driven" growth (Rehn 1996;
Steinbock 1998; Georghiou et al. 2003). Whereas investment-driven growth is

84   characterized by greater physical and human capital inputs into relatively stable production practices, innovation-driven growth is marked by innovations that create fundamentally new products and industries. The pronounced shift from the accumulation of productive resources to their creative recombination is manifested in rapidly rising R&D, which climbed from 1 percent of GDP in the early 1980s to 3.5 percent of GDP by 2002. Even excluding Nokia, which accounts for a third of total R&D expenditure, Finnish R&D remains well above the EU average (Asplund 2003, 10–11). Among EU members, Finland stands out with respect not only to per capita R&D, but also to per capita patent applications, innovation expenditure, and high-technology value added in manufacturing.

If a competitive economy is defined in terms of its ability to make the most efficient use of all available resources, however, it is nonetheless difficult to agree with the World Economic Forum's assessment that the Finnish economy is the "most competitive in the world" (World Economic Forum 2004). Finland's economic transformation was achieved at a heavy price, most visibly in the form of displaced workers. Unemployment peaked at 16.6 percent of the labor force in 1994 and has been slow to recover, averaging 9.0 percent in recent years. Meanwhile, Finland exhibits considerable weaknesses with respect to its business structure, ICT use, and commercialization of ICT. As described below, these weaknesses reflect the fundamentally political bargains that underpinned Finland's radical economic restructuring. This chapter begins, however, by focusing on Finland's status as a leading innovator and high-technology producer. What political and institutional developments made possible such a radical redeployment of economic resources during the 1990s?

FINLAND'S ECONOMIC TRANSFORMATION AS A LIBERAL REVOLUTION?

The Finnish experience easily lends itself to a classic liberal interpretation in which deregulation and liberalization fueled innovation, radical restructuring, and economic growth (Rehn 1996; Steinbock 1998). The center-conservative government's 1991 program to extend the ownership base of public enterprises most vividly captures the retreat of hitherto dominant, centralized actors such as the state and the subsequent rise of individual firms. This program was continued over the course of the decade under the auspices of a social democratic–conservative rainbow coalition government (Rehn 1996, 282–284). In fiscal policy, the center-conservative government reduced social security contributions and introduced a flat corporate tax of 25 percent in 1993 (26 percent as of 2004), thereby encouraging investment and promoting supply-side capacity. As described by Rouvinen, Ylä-Anttila, Hyytinen, and Paija in this volume, this represents but one of many liberal components in Finland's economic transformation. This section limits the focus to financial and trade liberalization because these policies most vividly capture the hypothesized shift from dense, national-level concertation to increased competition in financial and product markets.

Decentralization and Globalization: Financial and Trade Liberalization

During the 1990s Finland experienced a financial revolution characterized by the pronounced decline of universal banks and rapidly increasing access to foreign finance and equity. Financial reforms during the 1980s and poorly managed deregulation triggered an economic crisis and a wave of bankruptcies in the early 1990s (Honkapohja and Koskela 1999). Financial liberalization thus had the unintended consequence of shaking up the established network of universal banks and their respective industrial families as cross-ownership declined dramatically during the 1990s (Ali-Yrkkö and Ylä-Anttila 2002, 254). Simultaneously, the state retreated from the practice of accumulating massive public surpluses and channeling funds back to favored firms and strategic sectors at subsidized interest rates through the Ministry of Trade and Industry and the Finnish Central Bank (Rehn 1996, 238).

As the economy recovered, these developments enabled larger Finnish companies to secure direct access to equity finance and foreign capital and to become less dependent on universal banks and on the state. Nominal market capitalization rose from 23 percent to 150 percent of GDP during the 1990s, exceeding that in all Nordic nations except Sweden (Hyytinen and Pajarinen 2002, 27). Meanwhile, an influx of foreign capital turned the Helsinki stock exchange into "the most internationalized in the world" prior to its merger with the Swedish and Danish stock exchanges. This massive influx of foreign capital was critical in fueling the development of innovative young sectors.

Financial liberalization and internationalization not only liberated firms from traditional, hierarchical banking relationships and opened access to foreign capital, but also resulted in predictable changes in corporate governance. Traditionally dominant stakeholding institutions such as the supervisory board disappeared from corporate practice. Meanwhile, businesses displayed a greater sensitivity toward growth and shareholder value throughout the 1990s (Ylä-Anttila 2000, 12). These changes were reinforced by the corporate tax reform of 1993, which eliminated the taxation of dividends and favored

86   equity financing over debt finance. Corporate finance and governance, widely considered as the linchpin of the neo-corporatist or coordinated market economy, changed irrevocably as price signals and market relations supplanted associational or institutional affiliation, with predictable consequences for firm behavior and strategy.

Financial liberalization coincided with new forms of competition and new market opportunities associated with accession to the European Union in 1995. In addition to improving preexisting access to large, lucrative markets in Western Europe for Finland's developing ICT sector, contemporary policy makers in nations such as Austria, Finland, and Sweden viewed EU accession as a mechanism for exposing producers to greater levels of competition at home (Rehn 1996). Econometric analysis suggests that Finland derived particular benefit from EU accession, with an estimated annual contribution to GDP growth of 0.8 percent. In addition to increased exports to the EU and greater price competition at home, Finland benefited from concrete policy reforms in areas such as competition policy (Breuss 2003, 150–153). Liberalization started in 1989 with a corporate law reform that limited cartel formation and continued with the creation of a common cartel policy by the time of expansion. At the same time, financial integration allowed Finnish firms to make the most of these opportunities by opening up access to a wave of foreign direct investment during the 1990s. In other words, EU accession complemented financial liberalization by opening new resources to firms that sought to adapt to radically different market conditions.

Competition with Cooperation: The Limits of the Liberal Narrative

Though compelling, the narrative linking liberalization to economic change neglects important aspects of the Finnish experience. Rather than creating fiercely competitive fields of atomized firms, the retreat of universal banks and the dissolution of highly centralized industrial families coincided with the proliferation of interfirm networks. Indeed, liberalization could even be said to have increased cooperation by activating hitherto latent organizational institutions. Finnish business leaders and policy makers placed an increasing emphasis on interfirm collaboration and diffusion as a mechanism for adapting to new economic challenges. In so doing, they developed organizational linkages within peak and sectoral employer associations that were anathema to liberal principles. By the end of the decade, there was increasing discussion of a "network approach" to economic production and innovation in which "cooperation" received as much attention as "competition" (Schienstock and

Hämäläinen 2001; Castells and Himanen 2002; Rouvinen and Ylä-Anttila
2003). To more fully illustrate the limits associated with the liberal narrative, the next section considers the strengthening of centralized collective bargaining and policy making during the 1990s.

FINNISH ECONOMIC ADJUSTMENT AS NEO-CORPORATIST CONTINUITY?

Curiously, for a nation that is trumpeted as a paragon of successful liberalization, Finland presents compelling evidence for authors seeking to document neo-corporatist continuity in the face of global and postindustrial pressures (Wallerstein and Golden 1997; Wilensky 2002; Katzenstein 2003). Finland has witnessed neither a decentralization of collective bargaining nor aggressive efforts to disempower labor and other special interests. Organizationally, union density has been relatively stable in Finland, having fallen only gradually in the late 1990s. Finnish employers remain well organized, with little evidence of the "association flight" that has been documented in Germany (Asplund 2003). Indeed, in sharp contrast to literature on intersectoral cleavages, Finnish employer associations became more centralized during the 1990s. The wage-bargaining functions of the Confederation of Finnish Employers (Suomen Työnantajain Keskusliitto [STK]) and the lobbying functions of TKL were merged into the Confederation of Industry and Employers (Teollisuuden ja Työnantajain Keskusliitto [TT]) in 1993. Most recently, TT, which is composed of large export firms, merged with the small and medium-sized Employers' Confederation of Service Enterprises in Finland to form Finnish Industries in 2005.

Tripartite Concertation in Collective Bargaining and Fiscal Policy

Finland experienced not only relatively stable patterns of collective wage bargaining during the 1990s, but an active "centralization" toward the end of the decade. The Finnish experience thus deviates from the pronounced turn toward decentralization observed in nations such as Denmark and Sweden (Iversen 1996; Pontusson and Swenson 1996). The initial response to the recession was not decentralization but a hitherto unprecedented two-year wage freeze in 1992 and 1993. Concertation reflected Finland's opportunistic and inconsistent pattern of "fair-weather" corporatism. Intersectoral consensus was reached only in the wake of currency devaluations in 1991 and 1992, locking into place a redistribution of income between sheltered and exposed sec-

88   tors rather than preventing a costly devaluation (Rehn 1996, 262–263). Even
then, wage restraint did little to assuage growing concerns among employers
about wage flexibility and adjustment capacity at the level of the firm (Kauppinen 2001, 46). The STK adopted a radical stance toward decentralization in
the early 1990s, and collective bargaining devolved to the sectoral level.

Decentralization to the sectoral level, however, failed to secure wage moderation, making employers and employees alike amenable to a return in 1995
to centralized collective bargaining (Economist Intelligence Unit 2000, 43). A
national agreement was precipitated by the intervention of a newly elected
rainbow coalition government in 1995. In contrast to the previous center-
conservative government, which had adopted an ambivalent attitude toward
collective bargaining in the early 1990s, the rainbow coalition used targeted
concessions in fiscal and tax policy to bring trade unions to the table. The 1995
agreement was described as "the peak in terms of moderation and encom-
passment in the history of Finnish incomes policy," and the 1997 agreement
was even broader, with 98 percent coverage (European Commission 1999, 33;
Pekkanen, cited in Rehn 1996, 268). Most recently, the social partners signed
a collective agreement extending into 2007, underwritten by a government
pledge to increase funding on economic infrastructure.

The pronounced shift in employer attitudes toward centralized bargaining
was predicated on the latter's capacity to procure wage restraint. While economists remain skeptical about future developments, there is widespread
consensus that collective bargaining institutions were quite successful in securing wage moderation during the 1990s (Vartiainen 1998; Asplund 2003).
Real wage growth fell from 4.5 percent in 1994 and 7.1 percent in 1995 to average 3.4 percent between 1996 and 1998 before increasing in line with the
OECD average (Economist Intelligence Unit 2000, 43). The most recent agreement is estimated to generate overall wage increases of 3.2 percent, 2.6 percent,
and 1.8 percent between 2006 and 2007. In economic terms, wage restraint facilitated a dramatic shift from wages to profits during the 1990s that was instrumental in the pronounced increase in private sector R&D expenditure.

Centralized collective wage agreements reflected an equally dramatic
move toward tripartite concertation in fiscal policy during the 1990s. Government debt ballooned from 15 percent of GDP in 1990 to 59.6 percent of GDP
by 1994 as economic recession and soaring unemployment triggered automatic stabilizers (European Commission 1999, 27). Retrenchment efforts by a
center-conservative coalition government that was forced to adopt a go-it-
alone approach to fiscal retrenchment during the early 1990s met fierce resistance from societal actors. Government plans to cut unemployment benefits

faltered in the face of a threatened general strike in 1992, and the government   89
had to back down under similar circumstances when it attempted to lower
the wage threshold to employ a young person in 1993. Ultimately, the center-
conservative government was able to maintain fiscal discipline through a
broad array of expenditure cuts, but it paid a heavy electoral price in 1995
because of societal polarization and trade union resistance.

As a result, the subsequent rainbow coalition government turned to tripartite concertation to create an electorally sustainable austerity package.
Centralized, tripartite negotiation created tight linkages between otherwise
disparate policies, allowing the government to redistribute costs across different constituencies, policy areas, and time frames (Saari 2001). In so doing, the
government forged a consensual approach to fiscal retrenchment focusing on
cuts in pension expenditure, child allowances, and study grants. While the
dramatic 10 percent reduction in government expenditure as a share of GDP
during the latter half of the 1990s was as much about economic recovery as
about expenditure cuts, the absolute decline was a product of concrete policy
reforms (European Commission 1999, 28–29). In practice, trade unions
supported retrenchment with the understanding that unemployment benefits,
an important incentive for joining a trade union, would remain essentially
untouched (Rehn 1996, 269). Consequently, while unemployment benefits
were frozen between 1995 and 1999 and replacement levels were reduced by
2 percent in 1996, labor market reforms lagged behind Denmark and Sweden,
generating a nationally distinct pattern of fiscal adjustment (European
Commission 1999, 71–72).

Institutional Innovation and Its Contribution
to Neo-Corporatist "Continuity"

Although Finland experienced considerable continuity in collective bargaining and fiscal policy, established institutions fulfilled radically different functions during the 1990s. Neither policy making nor collective wage bargaining
in 1990s Finland approximated the type of positive-sum nominal-social wage
trade-off described in the traditional neo-corporatist literature (Cameron 1984).
Rather, neo-corporatist institutions contributed to economic restructuring
through wage restraint, a pronounced increase in profits, and macroeconomic
restraint. Concertation was originally provoked by Finland's economic and fiscal crisis, but its new role was solidified as European monetary union (EMU)
eliminated devaluation as an adjustment tool and placed new restraints on fiscal expenditure. These ostensibly liberal constraints on macroeconomic management triggered a conversion of traditional neo-corporatist concertation to-

90   ward more "competitive" ends (Rhodes 2001). The euro effectively reoriented neo-corporatist concertation away from full employment guarantees and increased social expenditure toward wage moderation and fiscal restraint. This pan-European trend acquired an innovation-intensive character in Finland as tripartite committees in the early 1990s increased public R&D while simultaneously reducing social expenditure. For example, the 2004 agreement traded wage restraint for government investment in infrastructure and R&D.

Moreover, neo-corporatist "continuity" masked fundamentally discontinuous, liberal innovations at the micro level. Thus, the share of firms reaching local agreements on issues such as pay, working hours, and organizational reforms within the framework of national agreements increased from 60 percent in 1992 to 90 percent by 1998 (Uhmavaara, Kairinen, and Niemelä 2000). Merit-based pay and profit-sharing arrangements emerged as especially popular mechanisms for achieving greater flexibility, while local negotiations increased the scope for flexible working hours, even as coordination at the sectoral and legislative level remained limited. The trade unions' willingness to accommodate greater firm-level flexibility in merit-based pay and working hours was instrumental in changing employer attitudes and sustaining centralized collective wage agreements during the 1990s (Asplund 2003, 107). Thus, although collective bargaining institutions appeared highly stable, continuity was predicated on institutional innovations at the firm level. Meanwhile, neo-corporatist institutions were embedded within and shaped by important liberal reforms such as financial liberalization, EU accession, and EMU.

TRIPARTITE CONCERTATION AND FINNISH TECHNOLOGY POLICY

Careful analysis of the Finnish experience thus reveals that both liberal reforms and neo-corporatist legacies played important roles in economic adjustment and restructuring during the 1990s. The Finnish experience is thus best characterized as an interplay between liberalization and preexisting neo-corporatist institutions, even within those policy areas considered the most liberal or the most corporatist. This section introduces Finnish technology policy to illustrate how ostensibly disparate or even contradictory institutions are combined in practice. "Technology policy" is of substantive interest because of its central role in promoting radical innovation and economic restructuring. Finnish technology policy has been repeatedly linked to increased private R&D, innovativeness, and/or productivity in multiple survey-based and econometric studies (see, e.g., Lehtoranta 2000; Maliranta 2000; Georghiou et al. 2003; Vuori and Vuorinen 1994).

Technology Policy: Actors and Structure

Technology policy is not new within the Finnish lexicon, but its salience increased greatly during the 1980s and 1990s as attention shifted from investment-driven to innovation-driven economic growth. Technology policy can be broadly defined as the constellation of institutions that prioritizes the redeployment of human and physical capital through process and production innovations. It was most concretely captured by the government's public commitment to raise aggregate R&D expenditure as a percentage of GDP to 2 percent by 1990, to 2.45 percent by 1995 and to 2.7 percent by 2000 (Asplund 2003, 26). The increase in aggregate R&D expenditure, which approached 3.5 percent of GDP by 2002, was facilitated by a continuous rise in public R&D. That said, Finnish technology policy adopted a "holistic" approach, focusing on a broader "national innovation system" (Lundvall 1992). In this view, innovation was not generated by individual state agencies or individual firms, but was instead embedded in a network of economic relationships, institutions, and policies. Technology policy thus aspired to coordinate a diverse array of actors and issues in order to promote innovation and restructuring (Schienstock and Hämäläinen 2001).

In describing the actors in Finnish technology policy, it is necessary to begin with private enterprise, which represented the primary engine of innovation and structural transformation. The private sector accounted for the largest increase in R&D expenditure over the course of the 1990s, even as public R&D expanded. Public R&D funding *as a percentage of national R&D* remained far below the EU and OECD average even though public R&D funding *as a percentage of GDP* was the highest in the EU (Asplund 2003, 33). In practice, this stemmed from an intense and intentional emphasis on private sector networking. If employer associations such as the Central Association of Finnish Forest Industries were legally prohibited from the traditional practice of price fixing, they acquired a new role in the 1990s in serving as fora for technological collaboration and diffusion. The prior existence of multiple, centralized, encompassing employer associations at the national and sectoral levels provided a foundation for interfirm cooperation, which was, in turn, aggressively linked to state agencies, conferences, committees, and universities.

The state played a critical role as a catalyst by channeling resources to facilitate the construction of these dense interfirm networks. Revenue from privatization was reallocated to long-term education and public R&D expenditure, which increased dramatically in real terms over the course of the decade (Asplund 2003). Public collaboration with industry was manifest in the in-

92 creasing salience of public organizations such as the Finnish Research Development Fund (Tekes) and the Finnish National Fund for Research and Development (Sitra). Meanwhile, the government sought to develop venture capital for emerging enterprises through the establishment of the public Start Fund of Kera in 1990, an expanded mandate in venture capital for Sitra in 1991, and the merger of the two bodies under the auspices of a broader government fund of funds (FII) in 1995. Rather than the specific regional and sector-targeted subsidies of an earlier era, this support was aimed at the horizontal promotion of R&D and technological diffusion (OECD 1999b, 54). In the end, this funding touched virtually all Finland's large high-technology firms and at least a third of its small and medium-sized enterprises (SMEs) (Castells and Himanen 2002; Georghiou et al. 2003).

Though it is seldom considered, trade union cooperation was an important and necessary component in economic adjustment. Finnish labor proved remarkably supportive of policies that promoted radical innovation, despite the fact that reorganization and restructuring threatened jobs. Labor cooperation stretched back to the late 1970s, when SAK supported increased investment in microelectronics (Vuori and Vuorinen 1994, 19). At the macro level, tripartite concertation secured macroeconomic restraint, a pronounced shift from wages to profits, while trade unions proved accommodating in negotiations over flexible work hours and other forms of workplace organization at the micro level (Asplund 2003). Meanwhile, trade unions were intimately involved in organizational and workplace innovations through a series of tripartite National Workplace Development Programs addressing ICT diffusion and telework arrangements (Schienstock and Hämäläinen 2001, 19; European Commission 2003b, 31).

Technology Policy: Origins and Development

Although new economic challenges triggered the redeployment of preexisting institutions and resources in the face of new economic challenges, Finnish technology policy did not develop automatically or functionally. Technology policy was a political construction, with state agencies assuming a leading role in provoking renegotiation through agenda setting. In response to growing concerns by business leaders and policy makers, new economic challenges, including structural rigidities and a comparatively narrow export base, were identified during the 1980s. Significant state initiatives such as financial liberalization and tax reform broke apart old bargains and forced the renegotiation of preexisting institutions. At the same time, the state responded to pressure that it actively assist firm restructuring. Given that traditional, region, sector, and firm-specific industrial policy instruments were being dismantled, the 1980s necessitated a search for creative new mechanisms for managing restructuring. This search was increasingly defined in terms of innovation-driven growth.

A tripartite technology committee convened in 1979 (subsequently institutionalized as the Science and Technology Policy Council [Valtion tiede-ja teknologianeuvosto]) accurately captures the form of active collaboration between business leaders, university administrators, policy makers, and trade union representatives that characterized Finnish political economy during this period. The influence of the committee (and of the council that succeeded it) was reflected in its capacity to assemble the most influential economic and political actors within Finland. As a result, the committee wielded significant power as an agenda setter and played a leading role in redefining national competitiveness in terms of technology innovation. More concretely, the committee provided the impetus for specific government initiatives to promote technological development, increase R&D expenditure, and facilitate interfirm coordination. It was a committee resolution to raise aggregate R&D expenditure to 2 percent of GDP by 1992, prompting the foundation of Tekes, that would eventually account for a third of public R&D expenditure (Georghiou et al. 2003, 88). A new discursive and institutional framework was thus already in place by the time of the 1991 recession, leading to an innovation-driven policy response to economic crisis rather than an emphasis on active macroeconomic management or increased investment.

Though the state clearly played a critical catalytic role as an agenda setter and as a facilitator for interfirm coordination, political bargains reflected the political interests and leverage of the social partners. For policy makers, technology policy provided a framework for packaging competitive reforms and advancing structural change without engaging in more aggressive deregulation or liberalization. Meanwhile, technology policy was relatively appealing from the perspective of large employers because it advanced financial market and capital taxation reform without threatening traditionally close collaboration between firms. Additionally, technology policy simultaneously supported adaptation by funding technological innovation and diffusion, giving large manufacturers direct input into the policy-making process through representation in bodies such as the Science and Technology Policy Council.

The position of labor offers an even more revealing illustration of how technology policy was shaped by the interests and influence of preexisting societal actors. As noted above, trade union enthusiasm for technology policy presents a puzzle, because rapid innovation can generate redundancies and aggravate unemployment. Labor support was partly a function of the trade unions' ability to shape policy and secure compensatory measures in other policy areas. At the national level, trade unions enjoyed formal representation within Tekes and on the Science and Technology Policy Council. Moreover, they were actively engaged with individual firms at the workplace level not only in collective bargaining, but also in government initiatives such as the 1995 National Telework Development Program and the 2002 Committee for E-Work Cooperation (Asplund 2003, 92; European Commission 2003b, 31). The 1996 Finnish National Workplace Program, the most ambitious of several such initiatives advanced by the rainbow coalition government elected in 1995, is especially indicative. This initiative on organizational innovation and diffusion engaged state R&D agencies, capital, and labor within a tripartite framework. Favorable reception by the social partners inspired a more expansive program, the 2004–2009 Workplace Development Program for the Improvement of Productivity and the Quality of Working Life (European Commission 2003b, 31). Though bipartite cooperation on technological innovation already existed in local-level collective bargaining, this initiative represented a more active effort by the government to engage societal actors in innovation and economic restructuring.

Technology Policy: Political Compromises and Economic Constraints

The development of Finnish technology policy may have been provoked by external economic challenges, but it is clear that it was defined by interaction and compromise among the state, businesses, and trade unions. As such, it represented a fundamentally *political* solution to economic crisis rather than a functionalist one. The distinction is important, for it reflects the fundamentally conflicting nature of the reform process as well as its suboptimal implications. Notwithstanding Finland's impressive achievements in technological innovation and economic restructuring, the economy remains hobbled by a dearth of SMEs and a high unemployment rate. These problems are not exogenous; they are directly linked to the particular political bargains that underpinned adjustment during the 1990s. To date, the Finnish pattern of adjustment reflects a highly centralized industrial structure organized around a comparatively activist state on the one hand and powerful, institutionally entrenched trade unions on the other.

Despite considerable success in diversifying production, Finland's underlying industrial structure remains more problematic. SMEs, an important source of flexibility, dynamism, and innovation, are conspicuously underrepresented in Finnish political economy. The amount of entrepreneurial activity and number of corporate spin-offs ranks among the lowest in the OECD, while the share of employment in newly established enterprises (0.5 percent) falls far short of that in comparable nations such as Sweden (1.8 percent) and Denmark (2.4 percent) (European Commission 2004, 63). Earlier statist and neo-corporatist developmental strategies were never especially attentive to the needs of SMEs, and their relative marginalization within highly centralized, national-level fora such as the Science and Technology Policy Council is not surprising. Somewhat paradoxically, the same level of centralization that facilitated the coordinated expansion of technology policies during the early 1990s limited their effectiveness at the end of the decade. Notwithstanding efforts to address the problem through the creation of regional employment and economic development centers in 1997 and a reorientation of Tekes funding, Finland's relatively homogenous and large-scale industrial base continues to be identified as a major weakness in performance evaluations (European Commission 2004).

Moreover, though Finland has demonstrated considerable prowess in reallocating capital into radically new industries, labor has been slow to follow. Flashy GDP growth rates in the late 1990s were not matched by equally impressive developments within the labor market. Unemployment fell only gradually and lingers at 9 percent despite robust economic growth. High unemployment reflects limited progress on critical policy reforms such as low-income-earner taxation and labor market activation. Despite progress in reducing the tax wedge on lower-income earners to 40.4 percent by 2002, it remains significantly higher than the EU15 average of 37.8 percent. At the same time, Finland's expenditure on the active labor market measures lags behind that of comparable nations such as Denmark, Sweden, and the Netherlands, giving it one of the highest ratios of passive- to active-labor-market expenditure in the EU (European Commission 2004, 98, 63).

High unemployment can be viewed as a product of the political bargains that underpinned Finnish adjustment. The decision to prioritize technological innovation and economic restructuring aggravated Finland's unemployment problem. At the same time, investment in technology policy and the pronounced shift from wages to profits was achieved in practice by giving trade unions veto power over controversial labor market reforms (Rehn 1996, 269). Hence, increases in public R&D expenditure became a politically attractive alternative to challenging trade unions on unemployment benefit reform. Despite modest restructuring in 1997, reforms were more limited in Finland than in other Nordic nations. Meanwhile, greater attention has been focused on workplace innovation than on work hours reform: the share of the labor

96 force engaged in part-time work (8.1 percent) is comparable only to that in Greece, Italy, Portugal, and Spain (European Commission 2004, 59). Finally, though collective bargaining has been perceived as quite effective in securing wage restraint, it has also inhibited labor market flexibility and wage differentials.

Slow progress on labor market reforms reflects broader problems associated with managerial reforms and Finland's inability to capitalize from ICT innovation. Despite Finland's status as a leading ICT producer, productivity gains are compromised by limited ICT use. In this sense, Finland contrasts sharply with nations such as the United States that derive most of their ICT-related productivity gains from use rather than production. For example, ICT-producing manufacturing's contribution to annual aggregate labor productivity growth (0.86 percent) was the third highest, behind that of Korea and Germany, among twenty-one OECD nations between 1996 and 2002, while the contribution of ICT-using services (0.12 percent) was the fifteenth lowest (OECD 2004, 96). An important limitation has been Finland's failure to implement the process-based and managerial reforms necessary to exploit new ICT technologies (Rehn 2003, 3–5). Finland's skewed industrial and institutional structure, as well as the specific political bargains that underpinned technology policy, effectively inhibit the translation of technological breakthroughs into organizational innovations.

Accordingly, attention has already shifted to a renegotiation and recombination of domestic institutions on order with the political bargains that were struck in the early 1980s and early 1990s. The highly publicized report of the Finland in the Global Economy Steering Committee has set the agenda by prioritizing innovation and knowledge-based adjustment in an increasingly open, internationalized, and flexible economy. To achieve these objectives, the report recommended targeted tax cuts to encourage SME creation, employment growth, labor market flexibility, and greater investment. The report also encourages increasing economic openness by increasing opportunities for foreign students, experts, and investors and by promoting networking abroad in growing markets such as Eastern Asia. Notwithstanding the increased emphasis on tax and regulatory reform, the report recommends increasing public R&D expenditure by 7 percent a year, with greater attention to strengthening business competence on the one hand and funding innovative start-up companies on the other. Meanwhile, a more active role for the state is envisioned in providing infrastructure to facilitate economic integration and also in active labor market policy, hitherto criticized for its limited coherence and efficacy (Finland in the Global Economy Steering Committee 2004).

The current center–social democratic government has embraced the report, and preliminary evidence suggests that significant movement has occurred along these lines. In addition to further reductions in personal income taxation and the tax wedge, the corporate tax rate was lowered from 29 percent to 26 percent, and capital income taxation was lowered from 29 percent to 28 percent in 2004. Some of these tax provisions, as well as other content directly related to the report, were included in the three-year collective bargaining agreement negotiated between employers and labor in 2004. Meanwhile, state agencies such as Tekes and Sitra have shifted toward a greater focus on business services and commercial application of new technological innovations on the one hand and the needs of SMEs on the other. Though the fate of these and other future reforms remains unclear, Finland's current institutional trajectory continues to defy descriptions of liberal change. Rather, we observe a continuing process of negotiation between key economic actors and policy makers that attempts to build on established, often latent, institutional capacities in facilitating economic adjustment.

CONCLUSIONS: CONCERTATION IN COMPARATIVE PERSPECTIVE

Viewed in relation to the traditional dichotomy between liberal change and neo-corporatist continuity, the Finnish experience illuminates an enduring role for concertation in the face of new technological innovations and shifting patterns of global competition. Concertation did not emerge as an automatic or functionalist response to new challenges. Rather, it was framed by extensive and ongoing liberal innovations in finance and international trade. Paradoxically, these liberal reforms triggered new forms of concertation, most notably in technology policy. Meanwhile, preexisting concertation in wage bargaining and fiscal policy was reoriented to advance ostensibly liberal ends such as radical economic restructuring. This active interplay between liberalization and concertation characterizes contemporary European political economy more generally, as demonstrated with reference to recent developments in countries such as Denmark and Ireland. This chapter uses Denmark and Ireland in order to generalize to two radically different institutional environments. The former country represents an archetypical example of decentralized coordination, while the latter is more generally characterized as a liberal market economy.

Given that institutional renegotiation is shaped by inherited institutional capacities on the one hand and influential societal actors on the other, Denmark and Ireland do not represent a mirror image of the Finnish case. Lacking a

98   pronounced tradition of statist intervention on the supply side of the economy, neither Denmark nor Ireland sought to improve competitiveness with recourse to radical technological innovation. Both countries, however, used
concertation to grapple with a rapidly changing global economy. As in Finland, innovative forms of concertation emerged from the interaction between
preexisting neo-corporatist institutions and liberalizing reforms. Finally, just
as trade union acceptance of Finnish technology policy was purchased at the
expense of more extensive labor market reforms, political bargains in Denmark
and Ireland simultaneously facilitated and constrained adjustment, driving a
process of "perpetual renegotiation" (Ornston 2004).

In contrast to Finland, Denmark illustrates how new forms of concertation
can successfully promote technological and process-based diffusion within existing industries. As in Finland, concertation emerged as a response to lack-
luster economic performance, growing unemployment, and soaring fiscal and
trade deficits during the late 1980s and early 1990s. In Denmark, policy makers simultaneously sought to leverage existing cooperatives between small and
medium-sized enterprises and historically encompassing forms of social protection in bolstering economic competitiveness. Rather than promoting radical technological innovation, however, policy makers used traditional cooperatives and active labor market measures to modernize existing low- and
medium-technology industries such as foodstuffs, clothing, and furniture.
This strategy reflected Denmark's rich tradition of interfirm cooperation, its
politically popular commitment to encompassing social protection, and its
historically limited supply-side interventions (Mjoset 1987, 427–428).

As in Finland, institutional adaptation was characterized by an interplay
between liberalizing reforms and preexisting institutions. In the wake of po-
larizing and fiercely contested efforts to cut social benefits during the 1980s,
policy makers turned to labor market "activation" during the 1990s. To increase labor market participation, the government reduced unemployment
benefit periods and replacement rates. At the same time, it expanded labor
market measures, adult education, and vocational training, while introducing
new paid-leave schemes for retraining and education (Jochem 2003, 132–134).
By 2001, Denmark was spending more on active labor market measures (1.6
percent of GDP) than Sweden (1.3 percent of GDP), despite Sweden's aggressive efforts to limit employment protection, cut unemployment benefits, and
promote labor market turnover (European Commission 2004, 63).

Danish flexicurity supported increasing emphasis on interfirm collaboration, technological diffusion, and process-based innovation within the context of existing employer cooperatives and networks. Explicit and consistent
linkages were drawn between training programs and employer associations,

while enhanced labor market and education policies were designed to fit the needs of industries more dependent on experience-based knowledge than on technological innovation (Nielsen and Kesting 2003, 375–376). As a result, small and medium-sized Danish enterprises could use collaboratively generated tacit or skill-based knowledge to generate productivity increases in traditional sectors. Absent more ambitious interventions in technology policy, however, Denmark lagged in high-technology output and overall R&D intensity (Innovation Council 2004, 61). Along with the expensive fiscal outlays stemming from Denmark's reliance on encompassing social protection, this tension is likely to drive ongoing institutional adaptation and renegotiation in the future.

If the Danish case illustrated how liberalization can trigger new forms of concertation in a different organized economy, the Irish case demonstrates that concertation can flourish within an overwhelmingly liberal context. Indeed, Ireland's underdeveloped industrial linkages, fragmented trade unions, and equity-based financial system reflected a strongly liberal British influence. Ireland appeared poised to follow the British lead toward still greater liberalization and deregulation as the 1980s witnessed increasing wage inflation, fiscal profligacy, soaring unemployment, and ongoing capital flight. Yet the government defied virtually all contemporary expectations by turning to the social partners and forging an encompassing tripartite three-year Program for National Recovery (PNR) in 1987.

In contrast to Finland, Ireland used concertation to promote innovation and economic structuring *from without* through the active recruitment of foreign direct investment. The PNR, which exchanged pay restraint and fiscal austerity for tax concessions, was explicitly designed to establish a cost-competitive foundation for multinational investment and economic recovery. Tripartite concertation was complemented by separate industrial policy initiatives to attract high-technology multinationals and link them to domestic suppliers and entrepreneurs (O'Riain 2000). The initial success of the PNR formed a foundation for ongoing three-year centralized wage agreements emphasizing wage moderation, fiscal restraint, and tax concessions (Hardiman 2002). Though it consciously drew upon earlier neo-corporatist experiments in centralized bargaining, Irish social partnership was framed by a hard currency regime and sharp limits on government spending. Moreover, tripartite concertation was focused on the active recruitment of footloose multinational capital as a strategy for economic restructuring and ongoing upmarket movement (Ornston 2004).

The reliance on multinational investment as an adjustment mechanism speaks to the political and economic limitations associated with Irish social partnership. Rapid economic growth and economic restructuring have provoked an onslaught of societal demands related to wage increases, taxation, and public infrastructure. These quickly escalating demands, which resulted in the breakdown of the three-year Program for Prosperity and Fairness in 2001, threaten Ireland's ongoing reliance on cost-sensitive high-technology assembly operations. Consequently, policy makers have sought other mechanisms for engaging multinational capital. The Irish state has aggressively promoted linkages between multinational investors and domestic entrepreneurs, and the 1990s witnessed a greater emphasis on supply-side investment in technical and vocational training, transportation infrastructure, and indigenous research (Ornston 2004). Contemporary concertation in Ireland, therefore, revolves around efforts to adapt the Irish social partnership, hitherto focused on pay determination and social policy, to facilitate supply-side investment and ongoing upmarket movement (O'Riain 2000; O'Donnell 2001).

5

TELECOMMUNICATIONS COMPETITION IN WORLD MARKETS

Understanding Japan's Decline

Robert E. Cole

Even as global demand for communications equipment exploded during the 1990s, Japan's trade position declined. The decline was pervasive and reflected a deteriorating competitive position. Total exports from OECD countries increased from US$49 billion in 1991 to $165 billion in 2001. While European and American exports grew steadily by over 10 percent per year, Japanese export growth was almost zero. At the beginning of the period, Japan accounted for 27 percent of total OECD communication equipment exports, but by 2001 its share had fallen to 8 percent (Organisation for Economic Cooperation and Development [OECD] 2003: 228–230, as seen in Figure 5.1. At the same time, Japan's telecom imports grew from roughly $2 billion in 1990 to almost $10 billion in 2001. This canceled out the modest rise in exports, leading to a sharp decline in Japan's overall telecommunications trade balance, from almost $10 billion in 1990 to $4 billion in 2001.

This dramatic aggregate decline hides a deeper problem. With some modest exceptions, Japan has been notably absent from the increasingly important Internet network equipment sector. Cisco and other American firms dominate this global market, estimated to be $39 billion in 2000 (Semilof 2000). Cisco in 2002 was estimated to have had 80 percent of the $6.6 billion market for routers, 69 percent of the $10.4 billion market for switches, 48 percent of the $2.1 billion market for IP telephones, and 30 percent of the market for network security equipment (Yamazaki 2003). Since this time, Cisco has lost market share in routers but mostly to Jupiter Networks. Long term, they seem to face a challenge from Chinese companies such as Huawei Technologies and ZTE rather than from Japanese companies.

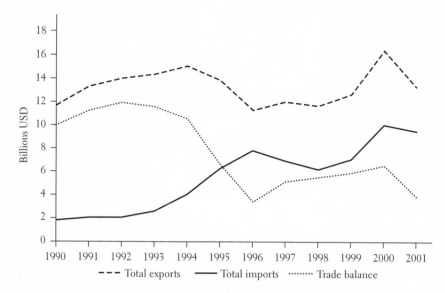

FIGURE 5.1   Japan's trade balance in telecommunications equipment
Source: OECD Telecommunications Database

Japan's market shares in mobile phones, in which one would have expected strong Japanese performance, also deteriorated. Japanese producers were fairly successful in export sales of first-generation analog phones and extremely successful in domestic markets with second-generation digital phones and mobile Internet services, such as i-mode. Nevertheless, Japanese service providers and handset manufacturers for second-generation phones had very little success in global markets, as shown in Figure 5.2.

In handsets, no purely Japanese firm has more than 2 percent of the global market; Sony's joint venture with Ericsson reaches only 5 percent. The failure to develop foreign markets in the 1990s led to extremely heavy — some would say excessive — competition among handset manufacturers for domestic market share. All things being equal, a very large number of domestic players limited the market share and profits of any one producer, thereby making it difficult to build a domestic base for the increased economies of scale necessary to compete in global markets. In provider services, NTT DoCoMo experienced staggering losses in 2002–2003 (some $17 billion at the then prevailing exchange rate) when the bursting of the IT bubble upended its efforts to spread i-mode globally (Nezu 2004).

There is a lot of hand wringing in Japan over the competitive malaise of the telecommunications sector. Some blame the United States for pressuring the

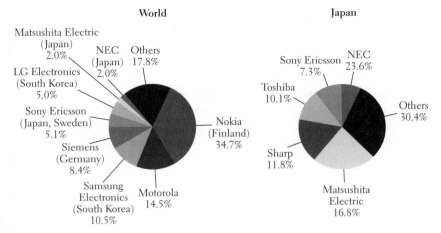

FIGURE 5.2   Global and Japanese domestic cell phone markets, 2003

Source: Nikkei Weekly 2004, 10

Japanese government to break up NTT, seeing its partial breakup as the source of Japan's problems with competitiveness in the telecommunications sector. How do we account for this dramatic shift in fortune? To address this puzzle, this chapter will first analyze Japan's decline and then focus on two key areas of global information and communications technology (ICT) competition: network equipment and mobile telephones.

EXPLAINING JAPAN'S PERFORMANCE LAG

Until 2003–2004, Japan was a consistent outlier in costs of accessing the Internet and prices of leased lines. Despite early public discussions about the coming importance of the convergence between communications and computers, Japan lagged in adopting and applying new Internet capabilities. Three factors explain this lag. They include institutional rigidities within NTT, NTT's relationship with equipment manufacturers, and the government's failure to deal creatively with NTT. These factors share a common explanatory thread: the failure to adapt to rapidly changing, often discontinuous technology.

First, powerful internal institutional rigidities slowed NTT's support for emerging networking technologies in the form of new products and services. Following their long-term mandate to provide domestic universal service, which requires a focus on high reliability, NTT researchers missed the potential of TCP/IP and the Ethernet. NTT's commitment to an alternative networking technology, asynchronous transfer mode (ATM) technology, involving huge sunk costs, further slowed its embrace of the Internet based on TCP/IP.

Institutional rigidities also led NTT to approach standard setting in a manner that damaged its global competitiveness. In networking equipment, NTT was slow to adopt TCP/IP, delaying the development of new networking equipment products, in part because it remained with the slow-moving International Telecommunications Union (ITU) standard-setting process rather than the more dynamic Internet Engineering Task Force (IETF) process. In second-generation digital mobile phones, NTT's adoption of its own operating standard created strong protection for their products domestically but isolated Japan from global competition.

Second, the traditional family of NTT equipment suppliers, NEC, Hitachi, Fujitsu, and Oki Electric, passively followed NTT's technological lead to their detriment, ignoring potential long-term benefits from reorienting product lines to compete in worldwide markets. They were opposed to the NTT breakup even despite potential benefits, since the breakup could be expected to accelerate the development of ICT infrastructure and associated products and services. Equipment suppliers seem to have feared the disruptions that a breakup might create for their up-to-then guaranteed NTT markets. Relationship contracting (long-term, trust-based relationships among upstream and downstream producers) in Japan, though lauded by some, can become a liability when rapid, often discontinuous change is required to take advantage of new opportunities (cf. Dore 1987).

Finally, the Japanese government failed to deal creatively with the NTT telephone monopoly in Japan. Powerful institutional forces and vested interests contributed to slow the process of change, shaping it in ways that preserved much of NTT's structure and pricing power (Kushida, 2006). NTT continued to be at the center of Japanese telecom R&D activity, capable of making important technological decisions and retaining its dominant relationship with equipment manufacturers.

INSTITUTIONAL RIGIDITIES AND DISCONTINUOUS TECHNOLOGY

Failure to adapt to rapidly changing, often discontinuous technology is a common problem for once-successful firms. A basic axiom of organizational analysis argues that once-successful firms often fail owing to the rooting of human capital in capabilities that enabled past success. Dosi (1982) argues that as a firm becomes more accomplished in a given set of capabilities, other capabilities that would allow it to pursue different directions decline. The expertise of successful corporate leaders is based on previously successful strategies and work routines, leading them to promote employees and favor strategies that will keep these tasks central (Burgelman and Grove 1996; Starbuck and Milliken 1988). For Japanese leaders in the private sector, the success formula was low-cost, high-quality hardware achieved through continuous improvement. For NTT, a public sector monopoly in the domestic market with a mandate for universal service, the success formula involved a carefully controlled slow rollout of highly reliable technology that protected its profits and made available large amounts of funds for R&D.

Specific institutional practices and arrangements heighten the problem in Japan. In the United States, sluggish responses by major firms to new challenges open up opportunities for new ventures. However, in Japan, the institutional hostility to start-ups greatly weakens this option. Japanese employment practices are further characterized by late screening, late promotion, and higher firm-specific human capital requirements for promotion (Kato 1993). These practices were held up by Kato and many others as a key ingredient in Japan's success during the 1980s. They gave Japanese managers, when they finally arrived in positions of leadership, a strong stake in the firm and an intimate knowledge of firm culture and practices and imbued them with a long-term perspective. However, when a firm faces the need for rapid and even discontinuous change and the old success formulas no longer work, organizational memory is no longer an asset but rather a barrier to innovation, since leadership is in the hands of those who most thoroughly internalized the old success formulas through long-term service.

INSTITUTIONAL RIGIDITIES WITHIN NTT AND SLOW SUPPORT FOR TCP/IP: THE TECHNOLOGICAL CHOICE OF "BEST EFFORT" VERSUS HIGH QUALITY OF SERVICE

The disruptive nature of the Internet as a "best effort" network was an anathema to NTT's focus on Quality of Service (QoS), making it inherently difficult for NTT to quickly support TCP/IP. The Internet maximized neither speed nor reliability. In the early 2000s, some 3 percent of all packets sent daily are dropped,[1] each packet is delayed by variable and unknown amounts, and the bandwidth available to each connection is unpredictable. The many low QoS and reliability features exhibited by the early Internet are a common feature of disruptive technologies (Christensen 1997). Many of these same technologies, however, incrementally add new features and improve reliability as one after another of its technical problems is resolved. The traditional "five nines" (99.999 percent) reliability target of telephone companies was simply not a design requirement for the Internet architects. All this was anathema to the QoS culture of NTT.

NTT'S INSTITUTIONAL COMMITMENTS TO ATM

ATM was consistent with the high-reliability culture of NTT. The ATM network, like the Internet, uses packet switching, except that its packages (called cells) are fixed in length and small. ATM is connection oriented, meaning that all host-to-host communications requests are provided a dedicated connection (fixed route) through the network. ATM emphasizes the active configuration of QoS parameters to ensure high quality and reliability (Messerschmitt 2000). As such, it is a very complex system. NTT was institutionally committed to ATM technology, into which it had poured extensive resources, even after doubts over its potential arose. NTT began research on ATM switching in the mid-1980s. Moreover, ATM was "competency-enhancing" (Anderson and Tushman 1997) in that it was a natural extension of the existing public telephone network, a network that relied on circuit switching, in which distance and duration of connection determine the cost of service (Yamashita 2004). Such competency-enhancing technologies are typically easier to incorporate into incumbent organizations and are thus more attractive. NTT, predictably, wanted to continue extracting high levels of profit from their existing fixed-line investments.

The development of ATM was designed to improve NTT's existing digital switches. NTT officials believed in the early 1990s that ATM, originally designed for voice, was the ultimate solution, mixing voice and data traffic over fiber. In 1991, Fujitsu became the first company worldwide to offer an ATM switching system that enabled high-speed two-way transmission and routing of voice, video, and data simultaneously. NTT anticipated replacing the existing narrow-band digital network with a large capacity broadband ATM trunk line network around the year 2000 (Fransman 1995). NTT officials were confident that they were building the new information highway.

NTT's technical and financial investments in ATM and associated ISDN services were enormous. Fransman (1995, 115) reports that NTT spent $2,547 million for R&D in 1993, 4.5 percent of its revenues; ATM was a top priority in that spending. Nezu (2002b, 17) reports that NTT's insistence on rolling out ISDN on a nationwide basis (only Germany made a similar mistake) led to a waste of roughly $9 billion of investment, all wiped out by the spread of DSL broadband access, which NTT had initially resisted (ISDN had roughly one-sixth the speed of DSL). NTT pursued this dead-end trajectory, but-tressed by tax incentives and public money, with the strong support and urging of the Ministry of Post and Telecommunications (MPT). The early decisions to support a "network build-out" of ISDN services made sense in terms of providing an interim technology before fiber optic networks became possible (Kushida 2004).

In 1995, NTT was experimenting with vBNS, a network built on the commercial ATM lines. It was built with a speed of 155 Mb/s and was expanded to 622 Mb/s in two years, with the aim of achieving 2.4 Gb/s. However, at the end of 1997, NTT and its collaborators concluded that a speed of 2.4 Gb/s was not feasible, and that it would be more effective to exclude ATM from the network and utilize IP directly over SONET (synchronous optical networking, a network technology for communication over optical fiber). Only at this point did NTT executives begin to realize that ATM was not the ultimate end-to-end solution (Oie et al. 2001). As it became clear that ATM was not the ideal technology, NTT's strong internal political commitments to ATM were revealed. NTT executives did not want to admit failure even after key engineers in their Basic Research Lab concluded that ATM could not provide the speed possible with TCP/IP. In particular, NTT's Network Service Systems group, which made telephone switching systems for ATM, pushed ATM and lobbied the NTT executives to continue supporting it throughout much of the 1990s. It was not until the late 1990s that NTT finally stopped their research on telephone switching units based on ATM.

SPEED AND ALTERNATIVE MECHANISMS OF STANDARD
SETTING: NTT'S COMMITMENT TO ITU

Although NTT's adaptation to discontinuous technology in the form of TCP/IP was rather quick by historical standards,[2] it was painfully slow from the perspective of coping effectively with the rapidly changing competitive environment. NTT's decision to have the entire ATM network go through the ITU (International Telecommunications Union) standardization process prior to implementation was a very time-consuming process.

ITU is the international organization within the United Nations system through which governments and private sector companies coordinate global telecom networks and services. The ITU has been known for its very slow standardization, with a long iterative process of negotiation and lengthy discussion before a common standard is finally selected. Moreover, to satisfy as many of these constituencies as possible, the solutions have tended to involve complex protocol solutions that are difficult to implement. NTT had long worked with

ITU on other projects, and it seemed only natural to managers looking for promotions at NTT that they would continue to do so with ATM.

The IETF, in which Americans have been the most active members, however, took a quite different approach. Comprised of volunteers, the IETF had its beginnings in 1986 at a meeting in San Diego attended by twenty-one individuals and has evolved into the principal body developing new Internet standard specifications. Much of the work in the IETF is done in working group meetings relying heavily on online communication; in these groups emergent policies are made by rough consensus rather than by unanimity in the ITU. The face-to-face working group meetings are much less important than the need to gain consensus on the working group mailing list. Every IETF standard is published as an RFC (request for comments), and every RFC starts out as an Internet draft.

After an Internet draft has been sufficiently discussed, and if there is rough consensus that it would be a useful standard, it is presented to the Internet Engineering Steering Group (IESG) for consideration. The IESG, after soliciting still more comments, has the authority to approve the draft as a proposed standard. Before this happens, however, it is "highly desirable" that there are independent interoperable implementations of each part of the standard. The greater the number of successful independent interoperable implementations, the more substantial the operational experience, and the more the candidate specification is used in increasingly demanding environments, the more likely the draft will be accepted and eventually become a draft or Internet standard. Most of the standards in common use are proposed or draft standards that have never moved forward to the status of "full" Internet standards, since people often found more important projects to work on or the specification was not considered all that important.

These working routines also make the IETF much more responsive to real-time market forces in its development of new Internet standard specifications than the ITU was in its development of ATM standards.[3] The Internet community's support for a given standard, among other things, is stronger for technology solutions that have been successfully deployed. Letting the marketplace decide the winners, though not without its problems, tends to be a faster process. This case is also instructive for researchers, who commonly distinguish between de jure standards created by committees and de facto standards created by markets (Besen and Farrell 1994; Funk 2002). The IETF process shows that a committee-based approach facilitated by online communication can be a powerful force that is quite in synch with market forces (cf. Shapiro and Varian 1999). Japanese researchers played only a minor role in setting

IETF's Internet standards in the critical early years (1986–1996). Though the number of Japanese attending IETF meetings has grown in recent years, they participate mostly as observers rather than as active members. That is telling in terms of their continuing status as followers rather than leaders.

NTT worked on standardizing the basic protocols of ATM for broadband network from 1985 to 1989. Yet, major problems remained. Because the paradigm of ATM involved setting up a dedicated connection from a sender to a receiver via a sequence of switches, a great burden was put on the functions of packet switches. Further demands were placed on the packet switches by the growing expectations for faster transmission speed. These technical problems combined with the slowness of the ITU committee-based approach to standardization enabled the Internet to become well established as a commercial entity before ATM could be effectively deployed (Messerschmitt 2000).

THE EVOLUTION OF INTERNET CAPABILITIES AT NTT

Of course, NTT engineers were not ignorant of the emergent Internet technology. A small informal group promoting Internet concepts emerged in NTT in the early 1980s.[4] Shigeki Goto, a research group leader (*kacho*) at NTT Research Laboratories, arranged to send a research group to Stanford University. In 1988, the Japanese team finally succeeded in connecting the NTT Laboratories computer network to the CSNet (computer science net) and ARPANET through CSNET in the United States. This was done informally because at the time NTT was forbidden to engage in overseas activities.

The Internet group operated initially as a "skunk works" (an informal group flying under the radar of the formal organization) within NTT. A key step in the process of formal recognition of this group came in 1992, when Shigeki Goto became a department head (*bucho*). This enabled him to start several Internet projects with the official support of the director of NTT software laboratories. Interestingly, in the late 1980s and early 1990s, despite the resistance to the Internet from most NTT researchers and executives, many of the NTT researchers engaged in ATM-related work, such as those writing software for ATM switching units, came to use TCP/IP as a tool to do their work. The initially small pro-Internet group promoted TCP/IP and Ethernet as desirable solutions, and gradually key researchers were won over. The turning point for their acceptance of TCP/IP was the arrival of Mosaic in 1993–1995.[5] One member of the Goto group, Ken Murakami, and his NTT colleagues developed a high-speed protocol for IP (MAPOS) in 1996–1997 and demonstrated its effectiveness by building a system between Kyoto and Tokyo. He offered it to others at NTT, but they were not interested since it "was not ATM." Ironically, he made it available to Cisco, which implemented a version of it on their routers.

LACK OF INITIATIVE BY EQUIPMENT FIRMS
TO REORIENT PRODUCT LINES

In keeping with NTT's long-standing view that its job was to lead equipment vendors in developing complex new technologies, NTT led its suppliers in developing ATM switches (Fransman 1995). In the same vein, the major Japanese electronic firms were accustomed to relying on NTT to set future technology directions in communications. With NTT slow to grasp the significance of TCP/IP and Ethernet, the major electronic firms, not surprisingly, lagged in the development of ICT products and services.

NTT took the lead in developing the software required for the broadband Integrated Services Digital Network (ISDN) services they expected to deploy over the ATM network. ISDN requires digital switches, and many researchers were consumed with developing hardware that would allow telephone lines to handle data. TCP/IP protocols, by contrast, are mostly implemented in software, running on both the routers and the user's computers. Software now drives most networking functions and allows new features to be added in the field.[6] Western telecommunication companies (e.g., Nortel, Alcatel) increasingly shifted their activities to software, outsourcing much of their hardware manufacturing. Japanese telecom companies have been slow to make this transition from hardware to software and therefore are, not surprisingly, almost completely absent from global markets for network products (International Data Corporation 2000).

Without a strong domestic market in the networking products associated with TCP/IP and Ethernet, Japanese electronic firms were unable to take advantage of economies of scale that could serve as a platform for competing in international markets. Moreover, Nezu (2002b) argues that the telephone and telecommunications equipment makers were initially reluctant to develop routers because they were much less profitable than the existing large-scale switching machines used by the telephone companies. By 1995, routers and the protocols they ran had become so complex that competition in this market required a steep learning curve. Even Bell Laboratories, which tried to build routers competitively in the mid-1990s, had difficulty.[7] All the major national telephone monopolies, not the least of which was AT&T, were slow in recognizing the significance of the Internet and associated products

(Naughton 1999). None of this would have mattered, however, if Japan had had a vibrant start-up sector that could take the initiative, as was the case in the United States.

JAPAN'S MISSED OPPORTUNITY IN SILICON VALLEY
AND CISCO'S FIRST-MOVER ADVANTAGE

The behavior of NEC and Fujitsu reveals their passive stance in following NTT's technological lead. In the early 1990s, Masao Hibino, president and CEO of NEC Magnus Communications, was general manager of modem development at NEC and stationed in Silicon Valley. He thought TCP/IP and Ethernet were important developments and sent information to NEC offices in Tokyo to that effect. They responded, however, that TCP/IP was not real communication because it was "connectionless," a view that stemmed from their belief in the necessity of a dedicated connection. At this time, he says, "NEC people thought ATM delivering ISDN services was the final solution to broadband. Everyone in Japan thought so and we worked with ITU-T (International Telecommunication Union–Telecommunication Standardization Sector) to get each standard approved for ATM." NEC also had mid-career industrial associates at the University of California, Berkeley, who were aware of the evolving Internet technology. Such individuals, however, were reluctant to take the risks entailed by strongly championing a non-mainstream technology.

The same was true of the many Japanese executives stationed in Silicon Valley, some of whom saw the Internet as an important development. Many had great difficulty penetrating the local knowledge networks. To succeed in Silicon Valley in knowledge acquisition, one needs to develop an exchange relationship in which one gives as much as one gets, labeled "informal know-how trading" by Eric von Hippel (1988, 76–92). In addition, as Annalee Saxenian, a knowledgeable observer of Silicon Valley puts it, one also needs to "marinate" in the Silicon Valley culture to be able to draw upon its knowledge (2003). The Japanese assigned to Silicon Valley appear to have been weak in both these areas.

By 1995–1996, NEC executives had begun to realize that the Internet was different from what they had thought, and by 1997–1998 they realized that ATM was not a solution to broadband. By then, however, they had made substantial financial and political investments in ATM switches and had developed related products, making a switch to TCP/IP difficult. Hibino believes that NEC might have caught up if they had shifted everybody working in

ATM to TCP/IP at that time, but they did not. In late 1999, the mobile phone market was growing rapidly, and the number of subscribers to DoCoMo's i-mode was exploding. NEC, accustomed to following NTT's lead, shifted resources, including personnel who had worked on ATM, over to second-generation phones, anticipating potential for exports to China and Europe. However, they were wrong, as we shall see.

Cisco already had a strong market position, and it was hard for NEC to differentiate their products and find a niche. At the same time, the hardware, and especially the software, had become quite complex. Cisco had proprietary IOS (internetwork operating system) intellectual property based on the TCP/IP protocol. Simply copying Cisco products would have led to legal action from Cisco. So in 1997, NEC made the decision to distribute Cisco routers, hubs, and switches. The problem, however, was not Cisco's intellectual property rights per se. Japanese vendors could develop their own router codes to get around that. The challenge was posed by the deployment of Cisco products by private sector enterprises around the world and those enterprises' familiarity with Cisco's router command line interface (CLI). Those customers wanted to ensure that any new hardware and software had interoperability and compatibility with Cisco products and operations. Cisco also drew competitive strength from the extensive versions of its software that accommodate different legacy systems.

Nonetheless, NEC continued to work on the technology and sought new combinations so that they were not using pure "routers." The idea was to develop new integrated products (e.g., products that included routing and bridging technologies) that would not run afoul of Cisco's IP. Despite these efforts, in the fall of 2003, NEC was engaged only in domestic production of routers, hubs, and switches. Cisco, Juniper Networks, and Redback Networks — all American producers — had an 88 percent market share of the world market for routers for telecom carriers ("Tsushin Seigyo Souchi: NEC-Hitachi Kyodo Kaihatsu [Telecommunications Controlling Equipment: NEC and Hitachi to Develop Jointly]" 2003). Cisco's share of the Japanese market, however, was only around 40–50 percent, for many Japanese companies stayed loyal to the products of domestic companies; these firms also sought to buy products from Cisco competitors such as Juniper Networks.[8]

Seeking to keep pace with the evolving technology, NEC and Hitachi announced plans in December 2003 to jointly develop next-generation routers designed for high-speed Internet connections to telecom service providers. The total development costs over a three-year period were estimated at about 20 billion yen ($180 million), with half the cost subsidized by the Ministry of

Economy, Trade and Industry (METI) ("Electronics Giants to Develop High-Speed Internet Routers" 2003). It is remarkable (or perhaps unremarkable) that despite strong foreign and domestic criticism of METI's old-style "industrial targeting," it continues to orchestrate and invest in such downstream product development activities.

Fujitsu's experience was broadly similar to that of NEC, but with a few differences. In the mid- and late 1990s, Fujitsu had two groups important to the adoption of Internet-related technologies: the Communications Systems Group, which focused on sales to the telecommunications sector and carriers such as NTT, and the Computer and Information Processing Group, which focused on sales to enterprises. The Communications Systems Group, like NEC, was accustomed to following the direction set by NTT therefore to seeing the future as ATM delivering ISDN services. By contrast, the Computer and Information Processing Group was more open to Internet technologies. However, investment decisions were made by top corporate leaders who, at the time, did not appreciate the potential of emergent Internet-related applications and businesses. As a result, Fujitsu ended up investing heavily in carrier routers but not enterprise routers, abandoning the enterprise global market to Cisco.[9]

In early 2004, Fujitsu announced it would stop developing routers themselves and instead distribute routers from others. Cisco and Juniper's overwhelming market shares and strong price competitiveness, combined with the continued weakness of Fujitsu's telecom equipment business (in the red 30 billion yen—$272 million—for fiscal year 2003–2004), were undoubtedly factors in this decision. The bankruptcy of WorldCom, its biggest U.S. customer, had badly hurt their performance in 2002 and 2003.[10] Fujitsu is seeking to rebuild its telecom business by strengthening the development of low-price servers with router functions, distributing products of other manufacturers, and exploring co-development with other large electronic companies ("Fujitsu, routa gaibu choutatsu [Fujitsu Will Procure Routers from Outside]" 2004).

JAPAN'S FAILURE TO DEAL WITH NTT CREATIVELY

As we have seen, NTT's role as the technological leader led equipment manufacturers to take passive strategies, to the detriment of the competitiveness of Japan's network equipment. In addition, NTT's power over prices and its reluctance to deploy DSL hampered competition and demand for the Internet, data networking, and other information services. Both of these problems stemmed from NTT's continuing status as an overwhelmingly powerful actor in the sector, in turn a result of the Japanese government's failure to deal with NTT creatively.

The privatization of NTT and the introduction of competition into telecommunication service sectors began in response to the AT&T breakup and in the context of Japan's push for administrative reform. In 1982, the Second Provisional Council on Administrative Reform (Rincho) proposed the breakup of NTT and allowing competition in all telecommunication service sectors. In 1984, the Japanese Diet passed three telecommunication reform laws setting the privatization process in motion and establishing MPT as the dominant regulator. Responsibility for price and service regulation shifted from the Diet to MPT. MPT chose to micromanage competition in the telecom sector, orchestrating the entry of new competitors by evaluating all price or service changes in terms of their potential impact on the competitive balance (S. Vogel 1996). In the early years, it favored new entrants over NTT and engaged in contentious struggles with NTT.

It was not until 1997—fifteen years after the AT&T breakup—that the final terms of the NTT breakup were announced. They stopped short of the AT&T model, using the newly legalized holding company structure to divide NTT into three companies—one long-distance and international company and two regional companies. NTT was forced to decrease the cross-subsidization of activities, but the holding company fell far short of the original goal of breaking up NTT. Furthermore, as Kushida (2006) shows, the government has continued to play an important role in NTT governance.

One effect of the breakup bargain was that the MPT switched from a pro-competition, anti-NTT stance to one more protective of NTT. They worked out the basic agreement on interconnection between competing carriers and NTT's infrastructure based initially on using NTT's historic costs to calculate interconnection charges. The net effect was to discourage new entrants by keeping charges high. Despite strong pressure from the U.S. and Japanese constituencies, which have been arguing for lower access charges, NTT has only grudgingly lowered its rates (they were still eight times higher than those in the United States as late as 1999) and interconnection charges still stood at twice the level of those in the United States, France, Germany, and the United Kingdom in early 2003 (Carbaugh 2002). In spring 2003, the Ministry of Public Management, Home Affairs, Posts and Telecommunications (MPHPT), formerly the MPT, announced that they planned to raise—not lower—the interconnection rates by 12 percent to compensate for NTT's lower revenue, the result of declines in fixed line usage, and to avoid the specter of a World-Com bankruptcy in Japan. It was widely rumored that they succumbed to the pressure from NTT-related diet members (Tilton 2004, 4–5).

Throughout the 1990s, cheap fixed-rate Internet service was simply not available to Japanese consumers. It was not until around 2000 that Japan's Internet access rose sharply. Government inaction delayed the rollout of high-speed Internet access through DSL. NTT was committed to fiber optics (fiber to the home or FTTH) as the optimal solution to solving the "last mile" problem and did its best to ignore DSL. When that failed, they acted to obstruct attempts by new companies to connect to their backbone infrastructure — easy for NTT to do because competitors had to install their own equipment in NTT's branch offices, and clear rules were lacking. In early 2000, politicians became concerned that Japan was falling behind not only the United States and Europe in making broadband service via DSL available to households, but Korea as well. As a result, the government clarified rules for competitors accessing NTT branch offices; the Fair Trade Commission (FTC) also intervened to pressure NTT to make its local loop available to entrants (Kushida, this volume).

With these changes, new subscribers to DSL exploded, reaching 4 million by September 2002 and 13 million by the end of 2004. Competition has led DSL rates to fall to 2000 yen or US$16.94 per month, the cheapest in the world. It is for this reason that some would claim real competition in Japan's telecommunication market has just begun (Nezu 2002b). By waiting so long, however, the Japanese gave their global competitors a huge head start. NTT had failed to realize the potential for DSL as an interim broadband technology before FTTH could be more fully deployed. Like other national monopolies, NTT had long been accustomed to being able to deploy new technologies that met its high standards at a pace it alone decided was appropriate. They had been reluctant to deploy DSL not only because of their commitment to fiber optic networks but because DSL undermined its ISDN services. Aggressive new competitors such as Yahoo!BB, however, were much more sensitive to potential customer demand and forced NTT's hand. Unlike FTTH, DSL did not require a large-scale infrastructure investment.

JAPAN'S FAILURE IN GLOBAL MARKETS FOR WIRELESS EQUIPMENT AND SECOND-GENERATION PHONES: THE PDC STANDARD

There is a striking dichotomy between Japan's impressive domestic mobile telecommunications market on the one hand, with its sophisticated handsets and highly developed mobile Internet services, and its weak performance in global markets on the other.

NTT DoCoMo's successful i-mode service has been the bright spot for Japan's telecommunications sector. As early as late 2001, the Japanese mobile

Internet market had 46 million subscribers, dwarfing equivalent services in the rest of the world. DoCoMo's i-mode in early 2003 had some 70 percent of the total mobile phone subscribers and an even higher proportion of the income from related services. Following Funk (2002, 221), we can characterize Japan's mobile Internet offering as providing great reach (the number of people participating in sharing information) but modest richness (the quality, depth, and bandwidth of information). DoCoMo's i-mode represented a successful improvisation in combining and reconfiguring existing technologies (e.g., using c-HTML) and a pioneering approach to the mass marketing of mobile Internet to a consumer market with a clever revenue-sharing model with content-providing partners.

Japan's weak performance in global markets, in large part, grows out of an important decision made in 1993—NTT's selection of a closed digital standard for its second-generation digital cellular phones, known as personal digital cellular (PDC). As Funk (2002, 13) points out, global mobile communication standards are relatively open in the sense that they were typically created in open standard-setting processes. This was not the case for PDC, especially from an international perspective. PDC was adopted only by Japan, and NTT dominates the standard-setting process. This is in stark contrast to the dominant communication standards for second-generation digital cellular phones, the European-driven GSM, adopted by 120 countries by the end of 1998 after broad participation in the standard-setting process (Funk 2002, 12). What foreign company would want to sign on to the PDC standard for its domestic market when one service provider in Japan (NTT) and its exclusive set of Japanese handset makers had considerable market advantage? Conversely, foreign handset makers were unlikely to make strong efforts to crack the Japanese domestic market since they would have to meet unique design requirements.

The PDC digital standard was largely developed by NTT and is still largely controlled by NTT DoCoMo.[11] The technology and the wireless R&D lab, along with a large number of the key engineers involved, were transferred to DoCoMo when it was established as an NTT subsidiary. NTT DoCoMo developed detailed specifications for the standard in cooperation with selected handset manufacturers. MPT required that the Association of Radio Industry Business ratify the new standard and that DoCoMo publish a set of specifications for the air interface. Other service providers then provided PDC services based on these specifications. However, despite this public disclosure, the four major handset manufacturers for DoCoMo—Matsushita, NEC, Mitsubishi, and Fujitsu—preferentially received more detailed information about the PDC standard. This gave them great advantages over their domestic competitors, enabling them to solve various air interface problems and to develop phones superior to those offered by other manufacturers. In return, they delayed the sale of their series of highly competitive phones to service providers other than NTT DoCoMo, giving DoCoMo significant advantages over its competitors (Funk 2002).

In this environment, DoCoMo has been able to dictate its terms to the handset manufacturers, who willingly accepted DoCoMo's leadership because they were guaranteed participation in DoCoMo's dominant share of the domestic market. Notable is the government's complicity in DoCoMo's dominance: the government failed to ensure that the public disclosure of the standard specifications for handsets provided by NTT DoCoMo be sufficiently detailed and available to all parties so as to create a level playing field. This process also can be contrasted to GSM standards in Europe, where detailed specifications were made available to all relevant parties in a timely manner.

JAPAN'S NONMODULAR HANDSET DESIGNS AND NTT'S DOMINATION OF MANUFACTURERS

European, American, and Korean handset makers rely on programmable processors using software for expressing features and other modes of differentiation. Their phones are more modular than Japanese handsets, with companies such as Nokia developing platform modules (called engines) that can then be used for different models, resulting in huge cost savings. Nokia was initially forced into the platform model by its need to serve multiple carriers across the European market. In a similar fashion, Samsung generally develops platforms for several phones to service diverse global markets. They develop a software-oriented core platform onto which they add many of the carrier requirements. They then use this core platform to meet other carrier's requirements. They can also differentiate using different hardware components for a given platform, so even if they use a common platform, the final product can be quite different.

In employing these strategies, Nokia and Samsung contrast sharply with Japanese handset manufacturers, who largely build every model "from the ground up," both hardware and software. Japanese carriers such as DoCoMo and Au (the KDDI brand) contract with their vendors to develop distinctive phones from the basic hardware and compensate them by guaranteeing high-volume sales or in some cases commit to paying a portion of the vendor's R&D costs. This model reflects the continuing power of NTT over the handset makers. Were the handset makers oriented toward servicing multiple carriers, they might have developed a platform model and thus become more competitive in world markets. After all, some Japanese firms have had very successful experiences in using the platform model in world markets, such as Sony's experience with the Walkman or global automobile manufacturers such as Toyota.[12]

It may seem strange that NTT chose to develop the PDC standard as its own unique solution to second-generation phone standards. NTT's failure to anticipate the negative impact of developing a proprietary standard on equipment exports (e.g., handsets, base stations) reflected, in part, NTT's insularity, resulting from its historical mandate to serve the domestic market. It also reflected its status as an engineering-driven rather than a market-driven company. Adoption of the PDC standard meant that the Japanese handset manufacturers were kept busy supplying unique and proprietary phones for the domestic market. NTT continued to upgrade features that kept the R&D staff of the handset manufacturers busy. As a result, handset makers did not have the resources to develop phones and other telecom equipment that could meet U.S. or European standards. Nor could they build economies of scale based on strong home market sales to penetrate foreign markets.

JAPAN VERSUS KOREA: DIVERGING APPROACHES TO GLOBAL STANDARDS

One can contrast the choices made by NTT with the choices made by operators in Korea. Both countries have highly regulated and protected wireless markets with three carriers, one of which is dominant with more than a 50 percent market share. Instead of the dominant carrier developing a proprietary standard for a captured domestic market, the Korean carriers, under pressure from the government and with a small domestic market, experienced a strong need to adapt to global standards and participate in global markets. They were early licensees of Qualcomm's proprietary code division multiple access (CDMA) technology. The CDMA standard has become increasingly accepted around the world, with some 164 million subscribers by mid-2003. The Korean carriers, as early licensees of Qualcomm, were able to exploit that technology and develop with it ahead of most others. As a result, companies such as Samsung and LG were able to leverage their CDMA experience along with their consumer electronics know-how (especially color LCD and camera technology) to take a large and growing share of the worldwide handset industry.

The contrast with the Japanese selection of its own proprietary PDC standard could not be starker.[13] The Japanese handset makers, in contrast to the Koreans, are vulnerable to the seduction of their large domestic market.[14] The business model adopted by the carriers and the handset makers, based on custom-built handsets, works admirably well for the large Japanese domestic market, yielding strong profits to major industry players. In other industries, the Japanese have been able to build on the economies of scale achieved in domestic markets to launch successful attacks on global markets. This success assumes that key users in the domestic market are sufficiently in tune with global user needs and that the domestic business models (including standards) can be applied to global markets. In the case of mobile phones, however, the business model underlying customized handsets and the advanced set of so-phisticated mobile services available in the domestic market have not been in tune with the capabilities needed to meet global demand. Similarly, the stan-dards adopted were not compatible with the emergent global GSM standards. Nor can individual Japanese handset makers easily shift to a modular phone strategy without sacrificing their profitable domestic market share to existing competitors.

Notable also is the pride that NTT engineers demonstrate in taking a lead-ership role, whether it be in developing ISDN, ATM, or their own PDC stan-dard for second-generation phones. As public employees (until the partial breakup in 1997), many felt they truly worked for the advancement of Japan and its citizens. That desire to play a leadership role building only on their own internal capabilities turned out in this case to be a significant liability. It is a great asset to have highly capable engineers and for those engineers to have great pride in their capabilities, but the decision to either make one's own technology and standards or cooperate with others in building an interna-tional standard is a business decision. Being capable of creating one's own technology and standard, however, does not necessarily mean that one should. Especially in setting worldwide telecommunication standards, open innova-tion is a more productive strategy (Chesbrough 2003). Certainly, the explosion of demand for European telecommunication equipment in the 1990s is to be understood in part as an outcome of European company success in collabo-ratively building and then leveraging the global GSM standard.

To add to the mix, Funk (2002, 71–72) speculates on what would have hap-pened if Japan had adopted the GSM rather than the PDC standard. Aside from benefiting Japanese consumers, since Japan's cellular service is much more expensive than that of European countries, Japanese mobile equipment manufacturers would have been better poised to compete in global markets. By working on GSM and creating European research labs such as those of Motorola, they might have developed their own patents and would have had access to the patent portfolio of other producers based on an exchange of patents agreement among GSM producers. Such agreements allowed the developers of GSM products and services to move the technology forward much more rapidly. Participating in these arrangements would have enabled

120 Japanese producers to achieve global economies of scale instead of retreating to the fortress of their far smaller domestic market, protected by the PDC standard. There is reason to think that Japanese producers would have done particularly well in global infrastructure competition because of their strong technology in frequency spectrum efficiency and low-power technology. On the domestic side, Funk (2002, 72) concludes, "if Japan had adopted an open standard like GSM, they would have been forced to develop a method of competing in the Japanese market that did not depend on control of the standard."

IS JAPAN LEARNING FROM ITS MISTAKES?

In more recent years, NTT DoCoMo seems to have learned something about the importance of open standards and of combining market pressures with a standards-setting committee system. NTT was held responsible for Japan's dramatic decline in its share of global equipment export markets by its adoption of the proprietary PDC standard. MPT pressured them to take a different approach to the development of the third-generation standard, and DoCoMo participated more fully in developing the global third-generation standard. DoCoMo forged an alliance with Nokia and Ericsson that includes acceptance of Japan's W-CDMA technology for outdoor applications in exchange for Japan's accepting the GSM network interface.

The Japanese handset manufacturers, however, are still a long way from building the global platform management and product line strategy based on modularity that has allowed Nokia and Samsung to become undisputed leaders in global mobile phone markets. To succeed in these markets requires a deft integration of committee-based standard-setting activity and global market competition that the Japanese firms have thus far found quite elusive. It is naive to expect that simply by developing a more intelligent standards policy (though surely that is needed) Japan would be able to restore its former share of communication equipment exports.

A lot has changed since 1990. Japan played a big role in first-generation analog phones when competition took place in intermediate goods and some of the Japanese handset makers successfully supplied phones to U.S. and other carriers. The industry has now moved toward a customer-driven market where brand plays a dominant role. Sales channels are different and more complex. Japanese cell phone technology has in certain respects evolved in isolation from mainstream developments of other global players. The reason is that the R&D and manufacturing expertise of Japanese handset makers is geared to customized Japanese carrier designs. In short, the phones of each Japanese carrier, such as DoCoMo and KDDI, have unique proprietary handset designs. As discussed, these handsets have a more integral architecture (less modular) than European- and American-produced phones and rely less on software to express desired features. Lacking modularity, they bear the heavy costs of designing every new product from the ground up. These factors combined constitute a significant though not insurmountable barrier to Japanese handset makers in their efforts to make a strong play for global markets.[15] At the same time, the approach of Japanese handset makers is not without its merits. Their integral designs allow them to optimize specific models to ensure more compact and more sophisticated phones with many advanced features. This is particularly appealing at the high end of the market.

Competition is made all the fiercer by the dominant position of Nokia and the rise to prominence of Samsung and LG Electronics. Samsung can now supply relatively high-end phones at prices that the Japanese find hard to match. Moreover, Samsung's brand is now far better known around the world than is that of any Japanese cell phone producer. All these conditions make it quite difficult for Japan to recover its previous share of mobile phone-related equipment exports. Ten years is a long time to be out of the global market. Many Japanese telecom firms are putting bets on the Chinese market, but there the challenges are as large as the opportunities.

CONCLUSION

The 1980s and 1990s saw Japan miss a huge opportunity because of its failure to effectively deregulate the telecommunications sector and break up NTT's monopoly power. As a result, NTT was able to keep prices high and to push its own proprietary technologies. This severely hindered the development of ICT infrastructure, and that, in turn, retarded the development of ICT products and services. Institutional rigidity and ill-conceived decisions regarding standard setting have clearly slowed the growth of Japan's ICT sector. The absence of a hospitable environment for new ventures and the constraints imposed by relationship contracting on key electronic firms by their ties to an NTT committed to ATM technologies slowed the private sector's embrace of the Internet and related networking technologies.

As a result, U.S., European, and later Korean firms reaped huge first-mover advantages in the global network equipment and mobile phone markets. In the case of handsets, this is ironic, for the Japanese have arguably the most sophisticated and content-rich handsets, along with a pioneering revenue-sharing model in their domestic market, and the highest number of subscribers accessing the Internet through wireless connections. Yet, their commitment to a proprietary PDC standard for second-generation phones proved a major barrier to translating these advantages into global market sales. Moreover, the dominant handset makers favored by DoCoMo were so subordinate to DoCoMo that they had to devote most of their resources to building customized models. This left them unable to move toward a more modular platform strategy that would yield greater economies of scale and allow them to better compete in global markets. As firms and customers move into third-generation phones, it is not clear that the dynamics have sufficiently changed so as to alter the current outcomes in handset global competition in which Japanese firms are weak players. Vodafone's Japan subsidiary has taken some steps toward a platform model, which makes sense when viewed from Vodafone's efforts to make Japan's technology available in its worldwide markets. NTT, however, has shown no signs of moving in this direction.

To be sure, the advantages held by Japan's competitors are by no means unassailable. A smaller amount of national real estate, high population density, and a small number of major metropolitan areas favors lower rates in Japan than in the United States over the long term. For Japanese firms to become more competitive will require, however, a further weakening of NTT power. It also will require a willingness of regulators to leave NTT firms to some combination of market competition and international committee-based standard setting. Japanese firms, especially handset manufacturers and infrastructure providers, need to learn to participate more effectively in the international standard-setting process. The weakness they have demonstrated in the past in setting global standards seems to be due to some combination of factors, including poor English-language capabilities; an engineering rather than a market culture, especially on the part of NTT; their insularity; a desire to protect domestic markets at all costs; and a failure to join the shift from a traditional committee-based approach to the more dynamic IETF model.

On a different level, the Japanese government also needs to further bolster market forces, something they have shown little taste for since their rapprochement with NTT. In 2003, as mentioned earlier, they raised interconnection rates. In so doing, they of course made it harder for new entrants to compete with NTT. However, this is part of a larger problem faced by all advanced countries with a landline infrastructure relying on the public switched telephone network (PSTN). It is the political problem of how to continue delivering universal service without weighing down the leading part of the industry. The pattern has been to have the leading edge of the industry subsidize the lagging part, hence the rise in interconnection rates.

Nobuo Ikeda (2003) has proposed that in Japan the two parts of NTT be separated, with PSTN being run as a government-owned universal service company to be liquidated in the long run. This would leave the IP part of NTT free to innovate and grow without worrying about cannibalizing land lines or having to subsidize the provision of universal service for land lines. This is a novel solution and was predictably met by political opposition from vested interests. The Americans face the same problem, and a good solution is not in sight. To date, the Federal Communications Commission has responded to the declining revenue base for universal service (resulting from the loss of revenue by long-distance carriers) by increasing the universal service charge for wireless. Again, the problem is one of burdening the innovative elements of the industry with the need to subsidize the declining portion. Whoever finds a workable solution that frees up the IP part of the industry to innovate without weighing it down with a need to subsidize universal service may indeed create, among other things, a stronger competitive field for its players.

Japan's problems are attributable to a complex set of factors around the management of technology, ranging from NTT management's making the wrong technology bets on ATM and ISDN and the PDC standard, institutional rigidities within NTT, and weak approaches to setting global standards, to a failure by government to make a stronger move toward restructuring the slow-moving, monopolistic NTT. There are those, however, who attributed Japan's decline in telecommunication global markets to the weakening of NTT, brought about by pressure from the United States (Fujii 2003). Such analyses fuel nationalistic sentiments but, as I have shown, are far off the mark. Of those proponents, we can ask the following simple question. Would Japanese consumers have experienced the recent rapid spread of DSL and rapidly falling prices if NTT had been stronger? The answer is clearly no! Indeed, the consequences of a more rapid restructuring of NTT accompanied by the strong encouragement of new entrants and the rapid introduction of new technology might have led to more positive outcomes for the Japanese telecom sector. Such initiatives would have produced broad benefits for Japan's whole economy and society. In terms of its standing in the United States, AT&T is now but a shell of its former formidable organizational body. Who can deny, however, that the United States, despite the excesses of the 1990s and resultant overcapacity in fiber networks, has emerged much stronger in global telecommunications as a result of the AT&T breakup? Such is the power of "creative destruction" as envisioned by Joseph Schumpeter. It is one of the features of industrial evolution that market forces sometimes may lead to better management of technology than government's efforts to protect incumbents and micromanage business decisions.

Initially prepared for Conference on Institutional Change in East Asian Economics, Harvard University November 7–8, 2003. The author is indebted to the Doshisha Business School's Institute for Technology, Employment and Competitiveness (ITEC) for financial support for this research as well as to a Ford Motor Company IT Research Grant to the Management of Technology Program, Haas School of Business, University of California, Berkeley. The author also benefited from the research assistance of Yasuyuki Motoyama and Toru Ebata.

1. Lecture by John Chuang, September 10, 2001. Chuang is professor of information management systems, University of California, Berkeley.

2. Interview with Shigeki Goto, Tokyo, June 11, 2003.

3. I am indebted to Ye Xia at University of Florida, Gainesville, and to Ken Murakami, senior research scientist, NTT Laboratories, for their description of the workings of the IETF and ITU. This section also relies heavily on two IETF documents: IETF 1996, 1–31; IETF 2001, 1–27.

4. I am indebted above all for this account of Internet development at NTT to Dr. Shigeki Goto of the School of Science and Engineering, Waseda University, and formerly of NTT, and to Ken Murakami, senior research scientist at NTT Laboratories.

5. This section is based on an interview with Shigeki Goto, Tokyo, June 11, 2003.

6. Kevin Delaney, "Telecom-Equipment Concerns Focus on Software," *Wall Street Journal*, October 18, 1999.

7. I am indebted to Ye Xia, formerly of Bell Labs, for this detail.

8. This section draws heavily from an interview with Masao Hibino, president and CEO of NEC Magnus Communications, October 23, 2003, Tokyo.

9. I am indebted to Haruki Koretomo, chief scientist, Network Systems Group of Fujitsu, for this account; interview by the author, October 22, 2003.

10. Yoshifumi Takemoto, "Fujitsu Expects 16% Rise in U.S. Sales," *International Herald Tribune*, May 13, 2004.

11. This section on PDC, PHS, and the third-generation global standard draws very heavily from Funk 2002, 70–82, 183–194; an interview with David Hytha, an American executive with long experience in Japan and director of New Wave Networks, September 4, 2003; and feedback from Kenji Kushida, a University of California, Berkeley, Ph.D. student studying Japan's IT sector.

12. This section drew heavily from the observations of Kimio Inagaki, president of Jabil Circuit, Japan. I also benefited from conversations with Jan Rabaey, scientific co-director of the Berkeley Wireless Research Center.

13. I am indebted to Reza Moazzami, a telecom consultant, for his observations on the Koreans; interview by the author, July 7, 2003.

14. This discussion was stimulated by feedback from Dimitry Rtischev.

15. These observations benefited from a dialogue with Kimio Inagaki, September 11, 2004, Tokyo.

6

Japan's Telecommunications Regime Shift

Understanding Japan's Potential Resurgence

Kenji Kushida

INTRODUCTION: JAPAN IN THE DIGITAL ERA

In the 1990s, Japan's telecom sector was blindsided by new rules of competition that embedded value in standards. As a result, Japanese firms' manufacturing prowess became less relevant in both domestic and international markets. Up until then, institutions, corporate structures, and the dynamics of competition were channeled toward infrastructure investment using Japanese equipment. The domestic telecom sector provided funds and expertise to other sectors and acted as a springboard for exports.

Dynamics of competition in the domestic sector had led firms to acquire technological expertise and master production processes that yielded sophisticated, high-quality equipment at low prices, which translated into international competitiveness.[1] However, with the advent of the Internet and related technologies, Japanese market success was undermined by foreign innovations outside the realm of manufacturing; Cisco's control of the de facto software standard for routers is a prime example. In addition, Japan adopted a proprietary cellular standard, isolating its domestic wireless market and preventing Japanese firms from exporting their sophisticated cellular equipment (Cole, this volume).[2]

By the late 1990s, it was clear to all that Japan's domestic telecom sector was well behind cutting-edge developments in information technology (IT). The government was becoming increasingly aware that the existing regulatory structure and institutional environment hindered firms from using new technologies and pursuing new business strategies, and that the isolation of its domestic cellular market had cost Japanese firms dearly in global markets. Over several years, and in several steps, the government changed the logic of telecom policy making, dramatically revamping the regulatory structure and creating new institutions for coordination. This new regulatory and institutional framework altered the dynamics of market competition, leading Japan to unexpectedly develop fast, cheap, and innovative landline and wireless telecom services.

By 2005, Japan's telecom sector had shifted from a carrier-led, equipment manufacturer–driven, hardware-oriented platform for producer-oriented investment and exports to a carrier- and service provider–driven, service-oriented arena focused on domestic consumption. The new regulatory structure and policy-making logic, on the one hand, and the resulting new dynamics of competition, on the other, constitute the latest regime shift in Japan's telecom sector.

The future of Japan's development in the digital era may involve the discovery of a new national comparative institutional advantage. As the market dynamics created by the high-performance, low-priced telecom services unfold, Japan may cultivate new strengths or discover new ways to add value to existing strengths in manufacturing. This future is uncertain, but in order to understand the realignment of Japan in the digital era and how it may reemerge in global markets, we must understand the domestic transformation that has taken place.

JAPAN'S NEW TRAJECTORY OF DEVELOPMENT IN TELECOM SERVICES

This chapter is motivated by the observation that Japan's telecom services have developed rapidly and unexpectedly since the late 1990s. Japan seems to be forging its own path of development, one that is not adequately captured by existing cross-national indicators of IT.

First, although Internet usage in Japan has surged, much of the growth has come from Internet subscriptions via cellular services. Overall Internet usage grew from 9 percent of the population in 1997 to over 60 percent in 2003. However, since 1999, more than 80 percent of Japan's Internet subscriptions have been through cellular services (Soumusho 2004). *How* people connect to the Internet affects *what* people can do with it—firms and individuals are likely to create and derive value from cellular Internet applications in different ways than through PC-based Internet applications. Japan was one of the first countries in which the level of Internet penetration needed to be disaggregated into PC-based and cellular-based Internet access in order to understand the network environment and its implications. The sudden development and rapid growth of Internet penetration via cellular services in Japan demands an explanation.

Second, Japan's broadband services grew abruptly after 2001, giving Japan the cheapest and fastest broadband services worldwide. In 1999, Japan was one of the most expensive countries in the Organisation for Economic Co-operation and Development (OECD) for consumer access to the Internet, but by 2002, Japan's monthly subscription rates for DSL (digital subscriber line) were the lowest (Soumusho 2003, 7). Japanese consumers also enjoyed the fastest broadband access in the world, through DSL reaching 50 Mpbs (1 to 3 Mbps are common speeds for DSL in the United States as of mid- 2005) and fiber-to-the-home (FTTH) services delivering 100 Mbps to 1 Gbps. By the end of 2001, 95 percent of metropolitan areas were covered by FTTH networks. According to the ITU, Japan had the highest price-performance for broadband Internet access (defined as the lowest price per 100 kbit/s of data) in the world by 2002 (Soumusho 2003).

Moreover, most Japanese broadband services were bundled with Internet protocol (IP) telephony subscriptions after 2001. IP telephony sends voice signals as packets of data over the Internet (voice over IP, or VoIP), bypassing conventional telephone circuits and switches. Its most well-known form is PC to PC, through software such as Skype. The distinguishing feature of Japanese IP telephony subscriptions bundled with DSL and FTTH services is that they can be used without PCs, connecting to conventional telephone sets directly. Moreover, the government allocated a set of telephone numbers dedicated to IP telephones, also allowing IP telephones that meet particular performance requirements to receive orthodox telephone numbers. Such IP telephones in Japan are essentially functional substitutes for conventional household telephones. The government estimates that by the end of 2004, Japan had close to 8 million IP telephony subscriptions.[3] Estimates vary, but a survey by the *Nihon Keizai Shimbun* (Japan's preeminent economic newspaper) found that about 40 percent of major firms had installed IP telephony and another 40 percent were seriously considering doing so.[4]

It is obvious that a wider range of content and applications can be delivered via a high-speed, low-cost broadband environment than by a slower, higher-cost network environment, giving Japan the potential to pioneer in a new array of applications. Perhaps less obvious is the potential of IP telephony. Let us recall that cheap long-distance telephone service in the United States allowed American firms, led by the financial sector, to move some offices to states with cheap labor. This sparked a broader reorganization of business structures, allowing firms to modularize their business processes, which subsequently facilitated outsourcing as well as offshoring. It is quite possible that creative

128 firms that initially adopt IP telephony to solve a particular problem — cutting telecommunications costs — may find that the technology can be used to solve an entirely different set of problems.[5] Japanese firms may not be the first ones to discover new uses for IP telephony, but the political implications of an increasing number of firms and households bypassing NTT's conventional telephone network are emerging in the form of calls to reorganize NTT. The puzzle is to understand how the market for high-speed infrastructure, low-cost broadband, and rapid IP telephony diffusion emerged.

Third, Japan developed a cellular environment with the highest price-performance for data transmission and Internet access. By 2003, Japan had three separate third-generation (3G) cellular networks that offered data transmission speeds comparable to DSL speeds for some services, several times faster than landline dial-up connections.[6] In 2003 and 2004, carriers began offering flat-rate subscriptions to 3G cellular Internet services, leading Japanese firms to begin searching for ways to add value or increase efficiency through the intensive use of high-speed cellular Internet access and data transmissions. Perhaps even more than IP telephony, flat-rate, high-speed cellular Internet access allows firms the potential to create new applications and discover new uses for the technology.

Thus, by 2004, Japan's landline and wireless network environment was vastly different from what it had been in 1997, and its characteristics were no longer adequately captured by common indicators of performance in IT.[7] Moreover, the trajectory of development seemed to be a departure from Japan's producer-oriented economic development of the past — postwar consumers had rarely enjoyed low prices for services, regardless of the sector. Taking these unexpected developments as a starting point and investigating their origin leads us to the story of Japan's latest regime shift in the telecom sector.

THE REGIMES

Telecommunications is a particularly appropriate sector to conceptualize in terms of policies and regulatory structures that shape market dynamics. It is a sector in which Karl Polanyi's view — that sustained government intervention is necessary for competitive markets to function — is particularly salient (K. Polanyi 1944).

Telecommunications sectors in most countries began with monopoly incumbent carriers. The rules created after their privatization have been critical in shaping market dynamics (S. Vogel 1996), and differences in regulatory structures across countries have led to a variety of competitive dynamics. Some countries went further than others toward privileging entrants at the expense of incumbents; the terms under which competitors could lease the last one mile of infrastructure owned by the incumbent can determine the scope of new firms' activities, as well as the technologies they employ. Though policies do not necessarily determine market outcomes, the regulatory structure powerfully shapes market dynamics in the sector by creating the incentives and constraints facing firms.

Yet, we cannot stop here. Policies cannot simply be treated as though they are exogenous; they are driven by the logic of policy making in a particular country. Market outcomes and new technology do affect policy-making decisions, but they are not the sole drivers of telecom policy. Differences in policies and regulatory structures across countries are usually the result of political dynamics, bureaucratic interests, and the interaction between governmental and firm-level actors.

My conception of telecommunications regimes is built upon the notion that the logic of policy making drives policies and regulatory structures, which in turn shape market dynamics. I propose a set of variables that can capture this conception of telecommunications regimes and be used in comparisons across countries or time. They include the principal government actors involved in telecom policy making, the orientation of policy making (ex ante or ex post), the state's method of coordination with firms, the principal industry actors, mechanisms of intra-industry interaction, and the source of standard setting. Applying this set of variables to Japan over time, we see that that there have been at least three regimes in Japan's telecom sector (see Table 6.1).

BUREAUCRACY-MONOPOLY

The first regime, which I label "bureaucracy-monopoly," needs no extensive elaboration for our purposes. The main outcome was the regulatory regime and market dynamics in which the state-owned monopoly Nippon Telegraph and Telephone (NTT) set national standards, determining the technological trajectory followed by equipment manufacturers.

The national Diet (parliament) approved NTT's considerable budget allocations, and the Ministry of Posts and Telecommunications (MPT), created from the prewar Ministry of Postal Affairs, was given formal oversight over NTT. However, NTT's extensive budget and its powerful R&D capabilities allowed it to essentially regulate itself in setting domestic standards and to play the role of technological leader of equipment manufacturers.[8] NTT's lucrative procurement contracts with its "family" of equipment manufacturers — NEC, Fujitsu, Oki, and Hitachi — effectively subsidized their R&D activities in other areas, such as consumer electronics.[9] Foreign firms were essentially

TABLE 6.1 Regimes in Japan's telecommunications sector

	Regime 1: "Bureaucratic Monopoly"	
	(Late 1800s to 1954)	(1954–1985)
Government actors	Ministry of Communications (MoC)	Nippon Telegraph and Telephone Public Company (NTTPC), National Diet, MPT
Industry Actors	MoC, cartel of family OEM equipment manufacturers (NEC, Hitachi, Fujitsu, Oki Electric)	NTT, KDD, NTT family equipment manufacturers
Government-industry coordination	MoC directly operates communications industry, equipment procured as OEM	Diet approves budget, formal oversight by MPT, but NTT with most real control
Intra-industry Interactions	Equipment manufacturers competing on the basis of quality for volume of OEM work	Same as before. NTT providing large procurement demand.
Standard setting	MoC	NTT (nominally MPT)

	Regime 2 "Controlled Competition" (1985–mid- to late 1990s)	Regime 3 "Strategic Liberalization" (mid- to late 1990s–present)
Mode of regulation	Ex ante	Ex post
Government actors	Ministry of Posts and Telecommunications (MPT)	MPT/Soumusho* (renamed and restructured in 1999), Cabinet Office, Dispute Resolution Commission, FTC (Fair Trade Commission), possibly judicial system
Industry actors	NTT, New Common Carriers (NCCs), NTT family firms	NTT, NCCs, startup firms (e.g., Softbank, Usen), NTT family firms, non-family equipment manufacturers, foreign firms (e.g., Vodafone, Ripplewood)
Government-industry coordination	MPT wields classic "Industrial Policy" tools (licensing authority, discretionary administrative guidance, etc); MPT interprets law and mediates conflict between carriers.	Increased legalization of rules. FTC, possibly judicial system play larger role in enforcement. New formal institution for dispute resolution between firms.
Intra-industry Interactions	NTT dominates competitors with R&D capabilities and interconnection rates. NCCs closely regulated by MPT. NTT provides stable demand for equipment manufacturers — not all on OEM basis, but advantages to NTT family firms. Standards set by NTT force NCCs to procure equipment from NTT family.	NTT has less control over interconnection rates. NCCs no longer tightly regulated by MPT. Shift to non-NTT standards allow NCCs to procure equipment from outside NTT family. New dynamics of interaction such as price wars, legal battles, use of Dispute Resolution Commission, and complaints to the FTC.
Standard setting	NTT, MPT	De facto (e.g., Cisco's routers), MPT, IETF, ITU

* The Ministry of Posts and Telecommunications (Yuseisho) was reorganized in the reshuffling of ministries in 1999 to create Soumusho. Soumusho initially had the English name Ministry of Public Management, Home Affairs, Posts and Telecommunications (MPHPT). In 2004 it was renamed the Ministry of Internal Affairs and Communications (MIC).

shut out of NTT's procurement by strict quality standards that NTT ensured its family of firms could pass. Overall, the regime focused on investing in domestic infrastructure using Japanese equipment.

THE REGIME SHIFT TO "CONTROLLED COMPETITION"

The privatization of NTT in 1985 was a critical juncture, enabling MPT to reshape itself as a powerful actor in Japan's logic of telecom policy making. Privatization was the culmination of a highly political struggle over power relations and the configuration of markets. Change was precipitated by technological developments that eliminated the economic rationale for a monopoly telecom carrier and by the ensuing breakup of AT&T in the United States, combined with a wave of liberalization and privatization of state-owned enterprises in Japan. Politicians, MPT, NTT, "family" equipment manufacturers, and other electronics firms were key actors in this political struggle. The final settlement resulted in a partially privatized NTT, a partially liberalized telecom market, and vastly increased regulatory powers for MPT.[10] However, NTT's R&D capabilities and much of its revenue structure were left intact, allowing NTT to retain its position as technological leader of the sector.

THE "CONTROLLED COMPETITION" REGIME

After 1985, MPT used its newly acquired regulatory powers to orchestrate the entry of competitors (new common carriers, or NCCs) into the sector and to micromanage competition, relying on classic tools of industrial policy such as licensing and administrative guidance. Steven Vogel (1996) has labeled this regulatory regime "controlled competition." A combination of MPT's micromanagement and NTT's continuing dominance in R&D shaped the market dynamics in this regime. Japan's decline in international competitiveness, as well as its development of cellular Internet services, resulted from these market dynamics.

JAPAN'S DECLINE IN INTERNATIONAL COMPETITIVENESS FOR TELECOM EQUIPMENT

Robert E. Cole's chapter in this volume provides a succinct analysis of why Japan's equipment manufacturers saw their exports and global market shares of landline and cellular telecom equipment decline. Equipment manufacturers, following NTT's lead in pursuing ATM technology and obtaining approval of standards from the International Telecommunications Union (ITU), were blindsided by TCP/IP and Cisco's software-based de facto standard for routers.

Discontinuous technology surrounding the Internet also shifted market dynamics in Japan's domestic market. NCCs had previously been hostage to NTT and its family of equipment manufacturers, since nobody else manufactured equipment that met the domestic standards. With the rise of TCP/IP and routers, however, Japanese ISPs found themselves in the position of purchasing their equipment from unfamiliar American firms, such as Cisco and Juniper. Some were unimpressed with the quality of the American equipment but had little choice since Japanese firms did not manufacture what they needed.[11]

In cellular markets, as Cole and others have argued, Japan's NTT-developed digital standard isolated the domestic market, preventing exports as well as imports (Funk 2002; Cole, this volume). However, the dynamics of competition within this isolated domestic cellular market were what led to Japan's development of cellular Internet services.

THE DEVELOPMENT OF CELLULAR INTERNET SERVICES

NTT DoCoMo, a subsidiary of NTT spun off in 1991 as a result of government policy, exerted a powerful role in shaping Japan's cellular market. DoCoMo had inherited NTT's extensive wireless R&D labs, allowing it to develop the domestic digital cellular standard, PDC, and to lead handset and equipment manufacturers in R&D. MPT adopted PDC as the domestic digital cellular standard in 1994, and DoCoMo was able to continually upgrade it. The upgrades were formally approved by the Association of Radio Industries and Businesses (ARIB) and made public, but DoCoMo privileged its "family" firms with specifications and information before public disclosure. As a result, "family" firms could roll out new handsets for DoCoMo's service before other manufacturers developed their new handsets for competing carriers. In exchange for early information, "family" firms within the family were forced to delay shipping their new handsets to competitors for several months (Funk 2002; Kushida 2002).

These competitive dynamics tilted the playing field in favor of DoCoMo, pushing desperate competitors to look for alternative strategies. A technologically simple cellular service known as the PHS (personal handyphone system), a product of MPT's industrial policy initiative to promote a cheaper alternative to conventional cellular services, revealed significant latent demand for messaging services.[12] NTT's domination of PDC, the demand for messaging services revealed by PHS, and the general expectation shared by Europeans and Americans that cellular services would eventually connect to the Internet led Japanese cellular carriers to begin developing cellular Internet services.

After 1996, Japan's major cellular carriers, including DoCoMo, all raced to commercialize cellular Internet services. Each carrier took an independent path to develop the technology and business model, and DoCoMo actually lagged behind for much of the time.[13] DoCoMo's famous i-mode service, with its innovative business model, was rolled out in early 1999.[14] Yet, few observers recognize the significance of the fact that, later that year, all major Japanese carriers rolled out competing services with similar features but different underlying technologies and networks.[15] Rather than simply the success story of a single innovative firm, Japan's cellular Internet services were the product of competition in the entire Japanese cellular market as it shifted toward the commercialization of cellular Internet services.

Japanese carriers were able to rapidly commercialize Internet services, requiring implementation in the handsets as well as infrastructure, owing largely to the nature of handset manufacturer-carrier relations. In DoCoMo's case, its R&D labs could essentially give specifications to handset manufacturers, which would reliably implement and deliver the new models. The other carriers did not have DoCoMo's R&D capabilities, but they worked closely with manufacturers.[16] Carriers could guarantee demand for the new Internet-enabled handsets, since it was the norm for carriers to purchase handsets outright from manufacturers before selling them to retail outlets.[17] Thus, the domestic dynamics of competition focused cellular carriers on the task of developing and deploying cellular Internet services, and the nature of coordination between carriers and handset manufacturers facilitated the rapid deployment of fully implemented services. By May 2000, 11 million out of 51 million cellular users subscribed to cellular Internet services, growing to 74 million out of 86 million subscribers by February 2005.[18]

THE REGIME SHIFT TO STRATEGIC LIBERALIZATION

The regime shift from "controlled competition" to "strategic liberalization" was not a political struggle culminating in a radical break, as the regime shift to "controlled competition" had been. Rather, it took place over several years in a series of incremental changes adding up to a significant transformation.[19] These incremental changes were driven by adjustments in corporate and government behavior as it became increasingly clear that institutions, market dynamics, and corporate strategies formed during the "controlled competition" regime were slowing Japan's adjustment to the new rules of competition in the digital era.

Major Japanese corporations were forced to shift their strategies, since the new technological paradigm, with its modular architecture and de facto standards embedded in software, did not play to their strengths. At the same time, the technology opened up new opportunities for start-up firms and entrants.

For the government, Japan's lag in IT, especially in Internet penetration, created bureaucratic and political incentives to change the logic of policy making. Government actors became interested in restructuring the regulatory framework in order to facilitate broadband deployment and foster Internet penetration and to reduce NTT's influence in shaping the sector.

The regime shift was driven by the interaction between these corporate strategies and government actors. Changes in the logic of policy making led to a new regulatory framework, which reshaped market dynamics by altering incentives and constraints upon firms. Corporate strategies often informed policy makers, but the regime shift was not driven simply by corporate interests acting upon policy makers. Political and bureaucratic actors had their own interests and agendas as well, though market outcomes did take them by surprise on several occasions.

STRATEGIC LIBERALIZATION: THE CHANGING LOGIC OF POLICY MAKING FROM EX ANTE TO EX POST REGULATION

Japan's logic of policy making began to shift as new government actors became involved in telecom policy making, and the mode of regulation shifted from an emphasis on ex ante to ex post regulation. Ex ante regulation by MPT, the hallmark of the "controlled competition" regime, rested upon both formal and informal tools of regulation. Formal tools at MPT's disposal included licensing requirements controlling market entry, pricing changes, and the scope of carriers' business activities.[20] Informal tools giving MPT considerable discretionary authority were enabled by MPT's position as a nexus of information exchange and its role in mediating and coordinating settlements for disputes between carriers.[21]

A wave of deregulation in the late 1990s removed many of MPT's most important formal policy tools of ex ante regulation. In 1997, Japan signed the WTO Telecom Agreement removing most restrictions on foreign ownership of carriers and infrastructure. The following year, MPT abolished most licensing requirements for market entry and price changes while relaxing restrictions over the scope of carriers' business activities.[22] Overarching political support from the cabinet's "Three-Year Plan for Deregulation," which encompassed several industries including telecom, made it easier for MPT to engage in deregulation requiring amendments to the Telecommunications

Business Law. In 2003, the ministry went on to abolish most of the classification, registration, and notification requirements.[23]

These waves of deregulation affected MPT's informal policy tools. After 2003, streams of information about strategies and pricing schemes carriers were considering no longer flowed automatically through the ministry. It is common knowledge that administrative guidance and informal regulation are often exercised through the process of firms lining up to consult with bureaucrats about how to fill out appropriate forms. By 2004, stacks of documents no longer awaited approval on bureaucrats' desks, nor did lines of businessmen wait to see the officials in charge of pricing or registration.

In 2001, the ministry removed itself as the coordinator and mediator of disputes between carriers by establishing the Dispute Resolution Commission (DRC). The DRC was located within the ministry, but was, in principle, a neutral third-party deliberative organization. Complaint filings, deliberations, and results were made public, removing such decision making from the government's discretion and from potential political interference. In sum, Japan's mode of regulation in the sector shifted from attempting to avoid conflicts ex ante, by deciding what price changes and business activities could be permitted, toward an ex post model of allowing firms to freely engage in their strategies, bringing conflicts to the DRC when problems arose.

The Japanese Fair Trade Commission further strengthened ex post regulation by entering telecom regulation for the first time. In 1999, it issued a warning to NTT DoCoMo about its practice of forcing handset manufacturers to delay the shipment of their products to competitors, and in 2000, it issued a warning to NTT regarding its treatment of DSL providers.

THE NEW REGULATORY FRAMEWORK

As the logic of policy making shifted, a new regulatory framework emerged. This framework was designed to increase the level of competition in landline telecom markets — in other words, to strategically liberalize the market. The changes entailed creating new rules where none had existed before, a classic case of reregulation for the purpose of liberalization.[24]

First, MPT created a new set of interconnection policies governing the terms under which competitors could lease NTT's last mile of infrastructure. Until the late 1990s, few regulations governed interconnection. This gave MPT broad discretion in determining interconnection rules and the prices charged by NTT, thus providing opportunities for political intervention (Fuke 2000, 20).[25] In 1997, acting on recommendations by the Telecommunications Deliberation Council, MPT revised the Telecommunications Business Law to establish clear rules for interconnection. NTT was required to lease its last

136   one mile of infrastructure when requested, and MPT established a formula dictating the prices NTT was allowed to charge (Fuke 2000, 43–45). In 2000, this formula was revised to further favor competitors.

Second, actors new to telecom policy making were involved in creating a set of policies that promoted the deployment of DSL. DSL technology sends a high-frequency signal through existing copper lines on top of conventional telephone signals. This requires that equipment be installed on both the user's end and inside the facilities of the carrier. The new interconnection rules did not include provisions for competitors to place equipment within NTT's switching facilities, known as *collocation*. In the absence of rules for collocation, NTT was able to stonewall requests for information, delaying access to its facilities.

In 1999, Tokyo Metallic, a startup firm, became the first company to commence DSL services in Japan. However, NTT had little interest in the technology, preferring to rely on its existing ISDN services and to wait until it could deploy fiber-optic networks (ITU 2003).[26] Tokyo Metallic and other startups struggled to expand their DSL services. NTT took five to nine months to assess whether collocation space was available within a particular facility.[27]

In July 2000, the cabinet entered telecom policy making by establishing an "IT Strategy Headquarters," which produced the e-Japan strategy in September. The e-Japan strategy identified Japan as lagging behind other advanced industrial nations in the development of IT. The policy goal was to create "ultra high-speed network infrastructure and competition policies," setting a five-year timeline to establish "one of the world's most advanced Internet networks" and aiming to provide low-cost Internet access within a year.[28] The cabinet passed the "Basic IT Law on the Formation of an Advanced Information and Telecommunications Network Society (IT Basic Law)" that November. The Basic IT law strengthened the position of Soumusho (the new ministry created by merging MPT with a couple of others) by creating a broad framework within which many specifics could be determined by ministerial ordinances.[29]

In October 2000, the Japanese Fair Trade Commission entered landline telecom policy making by taking an unexpectedly strong stance against NTT over its treatment of the DSL providers. Citing antitrust concerns, the FTC issued a warning (a form of administrative guidance) to NTT, the first time it had ever done so, though the lack of specific rules precluded it from taking punitive actions. The FTC's warning brought NTT practices hampering the deployment of DSL into the public spotlight. It also sent a signal that the FTC was closely watching NTT and willing to act once rules were established.

That same month, Soumusho revised several ministerial ordinances, requiring NTT to clarify the terms under which it offered collocation and to publicize how it calculated fees. The ministry also required NTT to lease unused fiber-optic and copper capacity to all carriers that requested usage, a procedure known as unbundling (Fuke 2003, 180–181). In 2001, the newly established Dispute Resolution Commission ruled against NTT in disputes brought to it by Tokyo Metallic and eAccess, another DSL start-up.

In 2003, the judicial branch of the government was thrust into telecom policy making for the first time when five carriers, led by KDDI and including British-owned Cable & Wireless IDC, took the extraordinary step of filing a lawsuit against the government. The lawsuit was over Soumosho's approval of a 5 percent hike in NTT's interconnection rates—an apparent reversal in its stance of favoring competitors that led many to suspect pro-NTT political pressure behind the move. Although the legal proceedings are expected to take years, the lawsuit brings a new actor with policy-making clout into the telecom arena.[30]

Thus, by 2001, the new regulatory framework had substantially altered the opportunities and constraints facing firms in the sector. In 1996, carriers were compartmentalized and regulated, though lack of interconnection and collocation rules left them vulnerable to NTT. By 2001, carriers were free to enter or exit the market, and foreign firms were allowed to buy existing infrastructure-owning carriers. NTT's competitors had a set of rules about interconnection and collocation upon which to formulate business plans, and they could reasonably expect the FTC or the Dispute Resolution Commission to enforce those rules if NTT did not comply, with the legal system as a final recourse.

MARKET OUTCOMES

Softbank, DSL, and IP Telephony

Japan's emerging regulatory framework supported innovative corporate strategies, unleashing a new set of market dynamics. Softbank, a successful start-up from the 1980s, and its founder, Son Masayoshi, famous for his early investments in Yahoo, arguably played the most significant role in changing the terms of competition. In 2001, with the market for DSL growing, a subsidiary of Softbank launched a price war, setting monthly subscription prices at about half the rate of its competitors (Softbank charged about 2400 yen, approximately $22 at US$1 = JP¥110). It also mobilized aggressive sales agents who blanketed metropolitan areas, handing out free to new subscribers DSL modems costing over $100. Other DSL competitors were forced to match Softbank's prices, giving Japan the lowest monthly DSL prices worldwide. It is unlikely that MPT in the controlled-competition regime would have permitted Son to wage what many saw as a reckless price war that threatened carriers'

138   ability to invest in the next generation of infrastructure. Softbank then delivered a second price shock. It bundled free IP telephony subscriptions with its DSL service, allowing subscribers to call other Softbank subscribers without charge, and set flat rates for long-distance calls to non-subscribers. Flat-fee telephony was unprecedented in Japan, since NTT (and its competitors, which paid interconnection fees to NTT) charged by the minute, with higher fees for longer distances. Softbank was able to offer flat-rate service because it had constructed a completely IP-switched network on top of leased lines from NTT, avoiding any interconnection fees. Son also set IP telephony calls to the United States at below-cost prices, at 8 yen a minute (6–7 cents), a fraction of what international carriers charged (200–300 yen for three minutes). The low monthly fees and attractive IP telephony prices led to a sharp increase in the number of DSL subscribers, and a sudden public interest in VoIP.

The Japanese government was surprisingly quick to support IP telephony by assigning a dedicated array of numbers to IP telephones (a 050 prefix) and later allowing IP telephones to obtain telephone numbers within the existing numbering scheme if they met quality standards.[31] Major corporations, interested in cutting communications costs, began adopting IP telephony. Thus, Son Masayoshi arguably went furthest in taking advantage of the new regulatory framework in a conscious effort to create a new broadband environment in Japan.

FIBER TO THE HOME (FTTH)

Driving forces behind Japan's deployment of nationwide fiber-optic infrastructure date back to industrial policy initiatives in the mid-1990s. However, the market for high-speed fiber-optic broadband services was shaped by the DSL market — a product of the new regime.

NTT had been pouring money into a nationwide fiber-optic infrastructure, and MPT had actively promoted the development of such an infrastructure since the early 1990s. Using classic tools of industrial policy, MPT created incentives for competitors to deploy fiber networks. Through the Development Bank of Japan, it provided low-interest loans, using a semi-public organization (the Telecommunications Advancement Organization) to help subsidize interest payments. By 1999, over 75 billion yen worth of loans had been allocated.

Market competition in the DSL market — low-priced subscriptions that bundled IP telephony — shaped the corporate strategies of firms commercializing consumer FTTH services. Usen (pronounced *yu-sen*), a landline music broadcasting company that had deployed fiber-optic cables to its grid of electric poles, was the first firm to offer household FTTH services in March 2001. Usen decided to charge 6,000 yen (approx US$54) per month, only slightly more than double Softbank's DSL price. NTT had been expecting to charge more than double that price when it rolled out its own proprietary FTTH service. Other firms entering the FTTH services market, including subsidiaries of regional electric power companies and NCCs, offered prices similar to that of Usen. Seeing Softbank's success with IP telephony, the government's legitimization of it, and firms' interest in adopting IP telephony, FTTH service providers began bundling IP telephony subscriptions as well. They promoted their competitive advantage of better sound quality and fewer dropped calls owing to the higher throughput speeds. By late 2004, Japan had over 2 million FTTH subscribers.[32]

CHANGING THE FUNCTION OF THE DOMESTIC MARKET

After the deregulation of FDI into infrastructure in 1997, the Japanese domestic market took on new functions in global competition — especially after its cellular Internet services developed ahead of the rest of the world's. Foreign carriers such as British Cable & Wireless had been operating mostly corporate-oriented services on top of leased infrastructure in Japan for many years, but only after 1997 were foreign firms allowed to own and operate infrastructure as NCCs. A dramatic series of events starting in 2001 revealed the new Japanese telecom market to be a place from which foreign firms could derive profits and gain technology through mergers and acquisitions.

In 2001, the British cellular carrier Vodafone gained management control and majority ownership of Japan Telecom, one of the original NCCs. Japan Telecom's wireless subsidiary, J-Phone, was one of three nationwide cellular carriers and the pioneer of camera-phone services. Vodafone proceeded to reorganize the company under a holding company and sold off the landline businesses of Japan Telecom. It took full control of J-Phone, renaming it Vodafone and transferring the technology and know-how of J-Phone's cellular Internet and camera-phone services to its European operations. Vodafone quickly introduced Vodafone Live!, using technology taken from J-Phone's J-Sky service, in most of its European markets to quickly become the largest European cellular Internet service provider.[33]

Ripplewood, the U.S. venture fund that had shocked Japan by purchasing a failed but prestigious bank (the Long Term Credit Bank) in 2000, purchased Japan Telecom from Vodafone. Ripplewood then turned around and sold Japan Telecom to Softbank, making a substantial profit of about 90 billion yen in the process.[34]

Not only had Japan's domestic market provided value to foreign firms in ways unimaginable in the mid-1990s, but domestic infrastructure had undergone a remarkable change in ownership. Japan Telecom's major shareholders had been the former state-owned railway companies. In persuading them to sell, Vodafone had taken a careful diplomatic approach, promising to "preserve the Japanese character of Japan Telecom" and not to immediately split off the wireless business and sell off the landline business — though that was exactly what it ended up doing.[35] It is unlikely that the railway companies would have sold their stakes directly to Softbank, since many established firms viewed Softbank with mistrust, fearing it could launch another reckless price war undermining everyone's profitability. Indeed, immediately upon its acquisition, Son argued that cellular fees were too high and declared his intent to deploy a new cellular technology that would allow Softbank to add equipment on the ends of Japan Telecom's nationwide landline infrastructure, creating a dramatically cheaper cellular service. This attempt was delayed or thwarted by Soumusho, however, which, citing lack of available bandwidth, refused to allocate spectrum, in turn prompting Son to take them to court.

STRATEGIC INTEGRATION INTO INTERNATIONAL MARKETS AND 3G MARKET DYNAMICS

Beginning in the late 1990s, Japan's cellular market began to reconnect to international markets. Rapid 3G infrastructure deployment was facilitated by government policies, but the market dynamics for 3G services were substantially influenced by the DSL market. Unhappy with NTT DoCoMo's domination of the domestic cellular market, MPT allowed competitors to adopt a new standard in the latter half of the 1990s. IDO and DDI, suffering from DoCoMo's control of the PDC standard, opted for CDMAOne, a standard using U.S.-based Qualcomm technology with equipment provided by Motorola, which promised better performance than PDC. MPT fully supported the new standard, going so far as to mobilize the Development Bank of Japan to help fund the infrastructure, which enabled the carriers to commence services in 1998.

As international deliberations over 3G standards began in the ITU, MPT took the strategic action of announcing in 1996 that it would only issue three 3G licenses, based on MPT's assessment of the applications. The availability of only two licenses for NTT's competitors sparked a wave of consolidations, leaving Japan with three nationwide carriers by 2001: NTT DoCoMo, J-Phone, and KDDI. Though MPT's official rationale in issuing only three licenses was to conserve spectrum, the ministry also probably wanted to encourage consolidations in order to ensure that firms would have enough capital to deploy infrastructure — a classic stance of Japanese industrial policy. In hindsight, the contrast with Europe's spectrum license allocations could not be clearer: by 2003, Japan had three separate 3G networks, while many European countries, their carriers having depleted their cash by buying auctioned spectrum, had one or none.

KDDI launched price wars in 3G services in late 2003 by offering flat-rate 3G cellular Internet access. DoCoMo and Vodafone had no choice but to follow suit within a year. This move attracted the interest of firms that could count on flat-rate cost calculations and fresh opportunities to create new services and applications.

CONCLUSION: TOWARD A JAPANESE RESURGENCE?

A New Strategic Policy Thrust

Japan's telecom sector has undergone a regime shift in reaction to the new logic of competition in the digital era. Firms and government actors have adjusted their strategies, recognizing that the institutions and market dynamics of Japan's controlled-competition regime had led to Japan's declining international competitiveness in IT. In the late 1990s and early 2000s, interactions between corporate and government strategies of adjustment changed the logic of policy making and the regulatory structure, reshaping the market dynamics. I label the new regime strategic liberalization to capture the policy orientation aimed at liberalization not for the sake of liberalization itself, but for the strategic goal of rapidly deploying high-speed networks at low prices to a broad segment of the population.

The next phase of Japan's development may very well involve a change in Japan's comparative institutional advantage as firms working within the institutions, regulation, and market dynamics of the new regime attempt to augment existing competencies and search for new ways to add value. Indeed, the next thrust of strategic policy making explicitly aims at accelerating corporate attempts to utilize Japan's high-speed network environment.

In early 2004, the cabinet office announced the e-Japan II strategy. Satisfied with the faster-than-expected deployment of high-speed, low-cost broadband infrastructure, the government shifted its focus to indicating broad directions for future development. Health care, knowledge management, and facilitating labor mobility were among the areas to which it pledged policy support. At the end of the year, Soumusho unveiled its u-Japan vision, with goals similar to those of the e-Japan II strategy and focused on facilitating the development of "ubiquitous networks" integrating applications, hardware, and both landline and wireless networks.[36]

Japan is also learning the value of standards in competition of the digital era. Japan has been working on R&D projects with China and South Korea, at both government and firm levels, to implement the next-generation Internet protocol standard IPv6. The ostensible technical problem fixed by IPv6 is the shortage of possible IP addresses under the current IPv4 standard. Despite possible workarounds to enhance IPv4, the Asian countries seem to be betting that they can use their domestic markets (especially China's) to set and possibly control the de facto global Internet architecture standard. Their hope is to privilege their own firms in a variety of software and hardware applications for IPv6, thereby breaking Cisco and other U.S. firms' dominance of Internet infrastructure.[37]

Japanese telecom equipment firms have also been reorganizing their global manufacturing capabilities in preparation for another push toward global markets, especially in cellular equipment.[38] Since Japan was early in deploying 3G networks, handset manufacturers have already rolled out several generations of incrementally improved products. In a familiar pattern, made possible again as other countries deploy compatible 3G networks, they hope to take products developed in the sophisticated domestic market into international markets.

Rethinking the Fit between Institutional and Market Configurations and Technology

I would like to end by making a theoretical point regarding the fit between institutional configurations and technological paradigms. Three strains in the literature contend that firms enjoy a competitive advantage when institutional and market configurations in a particular country, region, or sector have a good fit with characteristics of the dominant technological paradigm.

One strain examines the fit between national-level institutions and technology. In this conception, Japan's institutional configuration was a good fit for lean production. Lean production became a dominant production paradigm, forcing the United States and much of Europe to adjust their political economies (Aoki 1988; Tyson and Zysman 1989). In the digital era, it became Japan's turn to adjust—the telecom regime shift is one piece of a broader adjustment in Japan's political economy (Yamamura 2003; Newman and Zysman, this volume).

A second perspective takes the idea of fit between technologies and institutional and market arrangements to the sectoral level. Variations in performance across sectors are explained as a function of the fit between sectoral institutional configurations and characteristics of technology that differ across sectors (Kitschelt 1991).

The third strand looks at the functional fit between institutions and technologies. Steven Vogel and John Zysman (2002) argue that when comparing the United States and Japan, national institutions in the United States are conducive to certain tasks, such as fostering labor mobility, financing new ventures via equity markets, producing entrepreneurs, and promoting competition. These fit the logic of competition in digital markets better than Japan's institutional advantage in functions such as forging labor-management cooperation, providing stable financing via the banking system, managing competition, and achieving incremental improvements in production.

My analysis of Japan's regime shift in telecom suggests that functional advantages can shift on a sectoral basis. Japan's new telecom regime is continuing to create opportunities for entrepreneurs, especially in providing services and content utilizing the new broadband environment, and searching for new uses of the networks. Labor markets are becoming more liquid, not only in start-ups and firms in services and applications, but increasingly in larger firms requiring particular types of expertise in IT. The nature of strategic R&D has changed. NTT retains much of its technological expertise, but the locus of decision making about which technologies to adopt has decentralized considerably. Strengths in manufacturing and the availability of stable funding, especially in strategic directions supported by the government's new policy thrusts, have been retained. Given that functional requirements of the dominant technological paradigm can shift over time (for example, Yamamura [2003] makes the distinction between the innovation phase and implementation phase), it remains to be seen whether Japan's new functional advantages in the telecom sector will become a good fit. They may not. But if they do, understanding Japan's telecom regime shift will be the key to analyzing Japan's surprising resurgence.

NOTES

1. The domestic telecom market was not a spectacular springboard for exports compared to autos, machine tools, or consumer electronics — especially to the U.S. market. Fransman examines why Japanese firms were not as successful in penetrating U.S. telecom equipment market as other sectors, despite enjoying high global market shares (Fransman 1995). Analog cellular and microwave equipment was quite successful in U.S. markets (Funk 2002; Steinbock 2003, 147).

2. Japan's loss of competitiveness in telecom equipment was not due to the same set of problems that led to Japan's economic problems in the 1990s. Difficulty in adjusting to the post–Bretton Woods international financial system, institutional rigidities, bad

144 macroeconomic policies, and political paralysis, are a few factors to which scholars have attributed Japan's broader problems (Pempel 1998; Posen 1998; Grimes 2001; Katz 2002).

3. "IP denwa: Katei nimo shintou [IP Telephony Penetration Extends to Households]," *Nihon keizai shimbun*, March 29, 2005.

4. "IP denwa 'Dounyu' 9 wari [Ninety Percent (of Large Firms) Have Installed IP Telephones]," *Nihon keizai shimbun*, March 3, 2005.

5. This logic is set out in Cohen, DeLong, and Zysman 2000.

6. As of early 2005, NTT DoCoMo's FOMA transmission speeds were 384 kbps, compared to 28.8 kbps for conventional telephone lines. KDDI's "au" service delivers up to 2.4 Mbps, compared to 384 kbps to 3 Mbps for typical American DSL services.

7. By 2005, Japan ranked eighth among the 104 countries included in the World Economic Forum's Global ICT Competitiveness rankings, up from twenty-first in 2002. In terms of infrastructure readiness, Japan rose from eleventh in 2004 to third in 2005.

8. MPT also lacked staff with technical expertise. Observers also point out that because NTT had been the Ministry of Communications in the prewar regime while MPT had been the Ministry of Postal Affairs, many in NTT viewed MPT bureaucrats as little more than postmasters. NTT's large budget and its extensive R&D labs also contributed to NTT's ability to essentially regulate itself. See C. Johnson 1989; Fransman 1995; S. Vogel 1996; Anchordoguy 2001.

9. Okimoto and others have also argued that close relations between personnel in NTT's research labs and those of its "family" firms boosted the competence and capacity of family firms. Aside from joint projects, close relations were facilitated by *amakudari* (descent from heaven) practices, in which NTT personnel were employed at high positions in family firms' R&D operations (Okimoto 1989, 74; Fransman 1995).

10. See Kushida 2006. Also see C. Johnson 1989; Takano 1992; S. Vogel 1996; Anchordoguy 2001.

11. I would like to express my gratitude to employees of TTNet and Nifty for these perspectives.

12. MPT had been driving R&D efforts to commercialize PHS, technologically similar to cordless phones, as a cheaper alternative to orthodox cellular services. PHS services commenced in 1995, but after explosive initial growth, the lack of interconnection rules and the improvements in coverage and handset performance it spurred among its cellular competitors, led to its decline and the shift of focus to text messaging and data applications.

13. Tokyo Digital Phone (later J-Phone, and then Vodafone) went to a research lab in Keio University known for their work on Internet-related technologies, IDO (later KDDI) joined the WAP forum assembled by the American firm Unwired Planet, and DoCoMo strengthened ties to Access, a Japanese start-up company. J-Phone's first cellular information service, "Sky-walker," in late 1997 took DoCoMo by surprise. DoCoMo's president at the time, Ohboshi Koji, was reportedly furious about being behind the game. When J-Phone rolled out an early text-only version of its cellular Internet service in December 1998, DoCoMo was behind in the race for the second time ("Kontentsu kakumei no kishu tachi: 'Kokusaikijun' de dokomo ni taikou [Flag-bearers

of the Contents Revolution: Battling DoCoMo with International Standards]," *Nihon keizai shimbun*, October 8, 2003; "Kontentsu kakumei no kishu tachi: Shanai benchaa, nankan toppa [Flag-bearers of the Contents Revolution: In-house Venture Moving beyond Challenges]," *Nihon keizai shimbun*, October 9, 2003; "Kontentsu kakumei no kishu tachi: Shameru de onnagokoro tsukamu [Flag-bearers of the Contents Revolution: Grabbing Womens' Hearts with Sha-mail]," *Nihon keizai shimbun*, October 10, 2003.)

14. In DoCoMo's business model, i-mode was a portal, modeled loosely on AOL, rather than a channel for providing in-house content (Natsuno 2001). The main innovation was that, for Web sites officially approved by DoCoMo, DoCoMo offered revenue sharing with content providers and billing services integrated with DoCoMo's own billing service (Funk 2002).

15. After DoCoMo began i-mode services in February of 1999, KDDI introduced its EZWeb service in April, and J-Phone its J-Sky service in December.

16. For examples, see the illuminating series in the newspaper *Nihon Keizai Shimbun* articles analyzing the development of Japan's Internet connection services though interviews with key participants in " Kontentsu kakumei no kishu tachi: 'Kokusaikijun' de dokomo ni taikou [Flag-bearer of the Content Revolution: Battling DoCoMo with International Standards], *Nihon keizai shimbun*, October 8, 2003.

17. For details on how the distribution system worked, see Kushida 2002, 58.

18. See Telecommunications Carrier Association of Japan (TCA), http://www.tca .or.jp/ (accessed November 12, 2005).

19. For an analysis of gradual transformative institutional change, see the introduction to Streeck and Thelen 2005.

20. MPT limited the scope of businesses in two ways. First, it categorized telecom carriers into three types. Type I carriers owned infrastructure and consisted of NTT and the NCCs. Type II carriers leased facilities from type I carriers. Special type II carriers could provide services across prefectures, while general type II carriers limited their operations to local areas. MPT controlled market entry, exit, and prices charged by type I carriers. A "Supply Demand Adjustment Clause" in the Telecommunications Business Law gave MPT broad discretion over the market entry of firms. This clause allowed MPT to cite factors such as "a mismatch between the business and existing demand in the proposed region of operation" to deny an application, without needing to cite any specific criteria (Fuke 2000, 16). Likewise, no specific criteria existed for MPT to approve price changes, allowing them broad discretion. Second, MPT compartmentalized competition by dividing the scope of business activities into long distance, local, and international service. There was no explicit legal basis for this division, but MPT created a required "business area" category on the application form for carrier businesses. This led to an unwritten understanding that carriers were not to cross business lines — for example, NCCs engaging in long-distance service were not to move into international service, and vice versa (Fuke 2000, 32).

21. The Telecommunications Business Law stipulated that when carriers could not reach an agreement, the complainant could appeal, and MPT could issue an order for the carriers to reach an agreement or engage in arbitration.

22. Specifically, MPT changed most requirements on type I carriers to "notify" rather than "require permission." See Fuke 2000 for an overview.

23. Soumusho removed the type I and type II carrier classifications altogether.

24. Steven Vogel (1996) divides liberalization into deregulation and reregulation.

25. Two issues are important in interconnection: point of interface (POI) and price. POI refers to the level of the incumbent's network—regional, prefectural, or national—that competitors connect to. Initially, there were essentially no regulations governing how NTT arranged POI contracts with NCCs. Competitors wanted NTT to charge end-to-end fees in order to duplicate the least costly national infrastructure and connect only to the most lucrative prefectures. NTT, of course, wanted to charge NCCs according to the level of the POI they connected to. In 1991, MPT stepped in to restrict POIs to one per prefecture for each NCC, creating a competitive structure that increased competition between prefectures but retained NTT's monopoly within each prefecture. In 1993, MPT promulgated regulations forcing NTT to charge NCCs on an end-to-end basis (Fuke 2000, 20, 35).

26. NTT's ISDN services also mostly charged by the minute, rather than having flat fees.

27. "Kotorii, NTT wo chousa, Kousoku tsushin senryaku misu utsusu: Shinki sannyuu gyousha to toraburu [Fair Trade Commission Investigates NTT, Reflects a Mistake in High-Speed Telecommunications Strategy: Disputes with New Entrants]," *Nihon keizai shimbun*, October 24, 2000.

28. "Information superhighway" initiatives by the United States in the early 1990s and South Korea's cyber-Korea strategy were among the international factors behind the political initiative. See Prime Minister's Office 2001. (Mark Tilton rightly identifies the e-Japan strategy as a classic example of industrial policy, characterized by an industrial catch-up orientation, a clearly identified problem [high price of broadband leading to low level of penetration], and a concrete goal [providing high-speed Internet access; Tilton 2004]).

29. "Juten go bunya no hyoka: IT senryaku honbu [Evaluation of Five Strategic Areas]," *Nikkei Communications*, July 1, 2002, 106–116.

30. "Five Carriers File Suits against NTT Fees," *Asahi shimbun*, July 18, 2003. A former MPT official asserts that the key figure leading KDDI in its lawsuit was an MPT official who took a post-retirement (*amakudari*) position in KDDI. This official had apparently long been an advocate of using the courts as a new avenue for policy making (Ichiya Nakamura, executive director, Stanford Japan Center Research, personal communication, December 19, 2003).

31. "IP denwa demo baugou ga kawaranai riyuu: mittsu no kufuu de soumushou no jouken wo kuria [The Reason Why Your Telephone Number Will Not Change Even If You Shift to an IP Telephone: Three Modifications Will Allow (IP Telephony) to Pass Soumushou's Conditions]," *Nikkei Communications*, November 24, 2003, 66–68.

32. Soumusho Web site at http://www.johotsusintokei.soumu.go.jp/ (accessed November 12, 2005).

33. "Kontentsu kakumei no kishu tachi: Nihonhatsu no jouhou saabisu kaishi [Flagbearers of the Content Revolution: Domestically Bred Information Services Companies]," *Nihon Keizai Shimbun*, October 6, 2003. By the end of 2004, Vodafone had introduced Vodafone Live! in twenty-one countries, mostly in Europe, with over 28 million subscribers (http://www.vodafone.com, accessed November 12, 2005). At

about that time, i-mode was offered in nine countries with only 3 million subscribers, through local carriers licensing the technology from DoCoMo (http://www.nttdocomo .com/ presscenter/facts/index.html, accessed November 12, 2005).

34. "Ripplewood Targets Big Bids for Hefty Investment Returns," *Nikkei Weekly*, June 21, 2004.

35. Michiyo Nakamoto, "Japan Telecom Succumbs to a Gentle Touch: Michiyo Nakamoto on the Success of Vodafone's Diplomatic Approach to the Takeover," *Financial Times*, September 21, 2001.

36. In an interesting departure from previous industrial catch-up-style thinking, both e-Japan II and u-Japan argue that the era of playing catch-up with the West in terms of IT infrastructure is over, and that cooperation with other Asian countries will be a key to future development.

37. I thank Bob Cole for sharing his insights on this point.

38. "Matsushita ga hokubei" 2004; "Beikeitai shijou ni NEC saisannyuu: Netto set-suzoku gijutsu ikasu [NEC to Reenter US Cellular Market: Taking Advantage of its Internet Connection Capabilities]," *Nihon keizai shimbun*, July 16, 2003.

7

The Emerging Economies in the Digital Era

Marketplaces, Market Players, and Market Makers

Naazneen Barma

The fast-changing global digital economy presents great prospects for emerging economies. They have the opportunity to reap the dramatic gains available in the world's fastest-growing markets and the chance to participate in cutting-edge technological activities through cross-national production networks. Many assume that economically developed countries, fueled by dynamic innovation, hold an incontrovertible economic and technological lead over the poorer parts of the world. This chapter, however, challenges the view that developing countries are merely passive markets for digital products innovated in the industrialized world and directs analytical focus to the roles that emerging economies can and do play in global digital innovation. It illustrates how explosive market potential in developing countries translates into new innovative forces there, sketches out some key patterns in the roles emerging economies play in the processes of global digital innovation, and examines their innovative potential by assessing their research and development (R&D) capacity. While recognizing the significance of the digital divide between industrialized and developing countries as one of the central features of the international political economy, this chapter's approach runs against the conventional academic view that the divide is about usage or access, issues to which much attention has already been devoted.[1]

It is essential to note from the outset that the developing world comprises a large and extremely varied group, individual members of which respond in very diverse ways to the digital economy. The optimal uses of information and communication technologies (ICTs) vary widely across developing countries, as does ICT-related government policy; consider, as an exaggerated example, the need to distinguish between the state of ICT use and access in sub-Saharan

Africa versus East Asia (United Nations Development Programme [UNDP] 2000; Digital Opportunities Task Force [DOT Force] 2001; United Nations Conference on Trade and Development [UNCTAD] 2002; World Bank 2002). Yet, for the purposes of considering the production possibilities represented by the digital era for the developing world, it makes sense to focus analytic attention on those newly industrializing countries, or emerging economies, that are increasingly able to break into digital production networks. Indubitably, modes of innovation and production profiles vary within this smaller subset of the developing world. Nevertheless, these emerging economies as a group adjust to the new economy in patterns that are different from those in advanced economies (Weber and Zysman 2002). This chapter seeks, therefore, to shed light on discernible patterns at the micro — firm or market — level, as well as considering the national and international dimensions of an innovative environment.

Although advanced countries are the main purveyors of radical, breakthrough digital innovation, emerging economies are likely to find that their strength in shaping global digital markets, at least in the short and medium term, lies in a different manner of innovation. In particular, emerging economies have begun to pursue two main avenues of non-breakthrough innovation that are increasingly significant in the digital economy. The first type of non-breakthrough innovation comes in the form of improvements to specific modular applications within a digital production chain that often come from on-the-job learning-by-doing. The second form of non-breakthrough innovation that emerging economy enterprises have successfully introduced into the global economy centers around modification of the production and distribution of modular applications to meet the unique needs of their home markets.

I characterize these non-radical forms of innovation as "modular innovation." The concept builds on the insight that the prevalence of networked production in digital sectors has enabled the producers of modular applications in global production chains to become the innovative center of the digital economy.[2] The modular innovations purveyed by emerging-economy firms can and have come in improvement of the product itself as well as in organizational and marketing modifications, particularly those that take into account the characteristics of new emerging-economy consumers and commercial infrastructure. Modular innovations can hence be both product- and process-oriented. In dynamic terms, they can cumulate over time into a trajectory that matches or even surpasses the impact of innovations on the technological frontier. The global economy comprises comparative advantages that map to different sources of innovative potential. Capital-rich advanced countries have the means to finance the expensive R&D necessary for radical innovation. Newly industrializing economies can rely on their rich human resources, track record of organizational innovation, and huge markets of increasingly sophisticated consumers to make technological advances through processes of learning-by-doing and user-driven innovation.

This chapter is organized to examine the different roles that emerging economies can and do play in the global digital economy and in ICT innovation. They are indeed marketplaces, but fast-growing ones with explosive potential; thus, rather than being passive recipients of ICTs innovated in the advanced world, they have the power to dictate the future of digital consumer products. In addition, they are increasingly relevant market players, particularly in terms of their niches in the cross-national networks of digital production and their role of producing and distributing modular applications for home market uses. Finally, and most recently, emerging economies also have great potential as market makers: they have the opportunity to shape future global digital markets as a result of their own prowess in digital innovation and the complementary resources they have to offer.

MARKETPLACES: THE NEXT 1 BILLION DIGITAL CONSUMERS

Although the digital economy continues to grow globally, poor countries represent the market potential of the future. And they are no longer simply the passive recipients of products and services innovated by and for the advanced world. They have their own very specific needs and tastes, and their buying power is sufficient across a number of different market segments to warrant the supply of customized products. Hence poor consumers are increasingly driving modular innovation in production technologies, business models, organizational management, and marketing and distributional strategies. These modular innovations are an essential type of the new value-creation patterns required in the global digital economy.

It is instructive to place the market power of the emerging economies within an international context. Global growth in ICT use has been robust over the last decade (Figure 7.1). Most strikingly, mobile cellular subscribers numbered 16 million in 1991 and shot to 1,329 million by 2003, overtaking the number of telephone landlines. While computer users have increased steadily, from 130 million in 1991 to 650 million in 2003, Internet connectivity has grown much faster, from 4.4 million users in 1991 to 665 million in 2003.

Yet growth in digital industries is far from even across the world. It has become almost axiomatic in ICT business strategy that the newly industrializing economies offer fast-growing and incompletely tapped markets. The thirty advanced, industrialized countries that make up the Organisation for Economic Co-operation and Development (OECD) count for less than one-fifth of the

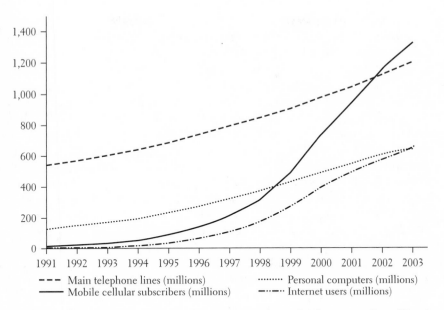

FIGURE 7.1   Global information and communications technology users (in millions)
Source: Data are from the International Telecommunications Union (ITU): World Telecommunication Indicators Database

world's population. On the other hand, China and India together make up more than one-third of the world's population, an ever-increasing share owing to population growth rates.[3] The emphasis on emerging markets comes from a pragmatic need: as the traditional markets of the digital era mature, companies must reach out to a new set of customers. During the last fifty years, about 1 billion people have come to use computers, the vast majority of them in North America, Western Europe, and Japan. Yet these markets have slowed in growth: computer industry sales in the United States are expected to increase on average only 6 percent per year for the next five years, while emerging-market demand is expected to increase at an average rate of 10 to 11 percent over the same time period (Hamm 2004). Thus, in order to continue to grow, digital industries must reach out to "the next 1 billion customers," who will come not from the industrialized world but rather from newly emerging markets. Digital-era growth opportunities for businesses in rich countries seem to be shifting inexorably to the developing world.

As Prahalad and Hart (2002) have argued convincingly, "Low-income markets present a prodigious opportunity for the world's wealthiest companies." Already tech companies are scrambling to make their mark in the emerging economies and cash in on the next big growth wave. In 2005, annual IT-related investments are expected to grow about 15 percent to $32 billion in China and 21 percent to $8.5 billion in India (Perez 2002). Emerging markets—led by China, India, Brazil, and Russia—are expected to see ICT sales surge 11 percent per year over the next five years, to about $230 billion. These markets are very appealing to rich-country companies not just because of their sheer population size, but particularly because of the growing ranks of the middle class—a new base of consumers for digital products, estimated at 60 million in China and 200 million in India. A. T. Kearney has estimated that the number of people with equivalent to $10,000 in annual income will double to 2 billion by 2015, with 900 million of these new consumers in emerging markets (Hamm 2004). Prahalad (2005) estimates the potential profits from serving the poorest 5 billion people in the world—a group he dubs the "bottom of the pyramid"—at $13 trillion per year globally (*Economist* 2004a). He values the purchasing power parity (PPP) of a fast-growing group of emerging economies—China, India, Brazil, Mexico, Russia, Indonesia, Turkey, South Africa, and Thailand, together representing 3 billion people or 70 percent of the developing world's population—at $12.5 trillion, or 90 percent of the PPP of the developing world. This is larger than the combined PPP of Japan, Germany, France, the United Kingdom, and Italy (Prahalad 2005).

In terms of emerging markets for digital products, more specifically, China had an installed base of 250 million cellular phones at the end of 2003. China Telecom is the largest mobile cellular operator in the world in terms of usage, with an annual growth rate of cellular subscribers in the past few years of more than 60 percent (International Telecommunications Union). India had an installed base of about 30 million cellular phones, growing at 1.5 million handsets per month, with the expectation that Indians will own 100 million handsets by 2005. Brazil already has 35 to 40 million cellular phones (Prahalad 2005). Table 7.1 demonstrates that while ICT usage per capita remains much lower than in the richer countries, growth in ICT usage has been torrid over the last decade in the countries Prahalad names as the emerging markets.

Digital industry giants have declared emerging markets a top priority and are pushing their products there aggressively, vying with each other for lucrative government contracts as well as for new middle-class consumers. For example, Sun Microsystems, Microsoft, and IBM have competed ferociously for deals with telecommunications and software firms in India, as well as for enormous state-by-state government contracts. IBM's revenues in Brazil recently surged past the $1 billion mark; the company plans on hiring two thousand people in Brazil and spending an additional $100 million on market development there (Hamm 2004). Microsoft famously got off on the wrong foot in

TABLE 7.1   Growth in information and communication technology use in select countries

	Telephone mainlines (per 1,000 people)		Cellular subscribers (per 1,000 people)		Internet users (per 1,000 people)		Personal computers (per 1,000 people)
	1990	2002	1990	2002	1990	2002	2003
United States	547	646	21	488	8	551.4	658.9
Japan	441	558	7	637	0.2	448.9	382.2
Finland	534	523	52	867	4	508.9	441.7
Mexico	65	147	1	255	0	98.5	82.0
Russian Federation	140	242	0	120	0	40.9	88.7
Brazil	65	223	—	201	0	82.2	74.8
Thailand	24	105	1	260	0	77.6	39.8
Turkey	121	281	1	347	0	72.8	44.6
China	6	167	—	161	0	46.0	27.6
Indonesia	6	37	—	55	0	37.7	11.9
South Africa	93	107	—	304	0	68.2	72.6
India	6	40	0	12	0	15.9	7.2
High income	420	584	13	653	3.1	445.8	
Middle income	49	168	—	176	0	59.5	
Low income	16	28	—	17	0	13.0	

Source: *United Nations Development Program (UNDP) Human Development Indicators 2004; calculated from the* International Telecommunications Union (ITU) World Telecommunications Database, *7th ed. Personal computer data are taken directly from ITU.*

China: although it owns the desktop market there, it earns little money because 97 percent of its software is illegally copied. Every time Microsoft pressures the government to crack down on piracy, however, the state makes a move to support Linux, the open-source operating system rival to Windows.[4] Yet Microsoft is pouring $750 million into China over the next three years to help develop a software industry infrastructure, on top of the $1 billion it spends there annually in running its business. Sun Microsystems has countered by signing a deal with the Chinese government to supply its Linux desktop operating system and office program to as many as a million personal computers (Leander 2004).

What makes the emerging economies crucial in terms of innovation, however, is not just their sheer market volume potential. In developing countries, the world's wealthiest companies find consumers with unique needs and varied tastes. These middle-class emerging-economy consumers may have lower incomes, but there is sufficient buying power across the huge numbers of people in these growing market segments to drive demand for products that are customized to their needs and tastes. These submarkets are thus significant enough to drive modular innovation, particularly in specific digital applications and in organizational form, to respond to the existing commercial infrastructure. The innovative challenge lies in tailoring new products to these consumers and taking advantage of their uniqueness. As an example, emerging-market consumers are younger and less loyal to brands than their Western counterparts. Brown and Hagel (2005) report that these new demographics and consumer patterns are forcing companies to rethink the manner in which they design and deliver their products, and a growing number of established digital vendors acknowledge that returning to the drawing board is the only option in the emerging markets. Furthermore, advanced-country companies are increasingly recognizing that if they are not competing in the growing emerging markets, they are not developing the capabilities they need to remain viable back home. Providing goods and services for poor consumers forces companies to innovate in ways that promote long-term success.

Prahalad points out that "if we stop thinking of the poor as victims or as a burden and start recognizing them as resilient and creative entrepreneurs and value-conscious consumers, a whole new world of opportunity will open up" (2005, 1). As he further argues, and as rich country companies have learned the hard way, firms cannot profitably serve emerging market consumers with the products designed for advanced-country consumers. Rather, Prahalad argues, they will need to thoroughly reengineer products in order to reflect the different customer needs and production and distribution economics at the bottom of the pyramid: the demand for small unit packages that can be paid for with poor consumers' limited cash in hand, and the necessity of a cost structure that can produce goods and services in high volume to compensate for the low margin per unit (Prahalad and Hart 2002; Prahalad 2005). (Note that lower prices in emerging markets will likely put pressure on prices worldwide, which may reduce ICT industry growth rates and profit margins [Hamm 2004]). In short, emerging markets are not implicitly stuck relying on commoditized, hand-me-down innovation from the developed world (Weber and Barma 2003). They have their own lead users who pull technology development toward applications that specifically fit their indigenous needs and demands.

In addition, selling to the world's poor requires investment in market development and, in some cases, the creation of a commercial infrastructure that can unlock the latent purchasing power in emerging markets (Prahalad and Hart 2002). For example, recognizing the enormous business and development opportunities in emerging economies, Hewlett Packard has articulated its "e-inclusion" initiative, which focuses on providing technology, products, and services appropriate for the world's poor. Intel has a team of ethnographers traveling the world to provide input into designing or redesigning products to fit different cultures and demographic groups. This, in turn, leads rich companies to develop innovative new strategies for allying with other stakeholders on the ground in the developing world — nongovernmental organizations, international financial institutions, and governments — as well as catering to local stakeholders and conditions and undertaking locally tailored research and development. Following this logic, IBM has developed a $12 microprocessor and simple network computer that it supplies to Chinese companies that then sell computers and Internet access services in rural parts of the country; Hewlett Packard has agreed to install Poland's new computerized driver's licensing system using a pay-as-you-go scheme (Hamm 2004). These are some of the ways poor consumers can and will drive modular innovation in production technologies, business models, organizational management, and marketing and distributional strategies.

MARKET PLAYERS: A NEW ECOLOGY OF COMPETITION IN THE EMERGING ECONOMIES AND THE WORLD

The demand of poor consumers for customized, low-cost products and services has created a new ecology of competition and innovation in emerging markets. The industrialized world's most successful companies are finding tough competition on the unfamiliar terrain of emerging markets in the form of home-grown companies who know their local markets intimately and have grown up supplying to them. Furthermore, these enterprises have been able to leverage their home market advantages into larger inroads into worldwide markets. Yet a number of questions arise in examining emerging-economy firms as market players. Are they actually competing directly with advanced-country companies in their home markets, or are they targeting different market segments? Are rich-country companies adequately addressing the evolving needs of lower-income middle-class consumers in developing countries, or are domestic companies successfully catering to their home markets in a vacuum of competition from overseas? From a survey of the anecdotal evidence available, it appears that emerging-economy companies are competing quite directly with their overseas competitors, and that the former appear to have the edge on the latter in successfully gauging what their consumers want and need. At the same time, however, examining where emerging-economy companies have been successful demonstrates that they may have specific skill sets that make their forms of competitiveness distinctive given the structures of the global digital economy.

First, emerging-economy enterprises seem to be competing successfully in their home markets and making inroads into global markets on the basis of cheaper pricing structures and lower production costs. In China, for example, the new networking company Huawei can charge 50 percent less for gear than Cisco. It has captured a 16 percent home market share in routers, second only to Cisco, and is starting to make inroads into global networking gear markets from Russia to Brazil, already ranking second worldwide in broadband networking gear. Domestic service companies in India provide stiff opposition to foreign challengers. I-Flex Solutions, an Indian company that provides banking and software services, has built the world's top-selling software suite for managing consumer, corporate, Internet, and investment banking needs; its revenues grew 26 percent in one financial quarter in 2004, in a slow-growth worldwide enterprise software industry.

Second, it appears that closeness to market allows emerging-economy enterprises to capitalize on the demands and increasing purchasing power of their home market consumers. The South Korean companies Samsung Group and LG have taken advantage of the advent of the wireless age in East Asia to make their move away from the personal computer-centric era, which has been dominated by U.S. companies. While 30 million computers are expected to sell in Asia in 2004, this figure is dwarfed by the 200 million Internet-enabled cellular phones expected to sell there. Samsung and LG are taking advantage of their cellular phone lines rather than their personal computer lines; in the past four years they have risen to become the third and sixth largest mobile phone makers in the world. TCL Mobile is one of the top two Chinese mobile handset makers, and its solid position in the largest cellular market in the world has given it an edge in other developing markets in Africa, Asia, and the former Soviet Union.[5]

For digital industry powerhouses, these different forms of competition in newly industrializing economies means that they will likely have to invest substantial sums of money to succeed in emerging markets. In addition, they will have to dramatically alter the very business strategies that made them so successful in the advanced world. Dell, for example, introduced a consumer PC in China, the SmartPC, which was different from anything it had sold before: "It came preconfigured rather than built to order, and it was manufactured not by Dell but by Taiwanese companies. At less than $600, the SmartPC has helped Dell become the top foreign supplier in China. Its share of the PC market there rose from less than 1 percent in 1998 to 7.4 percent today" (Hamm 2004). Yet two local Chinese companies, Lenovo Group and Founder Electronics, both rank ahead of Dell and other foreign hardware suppliers and remain the top PC sellers, with market shares of 25.7 percent and 11.3 percent respectively. They have an advantage in reaching Chinese customers through vast retailing operations; when Dell set up retailing kiosks for the SmartPC and other products, it faced competitors selling stripped-down PCs for about $360 and had to withdraw from the consumer market. IBM recognized Lenovo's potential when it sold its PC business to the Chinese PC maker in December 2004. The move signaled a recognition by IBM that its future in China depends on close partnership with a local market leader. The deal offers the Chinese the chance to tap into overseas management and technological expertise, reflecting "the rising global aspirations of corporate China" (Lohr 2004).

Emerging-economy companies have increasingly been able to beat out rich-country competitors on their home turf with intimate local knowledge and low-cost, low-margin products. At the same time, some domestic firms are finding their strengths lie in niches in cross-national production networks as they take advantage of the constantly shifting determinants of competitiveness in the global economy.[6] This is the strategy that East Asian manufacturing firms used with great success in the 1980s as the East Asian Tigers became the original newly industrializing economies (NIEs) of the postwar era. South Korea, Taiwan, Hong Kong, and Singapore pursued economic growth strategies that differed in important ways, but all were successful in responding to the major shifts that continue to determine competitiveness in the world economy today. Lall (1999) identifies these successful competitiveness adaptations as a new pattern of competition marked by knowledge- and technology-based advantages rather than on factor endowments; the emergence of new, less hierarchical organizational structures where firms are embedded in dense technological and productive networks; and the restructuring of old industries, driven by radical technological change.

What all the East Asian Tigers succeeded in doing was moving away from a reliance on low labor costs and hence from static sources of comparative or cost advantage by moving up the technological ladder and the economic-value chain. They diversified into complex technologies, not just adopting more capital-intensive technology but also moving into more advanced technological functions within activities. For example, they moved from being key nodes for simple assembly in cross-national electronics manufacturing networks to manufacturing their own goods with local content, and finally into design, innovation, and product development (Lall 1999). The challenge is structurally the same for economies hoping to make their mark in the digital era today: how to move from static advantages to dynamic innovation.

India has emerged as an important player in global digital markets as a result of its huge reserve of well-trained software engineers and one of the largest

158   pools of engineering and scientific manpower in the world. India's IT successes have come through "body-shopping," whereby programmers are sent
abroad on a contract basis,[7] and its large and growing business process outsourcing industry. But the longer-term innovative potential of both these activities is questionable. Some have argued that young Indian engineers benefit immensely through body shopping, learning technological, business, and
organizational management skills abroad. In turn, they represent an important source of knowledge and technical transfer back to India. Yet India's software industry has competed internationally on the basis of low-cost skilled professionals, which becomes less viable as the growing demand for programmers
increases their salaries, decreasing their advantage in shaping future markets.
Moreover, increased human capital in the form of returned body-shoppers
may yield very little in terms of innovation if there are no domestic outlets for
the skills with which the programmers return. In business process outsourcing,
as competition from other parts of the world heats up, leading companies in
India are fighting to win higher-value-added activities to continue to compete
in and innovate for global digital markets.[8] In this sense, the future of innovation in emerging economies lies exactly where it did in the past: moving away
from a static comparative advantage in cheap labor and toward building
dynamic comparative advantage higher up the value chain.

Borrus and Cohen (1998) discuss more specifically the structural changes
in the competitive dynamics in the global digital industry in the past decade.
First, the ICT industry has been increasingly characterized by the growth of
networked production, where a growing number of core functions are contracted out, including production and final assembly itself. This phenomenon
encapsulates the increasing modularization of digital production. It has commodified a growing range of advanced intermediary products, disaggregated
the organizational form of the major, integrated producers (beginning with
U.S. firms), and shifted the geography of production toward emerging economies, particularly centering many cross-national production networks in Asia.
Second, the ICT industry has seen a shift in power from integrated producers
to major users such as banks, insurance companies, and automobile manufacturers. These consumers have increasingly pushed the changes in ICT policy, such as telecommunications deregulation and a demand for interoperability of standards and no proprietary standards and systems. In addition,
these major users have pushed the development of new applications that have
become large new markets in data communications, including corporate private networks and intranets, for example. These new networked applications
have increasingly driven the personal computer industry and propelled
growth for hardware and software companies. Borrus and Cohen suggest that

emerging economies should focus on the markets for these applications in seeking to develop competitive domestic ICT industries. Third, there is new competition to set market standards in the ICT industry, which has shifted value-added, and hence power, in the production chain from integrated producers to holders of a standard located anywhere in the production chain. This means that new ICT markets are increasingly characterized by rivalry to set de facto market standards. Although U.S. companies have dominated this rivalry, it provides emerging-economy enterprises with remarkable opportunities in global ICT markets.

The implication of structural shifts in global digital production for emerging economies can be tied to the micro-political economy of innovation. Developing countries are not doomed to a lifetime of technological catch-up through the "stages of growth" of a single trajectory of industrialization and modernization (Rostow 1962). This chapter instead supports a perspective that is better able to account for and elaborate different trajectories of digital innovation in the developing world. The appropriate micro-institutional political economy model is captured in the "varieties of capitalism" approach, which emphasizes the importance of the set of relationships the firm is embedded within and the characteristics of those relationships.[9] A varieties of capitalism perspective yields the insight that there are indeed different mechanisms at work, at the firm level, in responding to various production and innovation challenges. In terms of innovation for the global digital economy, in particular, we see a wide array of experiments being carried out in the market place. Successful innovations in modular applications and user-driven product modifications come from these varieties of experimentation in emerging economies.

MARKET MAKERS: INNOVATIVE POTENTIAL
IN THE EMERGING ECONOMIES

It is worthwhile at this point to take a step back and consider again the different dimensions of innovation. The concept is conventionally associated with breakthrough or radical invention, financed by expensive research and development operations. Yet in emerging economies these characteristics are rarely found. This does not mean, however, that there is no innovation occurring in the developing world. Lall (1993) points out that the view of technological innovation as major breakthroughs, where a technological lead emerges from a completely new production or process, is misleading. Rather, the correct scope of technological activity is much wider, including what are characterized here as modular innovations. These are sometimes considered incremental improvements;

nonetheless, they account for the larger share of production increases even in the advanced, industrialized world. This form of innovation in the developing world includes gaining "technological mastery" over imported technologies; that is, it includes learning the tacit elements of foreign technologies and building the ability to modify technology for domestic applications, for example, through imitation and reverse engineering. The modular innovations in the global digital economy that have been discussed in this chapter represent this type of non-frontier technological innovation.

There is much to be learned about the processes of and potential for digital innovation in emerging economies today by examining the industrial-technological innovation paths followed by the original NIEs of East Asia. As Kim and Nelson (2000) point out, reverse engineering and imitation were the basis of the creative innovation that propelled the rapid industrialization of the East Asian NIEs in the 1960s and 1970s. Hobday (2000, 158) concurs that emerging and advanced countries have qualitatively different paths of innovation:

> The innovation paths of the NIEs make an interesting comparison with Western innovation models, which stress new product development, dominant designs, and R&D. . . . In contrast with normal Western models, the NIEs began with mature, standardized manufacturing processes and gradually moved to more advanced stages of technology. . . . Typically, firms graduated from mature to early stages of the product life cycle, from standard to experimental manufacturing processes, and from incremental production changes to R&D. In this sense, the NIEs progressed "backward" along the normal stages of the product life cycle.

The R&D efforts of latecomer South Korean electronics firms in the high-growth 1970s and 1980s, for example, were mostly applied, targeted at improving manufacturing technology and, to a lesser extent, developing new designs (Hobday 1995). Lall states, even more forcefully, "The process of technological change in developing countries is one of acquiring and improving on technological capabilities rather than of innovating at frontiers of knowledge" (2000, 13). The assimilation and adaptation of a given technology can involve just as much technological effort in developing countries as more radical innovation, and often requires formal R&D. It is this gaining of technological mastery, which often comes from on-the-job learning-by-doing and the production of modular applications that cater to users in home markets, that explains most innovation in and much of the dynamic comparative advantage of emerging economies.

The overall competitiveness of companies in terms of the scope for innovation in turn depends on a host of different factors. These can be thought of as constituting a national innovation system, the supporting resources and

TABLE 7.2   R&D potential

	Tertiary students in science, math, and engin. (% of tert. students)	Patents granted to residents (per million people)	Receipts of royalties and license fees (US$ per person)	Research and Development (R&D) expenditures (as % of GDP)	Researchers in R&D (per million people)	High-technology exports (% of merch. exports)	
	1994–1997[a]	2000	2002	1996–2002[a]	1990–2001[a]	1990	2002
United States	—	298	151.7	2.8	4,099	33	32
Japan	23	884	81.8	3.1	5,321	24	24
Finland	37	5	107.5	3.4	7,110	8	24
Mexico	31	1	0.5	0.4	225	8	21
Russian Fed.	49	99	1.0	1.2	3,494	—	13
Brazil	23	0	0.6	1.1	323	7	19
Thailand	21	3	0.1	0.1	74	21	31
Turkey	22	—	0.0	0.6	306	1	2
China	53	5	0.1	1.1	584	—	23
Indonesia	28	0	—	—	130	1	16
South Africa	18	0	1.0	—	992	—	5
India	25	0	—	—	157	2	5
High Income	—	350	82.9	2.6	3,449	18	23
Middle Income	—	5	0.5	0.7	751	—	9
Low Income	—	—	—	—	—	—	9

[a] Data refer to the most recent year available during the period specified.

Source: UNDP Human Development Indicators Calculated from: World Intellectual Property Organization (WIPO) 2004 Intellectual Property Statistics; UNESCO 1999 Statistical Yearbook; United Nations 2003 World Population Prospects 1950–2050; and World Bank 2004 World Development Indicators.

policies that increase national absorptive capacity for technological innovation (Mowery and Oxley 1995).[10] The core characteristics of a national innovation system are public agencies that support or perform R&D; universities, which perform both research and training; firms that invest in R&D and application of new technologies; public programs intended to support technological adoption; and laws and regulations defining intellectual property rights (IPRs). For the purposes of examining the innovative potential of a group of emerging economies, I focus on (1) the level of human capacity; (2) research and development activity and funding, both public and private; and (3) the enforcement of IPRs.

A few key emerging economies are gaining on core advanced-country innovators in terms of the elements for the research and development essential to innovation. Table 7.2 illustrates several of the core arguments of this chapter.

High-tech exports provide a measure of international competitiveness, and the figures in the last column show that the developing countries in question have indeed emerged on the global scene in the past decade. Patenting activity and royalty and licensing receipts are dramatically lower in the emerging economies than in the advanced countries represented. Thus, advanced countries are indeed the major purveyors of radical, breakthrough digital innovation. Yet these figures represent only the types of breakthrough innovation that developing countries rarely engage in. Emerging economies are instead likely to find that their strength in shaping global digital markets, at least in the short and medium term, lies in the modular innovation associated with improvements to specific applications through on-the-job learning-by-doing and user-driven product modifications. Emerging economies are indeed equipped with the resources necessary for these types of innovation. The comparative figures shown in Table 7.2 on tertiary science, math, and engineering students and R&D expenditures and researchers are far more encouraging in indicating the modular innovative potential of the emerging economies.

In assessing a country's research and development activity, however, it is not just the quantity that matters. The sector in which R&D is performed and whether it is linked to specific consumer demands or product development are also significant. Mowery and Oxley point out that public sector R&D investments have expanded to complement increases in private sector R&D, but, citing Thailand and Argentina as examples, they add: "Efforts in developing countries to build up public sector R&D programs in the absence of demand from the private sector often fail to produce results" (1995, 84). In Latin America, for example, the model of national councils of science and technology "underestimated the relationship between market and technology, and the importance of the management of innovation at the enterprise level" (Correa 1995, 833). Table 7.3 breaks down R&D performance between the productive sector and higher education, and the source of R&D financing between the private and public sector. The figures demonstrate that the countries we conventionally identify as important innovators—i.e., the advanced countries and the East Asian NIEs—perform and finance more of their R&D in the private sector than in the public sector. The slower-growth emerging economies, however, such as those in South Asia and Latin America, tend to rely more on government financing of R&D and conduct less R&D in the private sector than in the public sector.

Emerging-economy governments often favor basic research facilities that are oriented toward frontier technologies. Instead, it is important to link public labs with private funding in order to reorient the research agenda and activities such that public R&D has good linkages with private firms. For example,

TABLE 7.3 Sector and source of R&D performance

	Sector of R&D performance (%)		Source of R&D financing (% distribution)	
	Productive sector	Higher education	Productive enterprises	Government
Industrialized market economics (a)	53.7	22.9	53.5	38.0
Developing economics (b)	13.7	22.2	10.5	55.0
Sub-Saharan Africa (except S. Africa)	0.0	38.7	0.6	60.9
North Africa	N/A	N/A	N/A	N/A
Latin America and Caribbean	18.2	23.4	9.0	78.0
Asia (except Japan)	32.1	25.8	33.9	57.9
NIEs (c)	50.1	36.6	51.2	45.8
New NIEs (d)	27.7	15.0	38.7	46.5
South Asia (e)	13.3	10.5	7.7	91.8
Middle East	9.7	45.9	11.0	51.0
China	31.9	13.7	N/A	N/A
European transition countries (f)	35.7	21.4	37.3	47.8
World	36.6	24.4	34.5	53.2

Notes: (a) United States, Canada, Western Europe, Japan, Australia, and New Zealand; (b) Including Middle East oil states, Turkey, Israel, South Africa, and formerly socialist economies in Asia; (c) Hong Kong, Korea, Singapore, Taiwan; (d) Indonesia, Malaysia, Thailand, Philippines; (e) India, Pakistan, Bangladesh, Nepal; (f) including Russian Federation

Source: Lall and Pietrobelli, 2002 42. Calculated from UNESCO 1997.

business R&D only accounts for 13 percent of the total in India; the rest is conducted by the public sector and universities, where it may not be relevant to economic applications (Dedrick and Kraemer 1993). In an effort to combat this effect, the government has established "science cities" around prominent research institutions to create centers for high-technology industrial development through stronger ties between research and industry. Mowery and Oxley (1995) argue that the optimal sequence for public investment in research and development is initially to target technical schools and universities that emphasize training, rather than to encourage basic research. This basic frontier-technology research in public laboratories seems to hold promise for economic returns only at a later stage of economic development.

This logic holds at the micro or firm level as well. Hobday (1995) concurs that the key to competitiveness for latecomer firms runs contrary to theories which stress R&D or place R&D at the beginning of the innovation process. Rather than radical innovation, behind-the-frontier innovation through imitation and reverse engineering was essential in allowing catch-up development. He debunks conventional wisdom: "East Asian latecomers did not leapfrog from one vintage of technology to another. On the contrary, the evidence shows that firms engaged in a painstaking and cumulative process of technological learning: a hard slog rather than a leapfrog" (1995, 1188). He also emphasizes the importance of home-market consumer-driven innovation in analyzing the success of latecomer electronics firms in East Asia. He points out that latecomer firms located in developing countries have two major disadvantages in terms of innovation: they are dislocated from the main international sources of technology and R&D, and they are dislocated from leading-edge markets and demanding users. In order to succeed, therefore, the latecomer firm must devise ways to overcome market barriers to entry and then forge the user-producer linkages that stimulate technological advance. With growing and increasingly sophisticated domestic consumer bases, emerging-economy enterprises may find that catering to their home market will further propel them onto global markets. These arguments reinforce this chapter's claim that experimental innovation in modular applications and user-driven product modifications is central in shaping economic success in emerging economies.

A closer examination of patenting data allows further analysis of whether emerging economies have built indigenous technological and entrepreneurial capabilities. Mahmood and Singh (2003) find that the original East Asian NIEs — Taiwan, South Korea, Hong Kong, and Singapore — have much higher U.S. patenting activity than other emerging economies, which they attribute to different sources of innovation in each country. Though it is important to bear in mind that patenting activity reflects bursts of innovation rather than modular innovation, Mahmood and Singh's data nonetheless demonstrate significant growth in innovative capability across the emerging economies over time (see Table 7.4).

Interestingly, the sources of innovation differ quite dramatically across the countries that Mahmood and Singh (2003) analyze. The relative contribution to innovation by multinational corporation (MNC) subsidiaries is highest in Singapore and India, minimal in Taiwan and South Korea, and in between for Hong Kong and China. Business groups contributed more than 80 percent of patenting from South Korea in the 1990s, compared with less than 4 percent in Taiwan. Individual inventors' importance is declining across all countries over time, but they still hold 59 percent of recent patents in Taiwan. Thus there is evidence to support the proposition that a country's industrial policy and profile shapes its innovative fabric. The predominant sector of innovation is business groups in South Korea versus other domestic firms or organizations in Taiwan; this maps to the well-documented difference in industrial profile between the two countries, with *chaebol*, conglomerates of many companies

TABLE 7.4   U.S. patents granted to emerging economies

Recipient countries	1970–1974	1975–1979	1980–1984	1985–1989	1990–1994	1995–1999
Newly industrialized economies						
Taiwan (ROC)	1	176	397	1,772	5,271	12,366
South Korea	24	43	91	424	2,890	11,366
Hong Kong	59	75	113	177	279	570
Singapore	21	9	20	47	148	499
Emerging Asian economies						
India	83	67	40	64	126	316
China	61	2	7	129	239	332
Indonesia	19	5	5	10	26	18
Malaysia	2	13	6	13	43	89
Thailand	4	3	7	11	15	56
Emerging Latin American economies						
Mexico	243	246	191	202	189	257
Brazil	86	100	110	156	260	353
Argentina	126	113	100	82	109	183
Chile	22	20	12	18	21	44
Venezuela	36	35	50	103	121	145

Source: Table 2 in Mahmood and Singh 2003, 1034. Data are from U.S. Patent Office.

clustered around a parent company, dominant in South Korea while small and medium enterprises are dominant in Taiwan. Further reflective of industrial profiles, the predominant sector of innovation is foreign MNCs or organizations in Singapore and a combination of domestic firms or organizations and foreign MNCs or organizations in Hong Kong, India, and China. The figures further demonstrate that research institutes appear to play an important role in all countries. In China and India, however, private sector R&D is not yet fully developed, as evidenced by a disproportionately high number of government-affiliated organizations among the top fifty patent holders.

It would be impossible to discuss the potential for digital innovation in emerging economies without considering in some way the relationship of intellectual property rights to innovation. Lax IPR enforcement in developing countries permits the learning-by-doing modular innovation that emerging economies have used in making their mark in global markets, namely imitation and reverse engineering. This has been true most recently with the East Asian Tigers, but, as Maskus and Reichman point out, "few now-developed economies underwent significant technological learning and industrial transformation without the benefit of weak intellectual property protection" (2004, 290). They cite Japan as an example: from the 1950s through the 1980s, Japan pursued an industrial property regime that favored incremental innovation and technology adaptation and diffusion. On the other hand, stricter IPRs may facilitate technology transfer to developing countries, as well as the local diffusion of that technology. Thus, stronger IPRs, since they promote local frontier-technology innovation, are most likely beneficial for leading newly industrializing countries that are launching serious R&D activity.

On balance, Lall (2003) argues that the effects of IPRs vary according to countries' levels of industrial, technological, and economic development, with the need for and benefits from stronger IPRs rising with income and technological sophistication.[11] As the World Bank points out: "Interests in encouraging low-cost imitation dominate policy until countries move into a middle-income-range with domestic innovative and absorptive capabilities. . . . Least-developed countries devote virtually no resources to innovation and have little intellectual property to protect" (2001, 131–132). Thus there is an inverted U-shaped relationship between the strength of IPRs and income levels: IPR intensity first falls with rising incomes as countries allow slack IPRs to build local capabilities through adaptive innovation, then rises as countries begin to engage in more innovative efforts. Lall (2003) concludes that the income per capita threshold at which innovative activity begins is fairly high: $7,750 in 1985 dollars.[12] Innovative capacity is the constraint here — if a country has little indigenous innovative capability, IPR strengthening cannot stimulate domestic innovation, and stronger IPRs have no stimulating effect on incremental innovation through absorptive and adaptive technological activity.

IPR enforcement also affects where emerging economies may position themselves in cross-national production networks. Global production networks have made it possible for countries to move up the ladder of technological complexity and value-added without necessarily building a local technology base. Lall (2003) argues that this is the case with many of the East Asian countries: although the global electronics production network encompasses only a few developing countries, almost all situated in Asia, few of these countries have strong domestic technology bases in electronics. The emergence of integrated cross-national production systems does not necessarily force emerging economies to better enforce IPRs: "Most TNC [trans-national corporation] assembly activity in the past has gone to countries that have isolated export-processing zones from the rest of the economy without having changed the IPR regime." In the longer term, however, stricter IPR enforcement may

be beneficial for countries hoping to locate themselves in cross-national production networks:

> IPRs in developing host countries may be growing in importance as, with technical progress, more complex technologies have to be deployed by high-tech systems even at the assembly level, raising the cost of technological leakages. Moreover, when competing host countries offer stronger IPRs it may be an essential prerequisite for all aspirants to offer similar protection. Countries that have high-tech assembly operations may need to strengthen IPRs to induce TNCs to move into more advanced functions like R&D and design. At the highest end of TNC activity, where developing countries compete directly with advanced industrial countries, the IPR regime would have to match the strongest in the developing world (Lall 2003, 1673).

Countries with stronger IPRs may indeed be able to attract those transnational corporations with higher-technology activity to be offshored. Yet, because integrated systems remain highly concentrated geographically, these considerations may not apply. Thus the optimal level of IPR enforcement varies by country, according to the specific income level, sectoral composition of economic activity, and production profile.

The global intellectual property regime, embodied in the World Trade Organization's Agreement on Trade-Related Aspects of Intellectual Property Rights (the TRIPS agreement), necessarily affects the prospects for technology transfer and innovation in developing countries. Maskus and Reichman (2004) point out that the global regime could, very simply, reduce the scope for emerging-economy enterprises to break into global digital markets by compounding technological backwardness and inhibiting innovation. This danger is heightened by the process of world market regulation in knowledge goods, which is driven by the lobbying of powerful private interests in advanced countries rather than by a global consensus on the public-good dimensions of knowledge. Product imitation and reverse engineering, along with temporary migration of students, scientists, managers, and technicians, are important non-market forms of international technology transfer. International IP standards can make the task of reverse engineering by honest means and the transfer of technology through people more costly, even impossible.

In this way, Maskus and Reichman (argue that private capture of the global process for IP regulatory-standard-setting "undermines the ability of governments in developing countries to devise and promote their own national systems of innovation" (2004, 304). They urge developing country governments to integrate international IP standards into their own national innovation systems in order to maximize the benefits. Emerging economies could, for example, become the promoters of a transnational innovation system in which properly balanced IPRs were not an end in themselves but rather the means of generating innovation in a healthy competitive environment; they could preserve the ability to reverse engineer routine innovations by honest means and foster exchange between innovators at work on common technologies.

The idea of national systems of innovation has been central to the logic of this section. It has become quite clear that there are country-specific drivers of technological activity and innovation; that is, technological specialization and modular innovation are heavily dependent on the resources embodied in national systems of innovation. In addition, there is wide variation across countries in the productive and innovative roles played by different economic stakeholders such as multinational corporations, business groups, small and medium enterprises, research institutions, and the public sector. Nevertheless, in examining the innovative potential of emerging economies as a group, a few broad patterns have also emerged: the centrality of experimental modular innovation in emerging economies as they attempt to close the digital production divide, the significance of having some proportion of R&D funded and conducted by the private sector, and the dual relationship of intellectual property rights to innovation.

CONCLUSION

This chapter has examined the different roles that emerging economies can and do play in the global digital economy and ICT innovation. They are fast growing and hence vitally important marketplaces, with their increasingly sophisticated users just beginning to exercise their power in dictating the future of digital consumer products. Emerging-economy enterprises are also ever more relevant market players, having leveraged their success in home markets into inroads in global markets through a number of distinctive competitive advantages. Finally, emerging economies also have great potential as market makers: they have the opportunity to make new and different marks on the future global digital economy with their distinct national innovation systems and advantages.

Although advanced countries are the main purveyors of radical, breakthrough digital innovation, emerging economies will continue to find that their strength lies in the experimental modular innovation that is achieved through improvements in specific applications driven by on-the-job learning-by-doing and user-driven product modifications. Modular innovation in the emerging economies adapts product characteristics, business processes, and the commercial infrastructure to yield dynamic digital innovation that is, at this point in time, fueled to a great degree by the growing consumer base of the developing world. At the same time, the changing structures of the global digital economy provide unique and varied opportunities for emerging economy enterprises to leverage their innovative potential.

The future of digital innovation promises to continue to hold varieties of experimentation. One particular area to watch for new advances is the nexus forged between local business ecosystems in emerging markets and the broader cross-national networks that are the bedrock of the global digital economy. Understanding the trajectories of modular innovation in emerging economies will continue to be central to an analysis of the role that these countries can and will play in the global digital economy. Although different forms of modular innovation in emerging economies may not necessarily pose a direct challenge to currently dominant digital producers, they do have the potential to alter the structure of future global digital markets. Thus, in terms of both their market power and their production and innovation possibilities, emerging economies are positioned to increase their presence in the digital era.

NOTES

1. The international "digital divide" can be conceived of as the gap between developed and developing countries in terms of information and communication technology (ICT) implementation, access, and usage rates (Bridges.org 2001; T. Dunning 2003). Conventional wisdom, as represented by development organizations (DOT Force 2001; UNDP 2000; UNCTAD 2002; World Bank 2002) and the scholarly literature (Kraemer and Dedrick 1994; Yue and Lin 2002; Braga, Daly, and Sareen 2003), tells us that ICT may have tremendous implications for economic development. The most Pollyannaish of such views are techno-determinist, treating ICTs as a silver bullet for slaying developing-country woes. In such formulations, emerging economies can "leapfrog" along developmental paths aided by the potential wealth of a growing information technology sector and its beneficial spillover effects. A gloomier mind-set has begun to emerge, however, as numerous attempts to enact IT-driven development strategies have stalled in implementation. This view emphasizes the fact that the digital divide between industrialized and developing countries is growing, further miring the latter in poverty as IT-driven productivity continues to spur economic growth in the former. In this sense, "fairly sophisticated information technology capabilities should be thought of now as prerequisite to effective interaction with the world economy" (Weber and Barma 2003, 17).

2. Borrus and Cohen (1998) argue that the growth of networked production and thereby the commodification of a growing range of advanced intermediary products is a major structural change in the competitive dynamics of digital industry.

3. Population data are from the U.S. Census Bureau (mid-2004 statistics) and the OECD (2003 statistics).

4. As the newly industrializing countries continue to modernize, their governments are becoming increasingly important information and communications technology

customers. For example, in India, over half of all ICT purchases are made by the public sector which is required to use indigenous sources when available. Since government is such a large consumer in proportion to private interests in many emerging economies, government purchasing decisions may tip a market toward one particular form of a product over another. See Weber and Barma 2003 for an extended discussion on the role of governments in the use of open-source and free software (OSFS) in the developing world, including a discussion of the multiple motivations surrounding the adoption and use of OSFS applications in the developing world and a catalog of such initiatives. Also see Weber 2004 and Weber and Barma 2003 for a definition of OSFS and a discussion of the economics and political implications of OSFS solutions.

5. Examples of specific emerging-economy companies in this section are from Hamm 2004.

6. Borrus and Zysman (1997b) identified the importance of cross-national production networks in the digital era, as the production organizational counterpart to "Wintelism," or the struggle over de facto product standards throughout the value chain.

7. In the 1990s, 70 percent of India's software export revenues came from body-shopping.

8. Although China's IT industry is much less organized and of patchier quality than India's, this may change in the near future, because China already churns out more IT engineers than India. Russian and Eastern European engineers are as well trained and cost about the same as their Indian counterparts. Data on the offshoring of IT services are from the *Economist* (2004b).

9. See, for example, Hall and Soskice 2001. The varieties-of-capitalism approach ties the multiple networks of micro relationships around the firm to the macro political economy and vice versa.

10. Mowery and Oxley (1995) distill the considerable literature on national innovation systems.

11. See Lall 2003 for an excellent discussion of technological differences among countries. He has developed sophisticated country classifications of domestic innovation and national technological activity based on R&D financed by public enterprises and the number of patents taken out in the United States, which he then maps against an index of competitive industrial performance.

12. Maskus and Reichman (2004) agree on the threshold effects of per capita income on IPRs.

II.   THE EXPERIMENTS: VISION AND EXECUTION

8

MISSED OPPORTUNITY

*Enron's Disastrous Refusal to Build
a Collaborative Market*

Andrew Schwartz

COLLABORATIVE MARKETS

Why did Enron fail? The simple answer is that Enron was a fraud, a Ponzi scheme designed to enrich scoundrels. But beneath the off-balance-sheet transactions and partnerships that have drawn such intense scrutiny, Enron's efforts to reduce complex products into tradable commodities represented one of the most promising ideas of the past twenty-five years. Enron's failure was due in part to a business strategy that regarded competitors as ruthless and uncompromising. That mentality led the company to reject the very real possibility that rivals could, working together, create the new markets that in turn would open up profit opportunities for all. Enron's brilliant vision of the New Economy did not go far enough; it required a New Economy business model that emphasized cooperation among competitors.

Business rivals working together to create new markets, or what we may label collaborative markets, represent an important trend in the corporate governance of international business. The new "tools for thought" that define the digital age (Cohen, DeLong, and Zysman 2000) and a shift in U.S. government policy toward the deregulation of finance, telecommunications, energy, and other industries have emboldened industry producers, suppliers, trading houses, financial institutions to organize, operate, and own markets jointly. Dramatic increases in computer processing speed allowed the development of complex financial modeling techniques, supply chain management software, and credit procedures. Encryption and intellectual property protections permitted the secure interchange of data — including price quotations, product information, and credit information — necessary to effect multiple and rapid

174   transactions. Advances in data networking democratized the spread of the new tools for thought beyond the richest institutions, bringing the new markets to smaller entities. The final piece that enabled collaborative markets was the policy emanating from Washington that signaled politicians were willing to tolerate new market forms, even if natural competitors would drive them.

Collaborative markets operate under the explicit governance of the founding interests, unlike traditional public markets, such as stock exchanges and commodity exchanges, which tend to be organized and operated by neutral third parties and regulated by the state. Today's collaborative marketplaces, which include business-to-business (b-to-b) exchanges and private commodity exchanges, transact currencies, metals, energy, automobile parts, chemicals, and dozens of other products.[1]

Collaborative markets offer industry powers a vehicle to procure and sell products in an institutional environment compatible with their interests, even as demand and supply determine the prices of individual transactions. Specific value propositions include price transparency, increased price competition, reduction of sales and marketing costs, control over product quality and reliability, and standardization of billing and credit terms and delivery mechanisms. Perhaps most important, collaborative markets may present potential barriers to entry from unwanted competitors to the market organizers and, in some cases, may permit the organizers leverage over suppliers regarding product quality and selection.

Collaborative markets represent a break with traditional economic theory by positing an institutional rationale for market participants, including market competitors, to collaborate. Unlike oligopolistic economic models, in which a small number of competitors collaborate to restrict price or output, partners in collaborative markets compete as aggressively as ever, but within a set of market rules of their own design. Collaborative markets are hybrid (and novel) institutions of industry corporate governance: part free market, part directed market.

Enron's attempt to create financial markets from differentiated products — which refer in this chapter to bandwidth — represents a case study in the shortcomings of a traditional competitive approach in light of recent technological advances and the apparent willingness of government to tolerate industry ownership of markets. Enron might have succeeded in creating a trading marketplace in bandwidth, and indeed in other product markets, had it collaborated with its competitors. The Enron case teaches modern managers an important institutional lesson: businesses should consider collaborating not only with suppliers and customers, but also with rivals, to develop new markets, protect market share, and create production efficiencies.

THE NEW ECONOMY OF ENRON

Enron's attempt to reduce telecommunications capacity to a tradable commodity represented one of the most promising and potentially profitable ideas of the New Economy. Bandwidth trading, as it came to be known, was the centerpiece of Enron's strategy to transform complex products into financial instruments that could be traded, hedged, and financed.[2] In 2000, Schroder's estimated the ultimate worth of Enron's bandwidth trading effort at $36 billion, more than the combined value of Enron's electricity and natural gas business.[3] Bandwidth trading was supposed to transform Enron from "the world's leading energy company" into "the world's leading company."[4] Enron's fate depended on its plan to transform financial markets, epitomized by bandwidth trading, far more than on its accounting practices.

However, bandwidth trading was a dismal failure. Although domestic energy trading brought Enron tremendous profits through the bankruptcy year of 2001,[5] bandwidth trading dragged the company into the financial mud. Enron Broadband Services (EBS), the telecommunications division, lost $357 million in the third quarter of 2001 and $494 million over the first nine months of that year.[6] By the end of 2001, EBS's $1 billion of investments in broadband assets were worth only pennies on the dollar.[7] It was little wonder that when Enron executives fabricated transactions, EBS often was involved.[8]

Why did Enron's bandwidth trading effort fail? Some technical experts contend that Enron's idea never made sense. Michael O'Dell, chief scientist of WorldCom's UUNet, called bandwidth trading a "largely absurd notion that could only be created by financiers . . . [who] have no notion of how bandwidth is actually used in building real networks."[9] Leo Hindery Jr., the former CEO of Global Crossing, added that Enron "was way out of [its] league in the telecommunications business. The [valuation] numbers they throw around are laughable."[10] In this view, Enron was just another dotcom with a "pie-in-the-sky," "know-it-all" business plan that was at best premature, at worst reckless and even criminal.

Some telecommunication analysts argue that the capacity glut doomed bandwidth trading from the start. The overhang in bandwidth was destroying prices throughout 2000 and 2001. Major telecommunications networks were unlikely to support bandwidth trading while prices were plunging, since bandwidth trading would expose to customers the extent of the "buyer's market" and thus potentially accelerate the bandwidth price plunge. Cheap, plentiful bandwidth also implied that users would transact with a network based primarily on reliability, value-added services (including commercial tie-ins), and network solvency, not on price, the differentiating purpose of a trading market. Enron and bandwidth trading never stood a chance.

The technical challenges and unfavorable market forces obstructing Enron's bandwidth initiative became insurmountable hurdles when executed by Enron's confrontational and impatient business approach. Two strategic choices were decisive:

- *Confrontation: "My way or else."* Enron chose to impose bandwidth trading on huge telecommunications companies that doubted the technical feasibility of bandwidth trading, the value of transparent and flexible pricing, and, critically, Enron's intentions. In the end, the industry rebelled against Enron's pushiness and perceived arrogance, refusing to physically connect with Enron's facilities and to transact with its traders. New financial markets can only emerge from the cooperation and trust of industry interests.
- *Impatience: "Too much, too soon."* Although Enron constructed a telecommunications infrastructure to trade bandwidth, the industry was unprepared to adopt Enron's model. Telecommunications networks and customers lacked the internal processes and infrastructure to sell and buy bandwidth flexibility. Most important, company managements lacked the trading mentality, and it would take time and education to convince them that trading in bandwidth was benevolent and inevitable. Enron rushed bandwidth trading, buying into the New Economy myth that major corporations now worked on Internet time, when in reality new financial markets needed time to gestate.

Enron might have remained a growing firm and bandwidth trading might be a fledgling industry rather than a meteoric failure if the company had followed a more moderate and less antagonistic approach. Enron's bungled strategy to commoditize bandwidth may have represented a missed opportunity of historic proportions. The illustration of Enron's strategic errors implies that a collaborative and evolutionary approach could revive bandwidth trading — it is obviously too late for Enron — and lend guidance to managers looking to better execute Enron's vision of the New Economy.

A NEW COMMODITY MARKET FOR BANDWIDTH: 1996 TO 1999

Deregulation of the telecommunications industry marked by the 1996 Telecommunications Act and the prospect of explosive Internet demand for bandwidth set in motion a revolution in telecommunications economics.[11] Seemingly overnight, dozens of backbone networks appeared to compete with what had been a market dominated by the three incumbent long-distance networks, AT&T, WorldCom, and Sprint. Joining these "next generation" networks were an even greater number and diversity of content providers, Web hosting companies, Internet service providers, and enterprises of all types and sizes thirsting for once unimaginable quantities of bandwidth. Bernard Ebbers, chairman of WorldCom, predicted in 2000 that in only three years UUNET would need to provide network capacity for a single day equal to their total annual capacity in 2000. The (wholesale) market was destined to undergo price uncertainty from the huge supply and demand coming on line. The key ingredients of a robust commodity market were in place: multiple and heterogeneous buyers and sellers, and price volatility.

The bandwidth commodity paradigm promised to solve problems across the telecommunications industry. For the natural bandwidth supplier (typically carrier networks), trading in bandwidth offered additional chances to sell excess capacity, to reduce provisioning times and expenses by extending interconnectivity with other networks, and to add capacity at peak times without having to build additional facilities (simplifying the build-versus-buy decision). For the purchaser, trading in bandwidth offered price transparency, a flexible choice of network suppliers, and the opportunity to buy bandwidth in short-term increments. From an industry vantage, the commodity paradigm promised efficient distribution of bandwidth derived from transparent price signals, faster provisioning times from pressure on networks to pre-provision bandwidth, and, more important from Enron's vantage, the possibility that telecommunications companies would be able to manage financial risks. The missing element was a catalyst to jumpstart bandwidth trading.

Enron could not help but notice that those same dynamics that were transforming telecommunications — government deregulation and technological change — had transformed the electricity and natural gas markets not long before, establishing the foundation for the company's wealth and prestige. In 1996, Order 888 of the Federal Energy Regulatory Commission (FERC) led to the deregulation of the power industry and the creation of a competitive wholesale market. New technology emerged in the 1980s, when natural gas grew from an exploration waste product to the fuel of choice for the power industry. Today, the underlying electricity and natural gas commodity market is over $180 billion; by 2015, it is expected to be over $800 billion. Many analysts believed, Enron as well, that telecommunications offered even greater promise (see Figure 8.1).

THE ENRON WAY TO COMMODITIZE BANDWIDTH, 1999

Enron's initial strategy to commoditize bandwidth addressed two realities. First, the market required a single catalyst willing to devote time and money to design and install the commodity infrastructure. Second, it required cooperation and support from the telecommunications industry in order to respond to

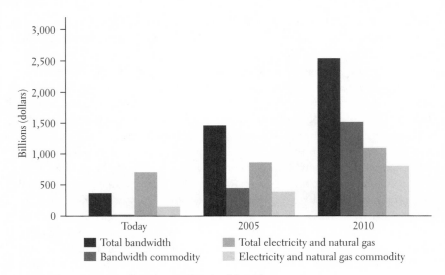

FIGURE 8.1   Forecasted markets in bandwidth and energy, 2000

skepticism about the entire bandwidth trading enterprise and to devise a master services agreement (MSA) between parties to establish standard guidelines.

To transform bandwidth into a tradable commodity, Enron set about creating the two elements fundamental to a liquid commodity market: (1) buyers and sellers with incentives and the will to trade, and (2) spot and forward markets. To satisfy the first element, Enron assumed the initiative by preparing the company to take either position — long or short — on a trade, depending on the purpose of the trade. Enron set itself up to be not only the architect of bandwidth trading, but the industry's first and largest client, a classic market maker.

On the supply or sell side, Enron set about constructing a huge network able to trade against a standard contract. Overall, the company would spend upward of $1.2 billion building a North American network of over 14,000 route miles while also operating in Latin America, Europe, and the Asia-Pacific region. As a network carrier, Enron would automatically adopt a natural long position, that is, it would be a supplier of bandwidth to sell into the market.

On the demand or buy side, Enron sought to create and distribute fabulously sexy applications that took huge amounts of bandwidth episodically. Through its content management business, Enron sought to link with the potentially highest bandwidth users (for instance, suppliers of movies such as the DVD and home video merchant Blockbuster) to create vast pools of bandwidth demand. As a content developer and distributor, Enron would automatically adopt a naturally short position, that is, it would be a consumer of bandwidth to buy from the market.

Enron was far more successful at adding bandwidth supply than demand. Unfortunately, Enron's supply added to what had become a huge overhang on the market, causing bandwidth prices to plunge and destroying the value of Enron's brand-new network. Perhaps demand will eventually catch up with supply, resulting in a viable two-way bandwidth trading market, but Enron's short-term failure to generate bandwidth demand left the company vulnerable to the slide in bandwidth prices.

Creation of a spot and forward market was a more complicated affair. As a commodity trader, Enron understood that commodity markets (and ultimately risk management) in bandwidth depend on the ability of a buyer to receive the contracted bandwidth and of a seller to deliver it. Consider the counterfactual situation if those conditions are unfulfilled. Assume that Company A sought to lock in a given price and quantity of bandwidth by buying bandwidth forward (in the future) from Company B. If Company B could not deliver the contracted bandwidth to Company A at the contracted time, then Company A would be guaranteed neither price nor quantity. Some markets (such as the S & P 500 stock index futures) address this problem by financially settling: the physical commodity does not change hands, only money does. But in the case of bandwidth, most participants will want to transact through the actual product, especially while reliable pricing data are unavailable. Markets that transact in the physical product at the present are called spot markets, and markets that transact physical product in the future are called forward markets. Both venues are fundamental to trade commodities and to managing risk.

The Master Services Agreement (MSA)

The link between the physical product and spot and forward markets is the standard contract or MSA. An MSA sets basic and mutually agreed upon terms of exchange for all market participants. In telecommunications, MSA specifications may include bandwidth quantity, time increment, delivery location(s), penalties for nonperformance (for example, outages), and quality of service (QoS). Previously, telecommunications companies swapped bandwidth bilaterally, executing a mutual service level agreement (SLA). In practice, bilateral MSAs may take months to negotiate. Clearly, market participants could trade actively and smoothly only if bandwidth transactions occurred under common, nonnegotiable terms, as spelled out in an MSA. The determining factor in whether or not the nascent bandwidth trading business would take off extended

180   from Enron's ability to craft an MSA acceptable to the telecommunications industry.

Enron's initial strategy for drafting the MSA signaled that the company would create a collaborative market in bandwidth. Enron would build a proposed standard contract with cooperation from the industry and offer to sell or buy against it, being fully willing to educate as one went and modify the contract to fit the changing industry. Enron's market builders understood that market integrity was crucial to convince the industry that bandwidth trading was good for all concerned. To this end, Enron constructed a market framework that allowed the company to monitor and deliver bandwidth but ensured oversight by neutral entities.[12]

Enron's first step was to distinguish the basic bandwidth commodity product among the many varieties of bandwidth. Minutes bandwidth, used to connect long-distance voice telephone calls, was traded in independent exchanges (Arbinet, Band-X) as well as by major carriers such as AT&T. However, telephone minutes are a slow-growth, low-margin business that is certain to be eclipsed by data transfer technologies in the near future. Enron also considered trading packets and specifically Internet protocol (IP) but found that it was technologically and commercially infeasible to deliver packets because of hardware and software complications related to measurement, billing, delivery, and standardization. To take advantage of the ever-increasing flow of Internet information while remaining aware of the state of technology, Enron chose the circuits that networks were using to transmit data over long distances. Specifically, Enron proposed city pair circuits, measured in bits per second, good for a fixed period of time, presented at a specified quality of service (QoS), and offered in intervals corresponding to the synchronous optical network (SONET) protocol.[13]

To affect bandwidth delivery, Enron planned to interconnect customers through a series of switching centers or hubs in metropolitan areas ("metros") with high bandwidth usage, including New York, Los Angeles, Washington, Miami, and San Francisco — twenty-three switches in all by January 2001.[14] Each metro hub was designed to collect bandwidth from or deliver it to carriers and large enterprises located in strategic buildings and once it was collected to transmit it to a distant metro according to the terms of the trade. Enron referred to the points of interconnection (or points of presence [POPs]) as pooling points. Once interconnected, bandwidth users could choose among the networks vying to supply long-haul capacity, in much the same way as travelers could go to the airline terminal and choose from among competing airlines. Like the airline traveler, the transmitted bandwidth passed through two hubs (terminals), one at the departing city and a second at the arriving city.

Enron initiated development of the independent Bandwidth Trading Organization (BTO) with the help of the Competitive Telecommunications Association (Comptel), an industry trade organization, to oversee the legal and financial development of bandwidth trading and assure its integrity. Composed of the major prospective participants in bandwidth trading, the BTO was empowered to "formulate, monitor, [and] enforce trading and market rules."[15] The BTO's crucial task was to organize and write an industry-acceptable MSA laying out the obligations and responsibilities of parties to a trade.

Enron proposed that PricewaterhouseCoopers (PwC) act as the neutral "pooling point administrator," working under the auspices of the BTO. PwC would act analogously to an auditor — recording and provisioning bandwidth trades, monitoring and verifying performance and delivery, overseeing BTO fee collection, verifying compliance by pooling point operators, and, in some cases, managing disputes and claims.[16] Although Williams expressed skepticism about Enron's pooling point concept, arguing that companies were already interconnected and had been exchanging bandwidth bilaterally for years, most networks agreed that Enron's pooling point architecture, combined with the monitoring of a neutral entity (PwC), facilitated the exchange of bandwidth according to standardized terms.

Bandwidth Trading Skepticism

Enron's bandwidth trading initiative met with a degree of skepticism and opposition within the telecommunications community, as the company had anticipated. Many doubted that bandwidth was a commodity on technical grounds:

- *Bandwidth could not be traded or even standardized.* Most networks manually provisioned SONET circuits, resulting in long provisioning times that precluded trading. Moreover, networks were increasingly swapping wavelengths (a broader and more flexible bandwidth variety) rather than SONET circuits. Most important, bandwidth could not be standardized since, as networks religiously reminded us, the reliability, quality, and bandwidth services varied widely across service providers.
- *Operations and billing software systems (OSS) were unable to manage bandwidth trades.* Although current OSS could switch circuits between networks, in theory, none exists in practice that can rapidly reserve a circuit and then tear down a circuit to correspond with a trade made and then unmade. In addition, network billing systems of carriers are notoriously muddled and unprepared to provision and procure bandwidth suitable for a commodity paradigm.

- *Bandwidth trading would not solve the "last-mile" problem.* Even with an elaborate pooling point system, bandwidth trading would not interconnect many large enterprises; they are typically located in buildings served predominantly by the Baby Bells (RBOCs).

Others questioned the feasibility of bandwidth trading on competitive and regulatory grounds:

- *Dominant long-distance incumbents, including AT&T and WorldCom, were likely to oppose bandwidth trading.* Bandwidth trading's main selling points — price transparency and ease of provisioning — threatened the incumbents' customer base and could accelerate the downward spiral of bandwidth prices.[17]
- *Competitors in energy markets, particularly Williams, might block Enron's grab for bandwidth trading.* While promoting bandwidth trading, the energy trading companies sought to limit initiatives that might permit Enron to dominate the business, such as Williams's attempt to discredit Enron's pooling point architecture.
- *Bandwidth trading was doomed if the competitive and regulatory landscape were in constant flux.* Established commodity markets tend to be underpinned by a known and stable set of actors. Long-established relationships have led to a measure of trust and routine that smoothes dispute resolution and clearing and credit procedures. In an industry in which major suppliers bankrupt occasionally and major purchasers simply disappear, as in telecommunications, establishing timeworn practices may be impossible. Stability in telecommunications is unlikely until the regulatory uncertainty of the RBOCs is resolved.

Despite these misgivings, Enron's initiative was too extensive for telecommunications companies to ignore, and most networks were willing to participate in the early stages. Besides, Enron was becoming a large customer of many networks (and vendors such as Lucent), and the networks were not anxious to alienate a potential revenue source. Level 3's CEO, James Crowe, typified this position when he remarked that he did not believe bandwidth could be treated like agricultural products, but the company would deal with commodity traders, anyway.[18]

Catalyzed by Enron, the bandwidth trading industry bubbled with hope in 2000. Within a year, telecommunications networks, energy traders, neutral bandwidth exchanges, investment banks, consulting firms, equipment vendors, pooling point operators, data services, and publishers devoted resources to the nascent business. One bandwidth exchange, RateXchange, grew to be worth nearly $1 billion on NASDAQ. Global networks such as Cable and Wireless, Sprint, France Telecom, PSINet, Global Crossing, Aerie Networks, Level 3, Qwest, and Teleglobe actively investigated the new markets. Start-up pooling point developers, such as LighTrade (reportedly financed in part by Lucent), entered the market. Enron and its usual competitors, Williams, Dynegy, El Paso, and Reliant, announced bandwidth trades. Support services, too, investment banks such as Morgan Stanley and Goldman Sachs, and consulting firms such as Accenture, KPMG, and Deloitte and Touche devoted resources to bandwidth trading. Well-attended conferences and publications with such titles as *The Bandwidth Desk, DJ Bandwidth Trading Alert,* and *Capacity* heralded the new industry. In December 1999, Enron and Global Crossing announced the first bandwidth trade. It was, of course, irrelevant that bandwidth trading barely existed. What mattered was that it might evolve into an industry worth billions.

ENRON TRIES TO DOMINATE BANDWIDTH TRADING, 2000

Enron's shift in 2000 away from a collaborative market approach to a more aggressive and confrontational strategy jeopardized industry support of bandwidth trading. Enron sought to impose obligations upon telecommunications carriers that would increase their costs, expand their network to connect with Enron's facilities, and direct business exclusively to Enron. It accompanied these demands with arrogance and intimidation, which insulted many industry executives. The company evidently believed that the industry had no option but to participate in bandwidth trading on Enron's terms. In shifting from catalyst to aspiring monopolist, Enron sabotaged bandwidth trading and compromised its corporate survival.

First, Enron pushed for a stern liquidated-damages provision in the BTO MSA, despite industry resistance. *Liquidated damages* refers to the reimbursement or penalties that a party to a trade would pay for nonperformance of the contract. Enron argued that bandwidth trades must be "firm," that is, "delivery dates and performance obligations are known and certain and backed by damages, rather than . . . 'best efforts.'"[19] If a bandwidth provider (seller) could not deliver the contracted bandwidth, it was obligated to pay the buyer a "settlement amount" based on the "replacement value" of the undelivered bandwidth as calculated by the buyer. Contrary to telecommunications industry practice, which was consistent with a "best efforts" type contract, the liquidated-damages provision obligated the network, as supplier, not only to the reimbursement of the original price of the bandwidth to the buyer, but also to potential (and hard to calculate) outlays over and above the original sale price. Given that networks partly saw the bandwidth trading market as a channel to sell excess capacity, the possibility that a sale could be negative to the income statement was unacceptable. In 2001, several opposing networks drafted the so-called BTO-2 MSA, which more closely resembled a "best efforts" contract. Few networks were willing to trade according to the Enron-backed BTO MSA.

Second, Enron insisted that industry participants make interconnection arrangements with the Enron pooling points. That was an expensive and time-consuming burden for potential users even when they shared co-location facilities with Enron. Networks still needed to set up equipment and to run fiber within the co-location building in order to connect to Enron; inside-plant provisioning is notoriously expensive and time-consuming.[20] Frequently, a prospective buyer or seller was located in a different building than Enron. In New York, for instance, Enron's pooling point was located not at 60 Hudson Street, the primary interconnection site for many networks, but at 111 Eighth Avenue.[21] A network taking delivery of bandwidth from Enron needed to arrange bandwidth connections in both the receiving and the delivering city (perhaps through a metro provider). Long-haul backbone networks were unwilling to commit the financial and human resources to connect with Enron, especially considering that bandwidth trading did not guarantee profitable sales.

Consider the economics of the following scenario. Carrier A is interested in buying bandwidth—let's say, an OC-12 (655 Mbps)—for one year from New York to Los Angeles on EnronOnline. In mid-2001, this circuit cost approximately $12,000 per month (0.4 cent per DS-0 mile). Unfortunately, Carrier A is not co-located in the same building in New York as Enron and therefore must purchase connectivity to Enron's building from a metro provider at $8,000 per month to buy the bandwidth.[22] Given this extra cost and the time and trouble required to connect with Enron, Carrier A may instead find it worthwhile to pay more for a circuit from a network with which it already interconnects. In addition, Carrier A may have a disincentive to transfer business to Enron since it may have existing agreements with networks to buy bandwidth bundled with value-added services or other commercial tie-ins. Consider another scenario, this time from the supplier side: Carrier B may be unwilling to devote the resources to connect with Enron unless it knows in advance that it will actually sell the bandwidth. Because it could not solve fully the issue of interconnectivity between networks, Enron's bandwidth trading initiative was destined to start slowly.

Third, Enron tried to force networks to connect with its pooling points and trade bandwidth exclusively with the company. It sought to cut out other traders, to become a monopolist. As a wedge against holdouts, Enron threatened to buy bandwidth from competitor networks and withhold business. Although wholesale networks were desperate for revenue and could dearly use Enron's business, none were willing to cede the business entirely to Enron.

Most networks, which already regarded Enron's chances to dominate telecommunications skeptically, considered its attempt to dictate competitive terms a potential threat to their independence and so refused to participate in bandwidth trading.

Fourth, Enron acted with a calculated arrogance that alienated the industry. One Enron bandwidth trader reportedly tried to intimidate a vice president of a global carrier by threatening that either his network could go along with Enron and fare badly, or it could oppose Enron and be entirely marginalized. The enmity between Enron and other networks supposedly grew to such a point that a conference planned in Europe attracted only Enron; the other networks canceled when they learned Enron would attend. Enron had unwittingly persuaded the industry that bandwidth trading was a zero-sum game in which Enron would be a winner and established telecommunications companies would be losers.

Enron's tactics were explicable, even predictable, to some extent. Unforgiving and abrasive personalities are common to a trading culture, where a trader is judged on his latest transaction and may not have the time or patience for conversation. Aggression is an integral part of the game. A visit to a major commodity exchange, for instance, will reveal a locker room of burly young men trying to intimidate one another. Telecommunications carries a different flavor and skill set entirely. Industry executives tend to be more refined and certainly more technically oriented than traders. They work at a far slower pace and consider problems deliberately, frequently choosing the more conservative alternative solution. Transaction negotiations may take place over months. Ironically, it was Enron's inflexibility in adjusting to the new cultural environment that handicapped its bandwidth trading initiative.

Enron believed, wrongly, that it could develop and dominate bandwidth trading by giving the industry no reasonable choice but to go along. This judgment, however, turned into a catastrophic miscalculation: the industry rejected bandwidth trading. Few companies connected and transacted with Enron. Most of Enron's recorded trades were with other energy traders seeking to establish price transparency and market awareness.[23] Having few real buyers and sellers translated into minimal bandwidth trading.

REALITY BITES: TELECOMMUNICATIONS, EBS, AND BANDWIDTH TRADING COLLAPSE, 2000–2001

The collapse of the telecommunications sector in 2001 exposed the folly of Enron's confrontational strategy and the weaknesses inherent in the bandwidth trading paradigm. The influx of "next generation" competitors to the long-distance incumbents, lured by the 1996 deregulation and the easy cash from venture firms seeking to churn out billion-dollar businesses, swamped the bandwidth market, creating levels of oversupply that could be resolved only through an acceleration in Internet demand—the appearance of the so-called killer app(lication)—or the demise of the new competitors. As we now know, it was the latter shoe that fell; the killer app failed to materialize, and Internet demand continued to "only" grow at about the historic norm of 100 percent.[24] Bandwidth prices plummeted, eviscerating the revenues of the next-generation carriers as well as those of incumbent long-distance carriers. The market value of long-distance carriers followed bandwidth prices, enfeebling several to the point of capitulation.

In 2001, the new companies that were serving local markets, called CLECs (competitive local exchange carriers; they included metro providers, DSL providers, managed service providers, etc.) were faring as poorly as their long-distance cousins. The regulatory relief offered in the Telecommunications Act of 1996 was proving an insufficient foundation for the CLECs to compete against the embedded advantages of the RBOCs, which were already connected to homes and businesses and had operated a network infrastructure for decades, if a famously inefficient one. Many CLECs were unable to attract enough customers to recoup high capital outlays and to overcome the RBOCs' operational foot-dragging in providing legally mandated support.[25] The RBOCs, virtual monopolists in local safe havens, remained largely immune to the telecommunications wreckage. Most CLECs came apart in 2001.

The collapse of the new competitive telecommunications companies undermined a prime justification for bandwidth trading, namely the existence of multiple buyers and suppliers. Most American markets outside the great metros would be served by only a handful of companies that would not benefit from the intervention of a third party. In addition, the turmoil's damage to sector balance sheets jeopardized the ability of companies to commit to buying and delivering bandwidth in the future, a sine qua non of bandwidth trading. Moreover, the drop in bandwidth prices called into question the financial rationale for networks to expend the necessary time and money to connect with Enron's pooling points. Finally, dislocation in telecommunications caused the industry to preoccupy themselves with survival (with the notable exception of the RBOCs), rather than with poorly understood and potentially disruptive initiatives such as bandwidth trading.

At EBS, losses mounted through spring 2001. Enron corporate had already made significant efforts to reduce its bandwidth trading efforts by redeploying assets from EBS to other parts of the company. Ken Rice, president of EBS, resigned in May. Reportedly, Enron tried throughout the spring and summer to sell its network, with no luck.

The hemorrhaging at EBS was noticed by Wall Street. From May to August, Enron shares dropped by more than half. Jeffrey Skilling, Enron's former CEO, blamed the collapse of fiber optic networks, and by extension EBS's failure, as the number 1 reason for the market value plunge.[26] Still, Enron expected to wait out the turmoil at EBS.

Later that fall, however, Enron's credibility and credit collapsed as the company exposed the infamous off-balance-sheet partnerships and accounting irregularities. No one was willing to trust a company that had operated with supreme arrogance and had raised expectations among investors. A run on Enron shares ensued, and Enron declared bankruptcy in December.

Bandwidth trading, to the extent it ever existed, faded with Enron's demise. The other energy traders either closed their bandwidth trading departments entirely (Reliant) or reduced activity substantially (Williams, Dynegy, and El Paso). Global Crossing, the carrier with the most aggressive trading effort, could not satisfy its debt obligations and finally declared bankruptcy in February 2002. Other global carriers, such as Cable and Wireless, France Telecom, and Sprint, suspended bandwidth trading initiatives. Independent bandwidth marketplaces (RateXchange) abandoned bandwidth trading entirely. Pooling point developers (LighTrade) failed to attract business and folded. Acolytes spoke of a bandwidth trading revival in 2003 or 2004, but most industry analysts remained unmoved.

WHITHER BANDWIDTH TRADING? 2002 AND BEYOND

Thanks to Enron, bandwidth trading will be forever associated with the great Internet bubble, and it will be many years before it reemerges, if it ever does. The obstacles to a viable bandwidth trading market remain. Despite the appealing logic of Enron's pooling point architecture, it is a matter of debate whether a bandwidth trading market can develop without a re-creation of a central hub, as in Shreveport, Louisiana, for natural gas, or pooling points managed by a neutral entity. In addition, it is open to question whether an industry undergoing rapid and unpredictable technological transformation is appropriate for the financial statics of a trading environment. This issue is embodied in the discussion regarding the bandwidth MSA as well as in the ever-declining price curve for bandwidth. It also is manifestly unclear whether a commodity market can develop under conditions of disruptive competitive and regulatory flux. The collapse of the next-generation networks and the CLECs and the uncertainty surrounding the deregulation of telecommunication's current power brokers, the RBOCs, complicate the perceived interests of the relevant players with respect to the possibility of freely traded bandwidth. Resolving these issues will require time and concerted industry commitment.

188   Enron's timing for bandwidth trading was atrocious. At a time when most telecommunications companies were focused on survival, a collaborative market strategy may have been insufficient for Enron to launch bandwidth trading successfully. At the same time, certain factors bolstering the case for bandwidth trading are now coalescing that could revive the dead industry. Vendors increasingly are designing products that increase interoperability and lower provisioning times. New operations software systems are being developed that could provision bandwidth automatically. The metro interconnectivity or last-mile problem is easing due to falling metro prices and co-location costs. Plunging long-haul bandwidth prices are encouraging companies to buy short-term contracts of a type ideal for bandwidth trading, versus the multiyear contracts (indefeasible rights of use [IRUs]) that had been the industry norm. Broadband penetration is ramping up to the 15–20 percent threshold that analysts consider the critical mass necessary for a burst in Internet traffic. Most important, the demand for accounting openness and the trillions of dollars in wasted telecommunications investments justify the price transparency and risk management that bandwidth trading would enable. Ronald M. Banasek, a bandwidth broker, accurately observed that "even with Enron no longer involved in the telecommunications industry all of these benefits [of bandwidth trading] still exist."[27]

Bandwidth trading's disappointed expectations and wasted resources discredited advocates within telecommunications firms. It will be a while before they (or their successors) can make the case for bandwidth trading at a senior corporate level again. To be sure, a paradigm that emphasizes interconnection among carriers, bandwidth differentiation, and near-real-time provisioning may be more appropriate for the bandwidth market than Enron's inflexible formula for building financial markets. Eventually, bandwidth prices will stabilize and even begin rising over some routes, thus renewing pressure to buy and sell in a liquid forward market. By that time, one hopes the next catalyst will learn from Enron's mistakes. It will grow the bandwidth trading market and develop a set of new value propositions organically with industry cooperation. As any commodity trader will tell you, no single company is bigger than the market.

LESSONS FROM THE COLLAPSE OF ENRON:
COLLABORATIVE MARKETS REVISITED

Enron might have avoided the worst of the telecommunications collapse in 2001 had the company sought to invoke the support of many disparate players, including rival networks, in the creation of a collaborative market in bandwidth.

Firms in other businesses increasingly view collaborative markets as alternative transaction venues. For instance, Goldman Sachs, Morgan Stanley, and
several energy traders own the Intercontinental Exchange (ICE), which trans-
acts in financial commodities, energy, and metals. Money center banks and
other major financial institutions have pooled resources to create currency
exchanges for greater efficiency. Business-to-business (b-to-b) exchanges, variants and perhaps precursors of commodity exchanges, operate with the cooperation and typically the ownership of key players in several businesses, including automobiles, cement, trucking, chemicals, and airlines. Managers
considering building or participating in a collaborative market may draw
several lessons from the Enron bandwidth trading fiasco.

1. *Create value propositions for diverse industry players, including competitors.* Attract participation among a variety of interests and construct a collaborative market that at once lowers operating costs for all players, introduces new
technology and products, and writes rules that help existing incumbents sustain or grow market share. Enron could have lowered costs for telecommunications companies by facilitating interconnectivity, by writing an MSA that
eliminated liquidated-damages provisions, and by encouraging buying and
selling of bandwidth between relevant parties other than Enron.

2. *Structure the collaborative market as a third party with independent management.* Create a true neutral entity to manage the collaborative market, perhaps with diversified industry ownership. Third-party management will help
assuage the concerns of firms mistrustful of the participation of competitors
and may help persuade government antitrust regulators that competition will
be free, fair, and transparent. Enron's enlistment of PricewaterhouseCoopers
and Comptel and the creation of the supposedly neutral BTO did nothing to
convince telecommunications companies that bandwidth trading was anything more than a scam allowing Enron to dominate telecommunications.

3. *Expect that the products, terms of trade, and market structure of the collaborative market will evolve.* Design the initial product offerings to be easy to
standardize, easy to deliver, and easy to bill. Product offerings that are overly
complex from the very beginning will put off some executives, who are skeptical of the complex corporate governance of the collaborative market project.
Enron sought to create a trading market in bandwidth prematurely, before
the industry was convinced it was a commodity. Most experts argued that
bandwidth trading would never resemble a trading market with the rapid buy-
sell staccato common to today's commodity markets. Enron may have been
advised to first create a bandwidth clearinghouse offering interconnection
services between diverse carriers and large enterprises. With the collapse of

global networks such as WCOM and Global Crossing and the retreat of international networks such as Telecom Italia and France Telecom from the United States, a series of interconnection hubs between carriers of global regions is clearly necessary and perhaps a natural precursor of an active bandwidth market.

4. *Consider a collaborative market initiative a long-term project.* Budget and staff a collaborative market assuming that it will mature in no less than two to four years. Although the technical set-up required is likely to be straightforward, the competitive issues, negotiations, and legal arrangements are complex and time-consuming. As a point of reference, regulated commodity exchanges research candidate products over a number of years, painstakingly sounding out key suppliers and consumers for interest (trading and hedging) in the new product market and for crafting the terms of a standard agreement. Exchanges organize conferences and road shows as a matter of routine in order to recruit support from the industry and from financial speculators. Also, exchanges list products that may demonstrate very low trading volumes at first but may become liquid over time. By contrast, Enron sought to create the bandwidth market in a matter of months, forging ahead in the face of intense industry resistance. They sought to force the major telecommunications companies to adopt the Enron bandwidth trading model in short order before these companies were ready internally to sign on, a strategy that was certain both to alienate the industry and to strain the decision-making capabilities of multinational corporations. Ironically, Enron, as one of the world's largest commodity traders, was well aware of the complex negotiations and time commitment inherent in building a commodity market from scratch.

Enron will be forever associated with the unbridled greed, criminality, and folly of the New Economy bubble of the 1990s. Bandwidth trading may join Enron as a warning signal of the dangers of lapses in critical thinking. Yet, the raw materials represented by the digital age's "tools for thought" and the salutary U.S. government policy are still in place. Enron's vision of commodity markets covering many differentiated products, including telecommunications capacity, may yet emerge, especially as collaborative markets become an accepted and common means of organizing market activity across wide swaths of business.

NOTES

1. Industry interests have a long history as market makers, though it is perhaps more common for industry interests (e.g., OPEC) to seek to control prices or output rather than to operate the markets per se. Competing agricultural interests created

the Chicago Produce Exchange in 1874, which evolved into the Chicago Mercantile Exchange. Similarly, rival dairy merchants created the Butter and Cheese Exchange of New York in 1872, which eventually became the New York Mercantile Exchange. In the case of cotton, an entire system of private law has evolved dating from the mid-1800s whereby industry representatives both write and police the rules (Bernstein 2001).

2. Enron was involved in dozens of markets, including electricity, natural gas, oil, orange juice, plastics, gold, steel, advertising, and even weather.

3. Niles and Meyerhoeffer 2000. Jeffrey Skilling, Enron's former CEO, produced the same estimate at an investor meeting in 2001 (A. Levy 2001, 31).

4. Remarks widely attributed to Skilling. By 2001, *Fortune* had already chosen Enron "America's Most Innovative Company" for six consecutive years.

5. Gas and power market-making operations and merchant energy operations recorded $717 million in the third quarter and $1,960 million in the first nine months.

6. Broadband assets were recorded at $1,277 as of September 30, 2001. (Enron Corp. 10-Q, Nov. 19, 2001, 44, 54.)

7. Ibid., 39.

8. For instance, EBS sold dark fiber to the off-balance-sheet partnership LJM2 for $54 million in spring 2001 to meet second-quarter numbers after Enron's CFO, Andrew Fastow, intervened to worsen the transaction terms for EBS (Enron Corp. 8-K, Jan. 29, 2002, 142–144).

9. Quoted in e-mail message in David Farber, Interesting People Series, at http://lists.elistx.com/archives/interesting-people/200201/msg00123.html (accessed November 12, 2005).

10. A. Levy 2001, 31.

11. Bandwidth is the telecommunications capacity of a circuit to carry or transmit data measured in bits per second. A typical telephone connection capacity is 64 kilobits per second. Networks may move data at far higher speeds. Common increments include: 45 million bits per second (Mbps), or DS-3; 155 Mbps, or OC-3; 625 Mbps, or OC-12; 2.5 gigabits per second (Gbps), or OC-48; and 10 Gbps, or OC-192. Most business plans modeled long-haul bandwidth, that is, between cities or regions (versus, for example, metro bandwidth).

12. The following description draws heavily from the Enron Market Development Proposal for Bandwidth Trading 2000.

13. For a detailed taxonomy of bandwidth trading products see Joshua L. Mindel and Marvin A. Sirbu, April 13, 2001.

14. Michael Rieke, Dow Jones Newswires, November 15, 2000.

15. Enron Development Proposal for Bandwidth Trading 2000, 7.

16. Ibid., 8.

17. Many analysts believed that Sprint, the third major long-distance incumbent, would not be threatened by bandwidth trading (and might be an advocate) because the company intended to focus on high-value-added enterprise customers rather than compete in the wholesale market.

18. Cited in Buster Kantrow, "Level 3 CEO: One Company Will Dominate Broadband Market," January 30, 2001.

19. Cited in General Information of Bandwidth Trading Master Agreement, January 15, 2001.

20. LighTrade, a pooling point developer, met the same problem and folded in 2002. Dynegy, Enron's competitor and former suitor, sought to relieve the metro interconnection problem by creating an interconnected grid or ecosystem among networks.

21. This example is borrowed from Rieke 2000.

22. For simplicity we will assume that Carrier A is located in the same building as Enron in Los Angeles to receive the bandwidth.

23. Attorney William Lerach filed suit in Federal Court alleging that there were only "20 legitimate trades." Cited in C. Bryson Hall, "Enron's Broadband Venture Facing New Scrutiny," in Yahoo! News, December 10, 2001.

24. 2000 growth estimate. See K. G. Coffman and A. M. Odlyzko, "Internet Growth: Is There a "Moore's Law" For Data Traffic?" June 4, 2001.

25. Local infrastructure cost more to build than long distance, per mile. Laying fiber in the metro is roughly eight times more expensive than laying it between cities.

26. From Skilling's comments at Senate Bankruptcy Committee Hearing, February 26, 2002.

27. Banasek 2002, 7.

9

The Relocation of Service Provision to Developing Nations

The Case of India

Rafiq Dossani and Martin Kenney

In 2003 the cover of *Business Week* posed a stark question for U.S. white-collar workers: "Is Your Job Next?" Motivating this alarming headline is the much larger question of whether the next great wave of globalization will come in services. This is such a profound question because the general wisdom in developed nations has been that although manufacturing might relocate to the developing world, it would be replaced by service activities. This chapter treats the globalization of services as the broader context for an examination of the political economy of the relocation of service employment. It focuses on the situation in India in order to provide insight into what might become an important source of employment in nations such as China and the Philippines, and also, perhaps, Anglophone and Francophone Africa.

Today, employment in the economies of advanced, developed nations is increasingly concentrated in the processing of digitized information and not in the manufacture of physical objects. Put differently, an increasing percentage of the population works at computer screens or on telephones. Even work in "manufacturing" firms is increasingly not on the factory floor, but rather in design, marketing, after-sales service, and monitoring. This strongly suggests that whatever further erosion there is in manufacturing employment, it is unlikely to have as dramatic an effect on the U.S. political economy as would an acceleration in the offshoring of services. How significant service offshoring will be for employment patterns in developed countries is difficult to calculate. However, services now make up the preponderance of all developed nations' total workforce. For example, according to the U.S. Bureau of Labor Statistics, in the fourth quarter of 2003, 83 percent of U.S. nonfarm employment was in services and only 11 percent in manufacturing. During the 1990s, more than 97 percent of the jobs added to U.S. payrolls were in services (Goodman and Steadman 2002, 3). Of these, business services and health care accounted for more than half of the total growth. Moreover, business-oriented industries grew from 30 percent of the total service employment in 1988 to 36 percent of the total employment in 2001, while consumer-oriented services fell from 55 to 52 percent (Goodman and Steadman 2002, 8). One recent study estimated that call centers employ as much as 3 percent of the work force of the U.S., and one consulting organization estimated that this will increase to 5 percent by 2010 (Mosher and Gist 2002).

Given the growth of services in developed nations, the scope for transferring services offshore is most remarkable. One of the earliest significant transfers beginning in the early 1980s was in software programming (Schware 1987). Software production was easily movable, because it can be directly done on a computer and does not demand extremely sophisticated communications capability (Arora and Athreye 2002; D'Costa 2003). Though offshore software production will undoubtedly continue to expand, the prospective displacement of a far larger and more diverse category of activities that come under the general rubric of services is far more interesting. The potential dimensions of this relocation of employment is best captured in the extreme words of an Indian executive who stated, "If you do not need to physically see the person doing the work, then it can be moved."

Estimates of the number of service jobs that could be offshored vary dramatically and are predicated upon varying assumptions. The U.S. General Accountability Office (2004, 44–45) provides a list of the various estimates and assumptions. The estimates range from one by Goldman Sachs that up to 6 million jobs could be affected in the next decade to Forrester Research's estimate that 3.5 million jobs will be moved by 2015. Another study that simply examines which types of jobs might be amenable to offshoring calculates that there are 15 million jobs in those categories (Bardhan and Kroll 2003, 6). All these are estimates, and all are likely to overestimate in certain categories and do not include other categories that might be affected. For example, the GAO (2004, 40) lists occupations at risk but does not project how many of those jobs will be lost. The difficulty with projections is that in addition to the call center, medical transcription, claims processing, and data entry types of activities, much more may be possible, even though it may turn out that call center work is not as movable as was once thought. In other words, the types of work that it is possible to discharge offshore are not limited to low-wage unskilled activities and will impact wages in highly skilled occupations as well. The operative determinants for offshoring are the skills available in the low-cost environment and the necessity of spatial proximity for the function to be discharged.

This chapter examines the dimensions and growth trajectory of what in India is termed the "information technology-enabled services (ITES)."[1] Despite the fact that this offshoring is at an early stage, we aim to provide some understanding of the scale and scope of the phenomenon and consider its implications. Currently, ITES is treated as an industry; however, from a value-chain perspective, it is more plausible to understand it as a spatial reorganization of the location of service activities in a wide variety of value chains. In this sense, service offshoring resembles general-purpose technologies (Bresnahan and Trajtenberg 1995; Helpman 1998). As such, today nearly all existing firms, Indian and multinational alike, are considering how to utilize the lower-cost service labor in developing nations that is now an exploitable resource. New firms are being established to mobilize this newly available labor to package their services to offer them in the global marketplace.

We begin by providing a brief overview of the relocation of services in the past. In the next two sections, we describe the technologies and environment for the relocation of services currently under way and then the important policies and regulations that contributed to the relocation. The fourth section examines the value proposition for relocating services from the point of view of the firm. The fifth section describes the characteristics of the different types of firms providing services from developing nations. Because India is the largest destination for services, in the fifth section we explore the Indian experience. In the discussion and conclusion, we speculate on the implications of the offshoring of services for developed and developing nations and possible policy initiatives for developing nations interested in the possibility of entering the ITES sector.

SERVICES AND THEIR RELOCATION

The disaggregation of service activities into discrete functions that can be relocated spatially is relatively recent. The initial relocation that grew to noticeable proportions was intranational, within the United States and began in earnest in the 1960s. The earliest efforts to minimize wage costs for back-office business processes saw U.S. firms move their back-office operations to smaller Midwestern towns where labor costs were lower, education was adequate, accents were neutral, and, at that time, the labor relatively more reliable. The cost savings were likely in the 20–30 percent range. Beginning in the 1980s, some credit card processing and call center services for the U.S. market were relocated to Latin and Central America and the Caribbean (Posthuma 1987). Later, some components of back-office services, such as payroll and order fulfillment, and some front-office services, such as customer care, were relocated

196   to English-speaking developing nations — especially India, but also other nations such as the Philippines.

ITES offshoring directly targets the back office and administrative functions. These staff functions include marketing, human resources, accounting,
facilities management, purchasing, finance, customer relationship management, and a plethora of others. Until recently, these functions were treated as
a fixed cost and received little management attention. And yet, these can account for up to 15–20 percent of corporate expenses and headcount. The discharge of most customer services is the result of an entire chain of bureaucratic activities, or what collectively is termed a "business process" (BP).[2] Each
of these activities represents costs to a firm.

The separation of a process into different activities is illustrated in Figure 9.1,
which depicts an insurance claim settlement process. The settlement of an insurance claim is a complex chain requiring the completion of a large number
of discrete activities. A large insurance firm such as Aetna would employ thousands of persons to undertake these functions. Any of the activities could possibly be offshored, but presently the vast majority of these activities are conducted in the United States, whereas initially the offshore operation in a
country such as India might simply key data into a standard form from information on a digitized "image" of the claim. It may then be possible to transfer
certain of the "Investigation and Valuation" activities to the Indian operation.
With experience, Indian accountants or engineers can be trained to "Determine fraud/exaggeration of claims" or, at the least, flag unusual claims. Here,
the Indian employees would make decisions requiring greater judgment and
have a greater impact on the firm's bottom line. Ultimately, depending upon
corporate strategy, it might be possible to relocate all activities that do not need
direct face-to-face human interaction. Obviously, the offshoring and outsourcing process is eased if portions of the value chain can be modularized
through the development of highly standardized linkages (Baldwin and Clark
2000; Gereffi, Humphrey, and Sturgeon 2005).

ENABLING TECHNOLOGIES AND
THE BUSINESS ENVIRONMENT

The reengineering movement that swept management in the 1990s focused attention on the savings that could be achieved by reorganization. One part of
this reengineering was to break down, examine, and standardize the activities
necessary to complete a process (Hammer and Champy 1993; Cole 1994).
This was often accompanied by a digitization of some business activities.
Reengineering permitted detailed consideration of the most cost-effective way

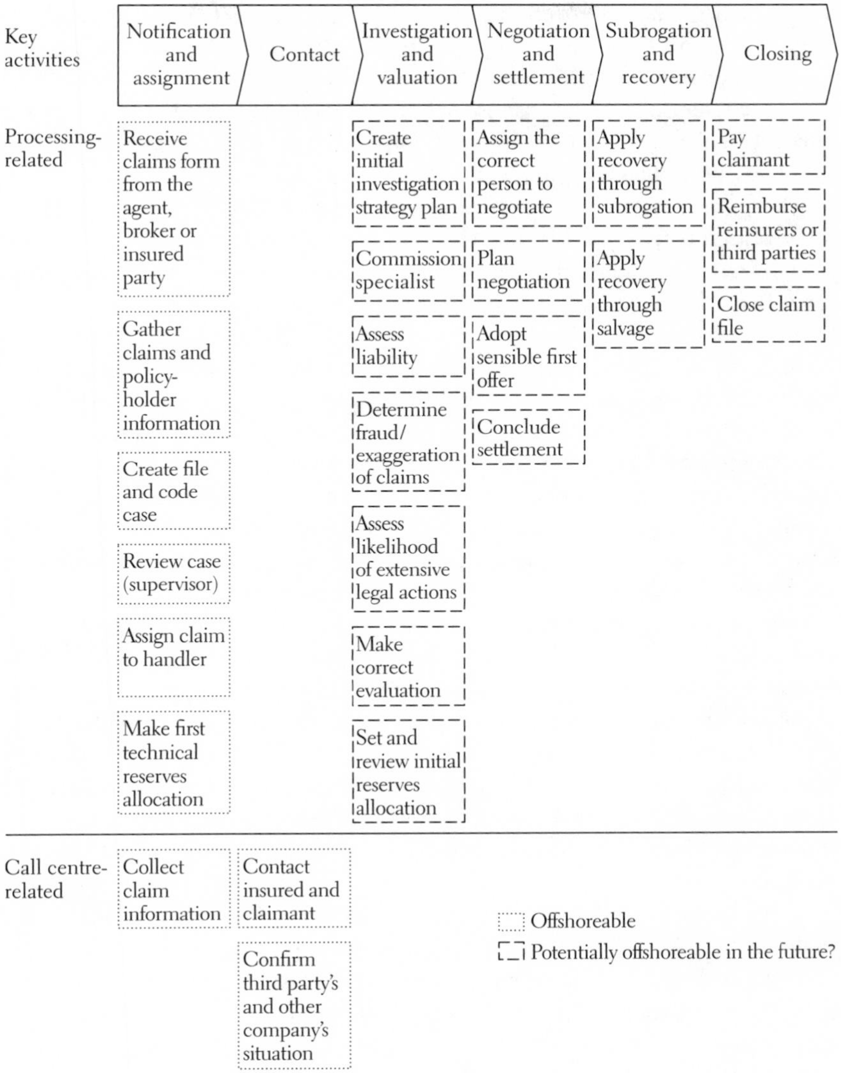

Key activities	Notification and assignment	Contact	Investigation and valuation	Negotiation and settlement	Subrogation and recovery	Closing
Processing-related	Receive claims form from the agent, broker or insured party Gather claims and policy-holder information Create file and code case Review case (supervisor) Assign claim to handler Make first technical reserves allocation		Create initial investigation strategy plan Commission specialist Assess liability Determine fraud/exaggeration of claims Assess likelihood of extensive legal actions Make correct evaluation Set and review initial reserves allocation	Assign the correct person to negotiate Plan negotiation Adopt sensible first offer Conclude settlement	Apply recovery through subrogation Apply recovery through salvage	Pay claimant Reimburse reinsurers or third parties Close claim file
Call centre-related	Collect claim information	Contact insured and claimant Confirm third party's and other company's situation				

Offshoreable

Potentially offshoreable in the future?

FIGURE 9.1   A typical claims-processing value chain

Source: NASSCOM-McKinsey 2002, modified by the authors

198   of completing each activity. In turn, it sensitized management to the ability to standardize and therefore the possibility of outsourcing activities that previously had been performed internally.

The current ability to offshore services is rooted in technological development in the 1970s. Engineers and corporate visionaries in Silicon Valley and a few other places in the world were designing the "office of the future," in which paper would be eliminated and replaced by digitized images on a screen (for a discussion, see Kearns and Nadler 1992). Though paper has not actually been banished, the information that was encoded on it has increasingly been digitized. Remarkably, the costs of transmitting bits of information have continued to drop exponentially for the last two decades.[3] Simultaneously, during the Internet bubble of the 1990s, telecommunications carriers installed a glut of new international fiber optic cable capacity, accelerating price declines. In cost terms, formerly distant locations such as India have become increasingly proximate, even as many of their other characteristics, such as labor costs, remain remarkably "distant." This provides the opportunity for organizations capable of spanning the physical distances and of mobilizing equivalent labor power in low-wage environments to undertake labor-cost arbitrage.

The increasing acceptance of standardized software platforms, such as IBM and Oracle for databases, Peoplesoft (which has been acquired by Oracle) for human resources management, Siebel for customer relations, and SAP for supply-chain management, facilitated offshoring. Adoption of these platforms meant that firms and employees had to make fewer asset-specific investments (Coase 1937; Williamson 1975, 1985). Employees in developing nations could learn a set of portable skills, lessening their risk. This encouraged investment in learning and facilitated the creation of a workforce compatible with the world market.

Technological advances were necessary but not sufficient to convince firms to move their service activities to India. The second important force was the conviction that such relocation could be undertaken with minimal disruption. For this, a degree of comfort concerning appropriate levels of security and assurances on business continuity was necessary. Two factors enabled this: the already successful offshore software operations of the MNCs and Indian software outsourcing firms that had a track record of satisfying international customers; and the fact that pioneers in BP offshoring were large multinationals, such as General Electric and American Express, that had established large Indian operations much earlier.

The final factor driving the offshoring of services is the pressure to increase profits. With revenues largely stagnant since 2000, firms are under intense pressure to cut costs while retaining service levels. Automation was one response, but many "routine" activities are not sufficiently routinized, and human intervention is still necessary. This last pressure even convinced firms that moving mission-critical, time-sensitive processes offshore was essential for satisfying stockholders.

GOVERNMENT POLICIES

In the last two decades, there have been a number of government policy changes that have affected the offshoring of services. The most significant of these was the global movement to deregulate telecommunications industries in both developed and developing nations, establishing competition in the market and contributing to a drop in prices, nationally and globally. International rates were formerly far higher than those for internal domestic calls, but this is no longer true on the whole. India deregulated very successfully; a number of other nations did so much less well, and prices did not drop as dramatically there.

The Indian government, from the 1980s onward, gradually deregulated business. Entrepreneurship flourished in those sectors that were deregulated or had never been regulated, such as software and business service provision. During this period, the Indian government gradually lowered tariffs on important information technology hardware and software. This allowed Indian firms access to the newest technology, which was a necessity for serving multinational clients (Heeks 1996). Indian tax policy was also important because profits on exports were not taxed, encouraging domestic firms to export.

Another important action by the Indian government was the establishment of the Software and Technology Parks of India (STPI). STPI was created as a freestanding entity with its own sources of funds, removing it both financially and in terms of governance from the Indian government bureaucracy, which had a well-deserved reputation for sluggishness and corruption. STPI branches were established throughout India to serve both domestic firms and MNCs. They built the physical infrastructure of the free trade zones and controlled and allocated telecommunications bandwidth.

In terms of intellectual property (IP) protection, the Indian government is bringing its policies into conformance with the demands of Western corporations and governments. Although India has had a more mature judicial system relative to its stage of economic development than many similar countries, enforcement was often delayed by overloaded courts. India's IP protection regime was always superior to China's, yet China has outperformed India in terms of economic growth. India has nonetheless agreed to come into full compliance with TRIPS by January 1, 2005, thus enabling product patent protection granted elsewhere to apply in India. This is of particular relevance to the pharmaceutical industry and will help the outsourcing and offshoring of clinical trials, which are becoming ever more expensive and regulated in developed nations.

The most important U.S. government policy facilitating the offshoring movement was the H1-B, J1, L1, and E1 visa programs, which permitted foreign white-collar professionals to come and work in the United States, either in pursuit of a permanent residency status or more temporarily. All of these contributed to the software body-shopping operations that gave the Indian firms their initial toehold in the U.S. market (Heeks 1996; Hira and Hira 2005). The program also made U.S. managers aware of the capability of Indian programmers and white-collar workers more generally. This convinced them that even greater savings could be achieved by employing Indians in their home market, where it was not necessary to pay for their stay in the United States. Visa programs also allowed Indian managers to come to the United States to facilitate the transfer. The Indian pioneer in the field of BP outsourcing was Tata Consultancy Services (TCS), the first software exporter from India, established in 1968. As with many such beginnings, chance mattered. TCS's capabilities as a software exporter were discovered by its hardware partner in India, Burroughs. Burroughs discovered that TCS's programmers did a good job installing and maintaining Burroughs's mainframe systems in India and thought that they could do so in the United States as well; therefore they invited TCS to send its programmers to install Burroughs's mainframes in the United States (see Dossani and Kenney 2004). In this way, U.S. visa policy has been very supportive of the development of the Indian ITES industry (Hira and Hira 2005).

OFFSHORING FROM THE PERSPECTIVE OF THE FIRM

The decision to offshore an activity is a strategic one involving the firm's consideration of its core competences and the trend toward specialization (Garner 2004). Although technology enables certain decisions, it does not determine them. In choosing which processes to undertake offshore, it may be thought that the simplest processes would be offshored first, since the skills for undertaking more complex processes might take a longer time to learn — and, in general, this is the case. However, as Bruce Kogut (2004) has shown, an MNC's foreign subsidiaries are capable of learning. Already many of the more mature MNC foreign operations, such as those of General Electric and Hewlett-Packard, have absorbed higher value-added activities.

The single greatest motivation for considering India or any other developing nation for outsourcing is, quite simply, that labor costs are significantly lower than those in developed nations. Yet savings in direct labor costs, though impressive, do not capture the entire calculation a firm undertakes prior to offshoring an activity. Only in a few instances are the offshored functions a set of skills that cannot be secured in the developed nation. Though hotly debated, the one vocation for which this generalization may be not true is software programmers, where there were labor shortages in the 1990s as demand increased rapidly.[4]

The wage differences between the United States and India are dramatic. For example, in 2004 a junior accountant at a large U.S. firm with less than one year of experience would earn between $35,750 and $42,500 per year (American Institute of Certified Public Accountants [AICPA] 2005), and approximately 10 percent more if he or she was a certified public accountant. In India, a newly graduated junior accountant would typically earn less than $9,000 a year. The differential for less skilled workers is even greater: the Indian wage rate for entry-level call center employees in metro areas is $2,400 a year. Moreover, even the more mundane jobs are considered attractive.

The infrastructural costs of siting an operation in India are approximately the same as they would be in a U.S. industrial park. Previously, on-site equipment and service may have been an issue, but during the last decade all major electronics vendors have established customer support operations in India, so maintenance and repair are no longer issues. Telecommunications capacity is critical for ITES, but in our interviews no firms expressed any difficulties with connectivity in terms of capacity or quality, though all had redundancy built into their systems. The most difficult remaining infrastructural issues appear to be the utility and transportation infrastructures. Corporations have developed private solutions, including multiple redundant back-up power and fleets of private buses to ferry employees to work. These do add overhead costs to the operations, but they appear to be manageable.[5]

Further savings can be had by taking advantage of the economies of scale derived from concentrating activities in fewer locations. In developed nations, activities such as credit card processing or customer service call centers are often scattered in a number of locations, because operations are limited in their ability to expand owing to the shallow local labor pools. Relocating to a large city in a developing nation can address the scalability problem because they have large labor pools and complementary services. For example, the Mumbai, Delhi, and Bangalore regions each have over 5 million inhabitants, and other cities such as Chennai, Hyderabad, and Kolkota have similarly sized workforces. The call centers in India we visited varied in size, but the median

202   size was 1,000 employees. The average size in the United States is under 400, and many employ between 150 and 300 employees.

The reengineering that occurs as part of a transfer process can uncover significant savings. The source of these savings is the study and planning necessary to transfer a business process. In the process of study, often aspects of the current methodology for discharging the process are discovered that do not add value. During the transfer process, it is easier to reform or abandon inefficient practices than it would be at an existing facility where they have become a "natural" part of the daily routine (see, for example, Adler et al. 1999; Florida and Kenney 1991; Kenney and Florida 1993). These reforms can be implemented without disrupting work patterns, since the workers in the new location are met with a fait accompli. Though difficult to quantify, significant savings can be achieved through this transfer process.

The expected savings from relocating the activity to India are at least 40 percent. One Fortune 500 firm that consolidated several global fulfillment operations to Bangalore reported overall cost savings as high as 80 percent.[6] The NASSCOM-McKinsey report (2002) found that General Electric (GE), one of the pioneers in offshoring service operations to India, in 2002 had achieved an annual savings of $340 million per year from its Indian operations. In 2005, GE sold the bulk of this operation to a group of private equity firms for approximately $500 million.

The ability of Indian operations to offer services in a fashion as timely as or more timely than would be available in a developed nation is an important attraction. Undertaking service activities in India permits firms to operate around the clock through, for example, global development teams or a division of labor in which Indian workers debug the day's software built in its developed nation's operation. This use of the entire day allows deadlines to be shortened. In the case of medical transcription, a doctor's notes for patients in intensive care can be completed in as little as two hours because Indian operations can afford greater slack resources to meet peak loads than their Western counterparts can.

Set against these cost and time benefits, there are significant strategic concerns. These concerns are usually not so pressing in the more highly commoditized and well-understood service activities. However, given the novelty of business service offshoring, even activities that might be considered routine in the developed country can be subject to quality slippages in the offshore destination owing to unexpected difficulties such as retaining staff, cultural misunderstandings, or employee dissatisfaction in the home country. For activities that have higher knowledge and creative inputs, the firm seeking to transfer an activity is often concerned about whether the service quality will decline. For example, the remote location may not be able to understand or match the quality needed. This is most likely for activities that have a large tacit component or where intimate market knowledge is necessary. Activities such as design and marketing usually create the highest added value and are not usually commoditizable; they are likely to be the most difficult to transfer.

MNCs may also be concerned about a loss of competencies in a certain location that would be costly (or even impossible) to reacquire in that location if ever again required. Overdependence on a single developing nation, could, if unique skills atrophy in the home nation, lead to a disruption of access. One way MNCs mitigate these concerns is by developing a "blended" strategy whereby the activity is shared between some domestic capacity, "near-shore" capacity in somewhat lower-cost labor nations such as Canada for the United States or Eastern Europe for Western Europe, and "offshore" locations such as China, India, and the Philippines.

A firm's decision on whether to offshore an activity is a complicated process. Implementation is also difficult, especially for firms without experience in the Indian environment. And yet increasing numbers of firms are deciding that their competitive situation compels them to respond to offshoring moves by their rivals. Whereas only five years ago offshoring services was not a high priority among most Fortune 500 firms, in 2003 it had become almost a mantra among corporate executives. For example, the senior vice president of Microsoft's Windows division, Brian Valentine, in a presentation to company managers advised them to "pick a project to outsource today" (Nachtigal 2003). The available evidence indicates that relocating service activities to developing nations, especially India, is compelling from a financial perspective and thus likely to continue.

THE INDUSTRIAL STRUCTURE OF THE OFFSHORE ITES INDUSTRY

There are two important dimensions for categorizing the ownership of firms in the ITES sector. First, are they domestically owned and operated or owned and operated by a multinational? Second, are they a captive or a firm that undertakes outsourced work? Because the potential market is so great and the economics so compelling, there has been a plethora of entrants from a large variety of backgrounds. Because ITES offshoring is only in its earliest stages, it is premature to predict which organizational forms will become dominant. It is not clear whether there will be a single ITES industry in India, nor whether the captives or independents will dominate or even compete. Further, there are niche areas such as medical transcription, geographical information system (GIS) data entry, and document conversion that may remain separate from the industry's mainstream.

Like the earlier movement of software programming offshore, the MNC captives led the way in the establishment of the first ITES offshoring operations. They are still the largest operations and, more important, the most sophisticated. This contrasts with software outsourcing, where the Indian firms soon became dominant in terms of the numbers of employees and earnings.

MNC Captives

The MNCs have ITES subsidiaries in a number of nations. Typical of large MNCs, for example, Siemens Business Services has a global help desk for products and IT support for hardware maintenance in Turkey; an Indian center for software programming and support, human resource and financial processes, application management, and services for SAP and legacy systems; a Vornazh, Russia, center for administrative and human resource processes and SAP support services; a global help desk for products in Cork, Ireland; and a Toronto, Canada, global help desk and remote services center. In the case of U.S. MNCs, however, India is becoming the preponderant location. Though they will have a number of locations undertaking an activity such as global back office financial operations and there will be 24-7 support with a European and North American office, the largest number of workers will be in India.

In the last four years, many firms that previously had the preponderance of their service functions in the United States, with some other operations in lower-cost developed-nation environments such as Ireland or Canada, have begun massive and rapid expansions in India. This would include firms such as Dell Computers, AOL, and SAP, which previously had had no Indian operations. Dell launched its Indian call center operation in June 2001. By June 2005 it had grown to nearly 10,000 employees in the original Bangalore facility, opened a second facility in Hyderabad, and was completing its own Bangalore campus. AOL's Indian operations experienced similarly dramatic growth. It commenced operations in July 2002 with 200 employees and as of July 2003 had grown to 1,500 persons (Ribeiro 2003). Such examples emboldened other firms to follow.

As internal operations, the captives have significant advantages. First and foremost, they have guaranteed markets for their services. Decisions on allocating volume are hierarchical and the information driving decisions is excellent. In the case of lower-value-added, routinized work, the advantages of captives may not be great and risks may be minimal, so the decision to outsource or do the work in house may be made almost solely on price, though even here there may be advantages to conducting these activities in house to allow the firm and its employees to gain experience. In the case of higher-value-added processes, it may be more prudent to retain them in a captive operation. Not surprisingly, the initial activities transferred were at the low end of the value-addition spectrum. However, this has not proved to be the end state for the more mature operations.

In a number of cases, higher-value-added activities have been transferred over time. For example, GE's Indian operation, which in 2005 was spun-out as an independent firm has moved up the value-added chain, adding employees doing actuarial support, data modeling, and portfolio risk management. In its health insurance operations, GE employs forty medical doctors to evaluate and classify medical claims. Leading firms such as GE, Intel, and Microsoft are hiring scientists and engineers with doctoral degrees. Some have already developed intellectual property for their employers. GE's India-based engineers have filed for ninety-five patents since 2000 (Kripalani and Engardio 2003). Microsoft's Beijing research laboratory is doing basic research that will contribute to its global operations. More recently, Microsoft announced that it was establishing a research laboratory in Bangalore that would augment its large global service operation in Hyderabad.

Financial institutions have been especially quick to open back-office operations in India. For example, two large investment banks, J. P. Morgan Chase and Morgan Stanley, have hired several dozen junior analysts in Bombay for analyzing U.S. markets. With their enormous need for white-collar talent, these firms can achieve enormous cost savings. For example, in 2002 the average salary for MBAs graduating from India's prestigious Indian Institutes of Management was $13,226, illustrating the possible cost savings (Saritha Rai, "As It Tries to Cut Costs, Wall Street Looks to India," *Wall Street Journal*, October 8, 2003). As a result, global money center banks have both outsourced activities to Indian firms and built large captives that undertake all manner of work including sophisticated data-mining and software development.

Operating a captive requires significant managerial talent. For those with long-established Indian operations (typically serving Indian markets), this is likely to be available internally, whereas the new MNC entrants are likely to experience significant learning costs. One dilemma they face is whether to staff the operation with expatriate executives or to hire domestically. During the initial ramp-up, new entrants had to send some expatriates despite the expense. For these firms, the expense of maintaining expatriates will become an issue, but at present the savings appear to be large enough to offset the expense.

Some MNCs are converting their developing nation operations into a global center of excellence. In many firms, business processes are nationally

206   based and were developed in different historical eras, so they vary in their performance of identical functions. Enforcing standard operating practices in the different national environments can be difficult because there is a constant tendency to "go native." This drift is endemic in even the best firms and may be most pronounced in the less intensively "managed" parts of the national unit's operations, such as the back offices. The transfer of these processes to a specialist organization dedicated to managing them not only creates economies of scope and expertise, but also provides an opportunity for standardization and the removal of the process from national drift. For global headquarters this can be a way of exerting control and improving monitoring. The risk in removing a particular process from the national operation is that the global operation will lose touch with the national environment.

The growth of the MNC subsidiaries in developing nations is best illustrated in India, where in 2004 these subsidiaries had become the largest sector of the ITES industry. There is every reason to expect this will continue for the foreseeable future. The advantages of a subsidiary are considerable in terms of reducing risk and possibly leaking knowledge, capturing profits internally, and using internal operations to benchmark outsourcing contracts. Since less than 10 percent of the Global Fortune 1000 firms currently operate in India, it seems likely that more firms will establish operations and those currently operating will expand.

Multinational Outsourcers

Service outsourcing has a long history and has grown rapidly during the last decade. Estimates of the total size of the BP outsourcing market vary widely. Different consulting reports have estimated the global BP outsourcing market would grow to $544 billion by 2004, $1.2 trillion by 2006, or $140 billion by 2008. The remarkable divergence in estimates is perhaps due to the fact that definitions differ, and because business service outsourcers are a varied category that includes data systems outsourcers such as EDS and IBM; payroll and accounting processors such as ADP; call center and customer relationship managers such as Convergys, Sitel, and Sykes; large consulting firms such as Accenture; and many others. Globalization is not new for these firms. Not only do the larger ones provide services internationally, but many of them already had cross-border operations prior to the current phase of offshoring to developing nations.

The international outsourcers established their Indian operations in 2001 or later as a response to competition from MNC subsidiaries and Indian independents. These MNC outsourcers have long-established customers and enormous domain knowledge, making them formidable entrants. These capabilities and existing customers have permitted them to scale up their developing-nation operations very rapidly. For example, in late 2001 Convergys opened its first Indian operation in New Delhi. By April 2003 this facility had more than 3,000 employees. It built a second facility in Bangalore that in 2005 employed 3,000 Indians and intended to increase its Indian workforce to 20,000 by 2007.

The ability to transfer customers to their Indian operations while providing backup in the United States and other locations allows global solutions and service level guarantees that firms operating only in India cannot provide. The conundrum for the MNC outsourcers will be how long their customers will support higher-cost U.S. facilities. It is a near certainty that the MNCs will continue to downsize their higher-cost U.S. operations as they undertake a global redistribution of their facilities.

MNC Specialists

India is also attracting smaller MNCs that perform various specialty services. These services are wide ranging but are based on specialized domain expertise. Though many of these are not strictly speaking business processes, they are included in the broader category of ITES. Examples of this type of work include medical transcription, tax preparation, patent application preparation, map digitization, cartoon animation, document entry and conversion, and other tasks. The sheer diversity of these services is remarkable.

Taken individually, these activities have limited employment potential. In aggregate, however, their total employment may be quite large. For example, there are approximately 270,000 medical transcriptionists scattered around the United States. The market is as decentralized as the location of the medical doctors, making sales and marketing difficult. Recently, there has been an effort to consolidate the industry. This consolidation might be hastened if it could be relocated offshore, where transcription can be done at much lower costs and with comparable quality. Transcription and map digitization are only two illustrations of a labor-intensive service activity that is being relocated to developing nations. Other areas include legal research using Lexis-Nexis, drawing of tables and figures, drawing and/or digitizing blueprints, etc. The variety of niches within which businesses could be built is remarkable given that transcription, paper-based document digitization, database-centric research, and many more activities exist in the pores of so many U.S. organizations and the economy as a whole. One drawback is that many niches may be too small to justify offshoring, yet cost pressures are encouraging an examination of its

208   feasibility. The MNC specialists are fascinating because of their sheer diversity and the likelihood that their decisions will be largely unnoticed by policy makers owing to each niche's relative insignificance. Yet if these myriad firms begin transferring activities and processes overseas, the aggregate impact could be great.

Domestic Specialists

Developing-nation specialty firms are also entering fields such as medical transcription, tax preparation, map digitization, and manuscript preparation. The difficulty for developing-nation entrants is their relative lack of domain knowledge. For those with deep enough domain expertise, it may be possible to transform their business proposition from offering simple labor cost arbitrage to providing significant value addition. For example, an Indian publishing firm that initially only prepared drawings for chemistry texts now offers a full range of back-office services, including copy editing, HTML formatting, and technical support. It has expanded its product list to include academic and professional journals and even time-sensitive publications such as newsletters. The enhanced capability not only allows the addition of greater value, but also provides the firm greater bargaining capacity with its customers. Developing domain expertise and specializing is difficult and has risks because the firm becomes dependent on a single industry or activity. Yet it also offers the potential to occupy niches that may not be drawn into the extremely ferocious competition found in the highly commodified sectors.

Domestic Independents

A large number of developing-nation firms have been established for the sole purpose of offering outsourcing services to foreign firms. Some of these are venture capital–supported and were formed during the Internet boom to provide back-office services to U.S. Internet firms such as Yahoo! and Amazon. Not surprisingly, the collapse of the dotcom boom forced these firms to rethink their corporate strategies. Since these firms were supplying back-office services, such as answering e-mails and Web-related questions, it was not difficult to switch their service offerings toward call centers. Newer independents have been funded by venture capitalists in an effort to take advantage of the outsourcing boom.

    The independents face significant strategic difficulties. Some independents have experienced rapid growth as they have found customers. Yet they often rely upon a few larger customers, making them vulnerable to contract termination. Because of the ferocious competition and the pressure to expand $\quad$ (often at the prodding of their venture financiers), the independents are pushed to pursue any and all business prospects. This militates against their expressed desire to develop domain expertise that would enable them to charge higher rates. Another difficulty is that the U.S. market is the largest in the world, but sizing a facility for that market means the facility is often idle during the day in Asia. The independents have been able to secure some business from Europe, especially England, which allows them to extend facility utilization; however, it is still difficult to utilize the entire facility for more than one and a half shifts. To increase capacity utilization, the independents bid aggressively for activities that do not require real-time processing. The MNC captives are at an advantage in this respect, because the parent firm can transfer a portfolio of activities so as to more fully utilize the facility.

The ultimate fate of the independents is difficult to predict, and for the smaller ones survival will be precarious. The larger ones should be able to strengthen their marketing in the United States. However, these independents might be acquired either by Indian firms or multinationals wishing to quickly enter the BP outsourcing field. For example, in 2003, the Indian software firm Wipro purchased a leading BPO firm, Spectramind. In May 2004, IBM purchased one of the largest Indian independents, Daksh. The strongest Indian independents may be able to remain independent and grow sufficiently to rival the multinational outsourcers, but survival as an independent may be difficult.

Developing-nation IT Industry Subsidiaries

When we consider ITES, the most important developing nation IT subsidiaries are Indian. They have grown remarkably fast over the last decade through the provision of outsourced programming and IT services to the global market (Arora and Athreye 2002; Singh 2002; D'Costa 2003). Because of their ability to use lower-cost Indian software talent, they have made significant global market share gains. Further, their interaction with the global economy has contributed to the development of executive and managerial talent capable of securing overseas contracts, managing the interface with foreign customers, and migrating activities across national and corporate boundaries. In the process, these firms have cultivated close connections with foreign customers, easing the burden of convincing them to trust Indian firms with other services.

Given the growth in ITES, the Indian IT firms believe that it is a sector in which they can expand. Their strategic question has been how to enter this new industry. The major firms have answered this question differently. Infosys and Satyam established subsidiaries, one of which, Progeon, has grown rapidly. It recently divided a five-year, $160 million contract from British Telecom with HCL BPO. In contrast, the Satyam subsidiary has experienced only limited growth. TCS, the largest Indian software firm, entered the outsourcing sector through a joint venture and has since made a small acquisition; it crossed 4,000 employees in March 2004. Finally, Wipro and HCL entered the industry through acquisitions. Wipro acquired Spectramind, and HCL acquired the Northern Ireland call center subsidiary of British Telecom, though the preponderance of HCL's outsourcing employment growth has been in India.

The Indian IT firms have significant advantages in terms of access to capital, linkages to customers, and experienced managers. But the ITES outsourcing business is quite different from IT. In terms of marketing, the customer's key decision maker for ITES is not the chief information officer or chief technical officer. These services must be sold directly to the various responsible divisions or departments, and the ultimate decision rests with the chief financial officer or chief executive officer. This means different marketing channels must be mastered.

The ITES workforce, from commerce and social science backgrounds, is also quite different from the engineers who comprise the workforce and managers of the IT sector. Since service-outsourcing work often requires direct interaction with customers, the salient workforce skills are interpersonal, rather than technical. Moreover, customer interaction can be extremely stressful, putting a premium on workforce management. In addition, many ITESs are undertaken in real time, so errors and mistakes have an immediate impact. Service level agreements are tightly written and monitored, and problems are exposed nearly immediately. In contrast, software bugs can be rectified later. The ability of Indian IT firms to manage nontechnical personnel in extremely price-competitive environments will be tested. There is also the possibility that the technical skills within the IT parent could be used to automate aspects of the BP outsourcing process, creating another level of value addition that would improve profitability and enabling the IT firm subsidiaries to create advantages beyond routine labor cost arbitrage. Although today rapid market growth ensures an appearance of success for many entrants, the ultimate success of the IT firms in the BPO space has not yet been settled.

THE INDIAN CONNECTION

At this moment, India is by far the most important location for the relocation of service activities, though other nations such as Russia, Turkey, China, the Philippines, Costa Rica, and Ghana have also attracted a certain amount of investment. India's attractiveness as a site for undertaking ITES is a combination of preexisting conditions and the result of a variety of policies. The preexisting conditions included a large pool of English speaking, college-educated persons, many of whom were unemployed or underemployed and willing to work for wages that were a fraction of those demanded in developed nations. Also, beginning in the mid-1980s, the Indian government began to liberalize its economy, and various states established policies aimed at attracting MNCs. The relocation of ITESs to India can be traced to the emergence in the mid-1980s of India as an offshore site for software production by both MNCs and a large number of Indian independents (Arora and Athreye 2002; D'Costa 2003). It was this experience in software that suggested to MNCs that other service needs might be fulfilled from India.

Nearly all the policies that enabled the development of the ITES sector were already being implemented as part of a general deregulation of the Indian economy and a policy of encouraging exports. The most significant policy reform for the ITES sector was the reform and deregulation of the communications infrastructure (Dossani 2002). Beginning in 1999, India liberalized its public monopoly telecommunications system and permitted Indian private providers to begin offering services. They could select their specializations, which ranged from the provision of niche services such as backbone and network management to full-service integrated voice and data operations. For larger cities, the result has been the creation of a telecommunications network with quality and cost levels approaching those of developed countries. This service is being extended to second-tier cities with populations in excess of 1 million.

By 2000, the conditions in India were prepared for the take-off of the ITES sector. According to NASSCOM (2005), employment growth was extremely rapid as the ITES grew at a compound annual rate of 52.6 percent from fiscal year 2000/2005. Software and software services employment grew more slowly at a compound annual rate of 23.6 percent. Sales of software and software services exports in fiscal year 2005 reached $12 billion (NASSCOM 2005). ITES employment had grown from 254,000 in fiscal year 2004 to 348,000 in fiscal year 2005. During the same period, export revenues in ITES grew from $3.6 billion in fiscal year 2004 to $5.2 billion in fiscal year 2005, a growth rate in revenues of 49 percent. (NASSCOM 2005). Whether such growth is sustainable is open to question; however, there is every reason to believe that growth, at some level, will continue. If ITES offshoring to India is restricted to call centers and financial data processing, the business may grow sufficiently large to overtake software outsourcing, but it will not have a dramatic impact on the employment situation in the developed and developing nations. On the other hand, if India can offer the entire spectrum of services, then the impact on both the United States and India could be enormous. What if India were to parallel in BP services the importance China has achieved as a manufacturing destination? As an indication of how fast the growth might be, as of March 2004, approximately 150 U.S. firms (almost all in the Fortune 500) had offshored service work to India, and on average they predicted their employment would increase by 50 percent during the next twelve months.

The initial activities relocated to India have been highly routinized and resemble the initial phase of software outsourcing. Experience, combined with the lower cost of more highly skilled personnel, may prompt a rethinking of earlier cost-benefit decisions. As discussed above, the cost of a trained accountant in India is so much lower than in the United States that it becomes possible to audit a greater number of cases and/or lower the threshold for universal auditing. The end result is a diminution of mistakes and fraud, leading to greater cost recovery. It was thus typical, in our experience, for the same process to employ a larger number of employees in India than in the United States, which the lower costs made possible. For example, in medical transcription, one person doing work in the United States was often replaced by at least two persons, both of whom transcribed the same material and compared notes. Sometimes, a third person, a supervisor, "arbitrated" the result. The cost differential makes such experimentation and back-up possible.

For developing-nation firms and policy makers, *where* the nation ends up in the value-addition process is critical. For example, even today, the Indian software industry operates in the low-value-added segments, typically in applications development, testing, and maintenance, while the high-end work, such as developing IT strategy, identifying software needs, systems design, and integrating the project with other packaged and custom components, is discharged by U.S. firms. If developing-nation ITES operations are not able to move up the value chain, offshoring may not prove to be so important to the development of the national economy.

There are several challenges to India's ability to maintain the current growth pace, although these are not likely to have a short-term impact. The first is a shortage of managerial talent. Particularly significant is locating managers capable of managing the migration of a business process from an overseas firm to Indian operations. The larger and apparently more successful ITES providers reported that it often took up to a year to make such a transfer for some of the more complex back-office operations, while the simpler ones, such as outbound call centers, could be transferred within a month. Another managerial task is the maintenance of a seamless relationship between the overseas entity receiving the work and the organization in the developed country. It is also necessary to have managers capable of maintaining and raising the productivity of operator-level staff. Though some firms, notably multinationals, have achieved productivity rates that match or even exceed those of their developed-country counterparts, productivity has been a problem for independent firms and is greatly exacerbated by high staff turnover levels caused by high demand and stress created by the unusual work hours in the call-center industry. An industry group, NFO World, estimates that 33 percent of Indian call center workers quit within a year of joining the industry (Ashok Bhattacharjee, "India's Call Centers Face Struggle to Keep Staff as Economy Revives," *Wall Street Journal*, October 29, 2003). The labor pool may also be shallower than statistics of overall graduation rates indicate: of the 1.1 million graduates each year in India, perhaps no more than 10 speak English well enough to work in a call center (ibid.), and quality may already be suffering. Although the turnover rates may be lower than in developed countries, some Indian firms we interviewed reported attrition rates of 7 percent per month, although 3.5 percent per month was the average rate. Wage pressures are also in evidence: our interviews showed that wages were rising at about 10 percent per annum.

Indian operations, particularly the independent firms, suffer from a shortage of expertise, especially in the fastest-growing vertical sectors such as finance, insurance, real estate, health care, and logistics.[7] Unfortunately, horizontal skills are also in short supply. According to the Outsourcing Institute, horizontal expertise is most needed for payroll, customer care, document processing, and benefits management. For this reason, though we expect growth to continue at a rapid pace, it might also slow as the best-qualified labor is absorbed and high turnover rates continue.

DISCUSSION AND CONCLUSION

The implications of the offshoring of service work are significant for both developed and developing nations. Service jobs, which formerly were rooted relatively close to where they were generated owing both to the sheer logistics of moving paper documents and to formerly high telecommunications costs, have now been made mobile by technological improvements and a new willingness on the part of management to consider offshore service processing. During the next decade, it is likely that globalization will sweep through the ranks of service workers, who until now have been largely immune to competition by lower-wage foreign workers. As enterprises seek to drive down their costs, a new round of globalization will occur within which a complicated multinational and likely multicorporate chain for data capture and processing

214   will emerge. The old image of the developed nations concentrating on information services, data processing, and knowledge creation may give way to a world in which data and information will simply be commodities processed in developing-world factories.

The relocation of services offshore and especially to India has the potential to reorganize the global economy in the same way as the movement of manufacturing to China has been emblematic of a reorganization of goods production. For the developed nations, already reeling from the continuing loss of manufacturing jobs, the emergence of India as an option for firms aiming to lower the costs of providing services creates significant policy dilemmas. Economic development policies will also need to be rethought. For example, over the past two decades many American states have devoted resources to creating service clusters, typically around low-end services such as call centers in smaller and more remote towns. These strategies may be at significant risk because such work may be offshored (Kenney and Dossani 2004).

For India and other developing nations, the offshoring of services may provide a large flow of new employment opportunities. In the case of India, current estimates are that ITES employment may increase to as much as 1 million by 2008. Obviously, all estimates should be treated cautiously, but given the rapidly changing technologies, they may prove to be conservative. What is clear is that the number of service jobs being relocated is increasing rapidly. Limits may, in fact, be elastic so that what is impossible to relocate today may become amenable to relocation tomorrow, especially because IT is evolving so rapidly.

A remarkable aspect of service offshoring is the rapidity with which it can occur. Manufacturing's movement offshore was a gradual migration that began in the early 1960s. Though punctuated by dramatic factory closings, there was an opportunity for the U.S. economy to adjust. This may not be true in services, where the objects are pixels and electronic pulses that can be transmitted by photons and radio waves (Kenney 1997; Cohen, DeLong, and Zysman 2000). A number of the firms we studied in India experienced vertiginous growth as they expanded from start-up to 5,000 employees in less than three years. When such growth rates are experienced by a large number and variety of firms, the cumulative effect can be enormous indeed.

Policy had an important role in India's ability to be a lead location for business process offshoring, including, as discussed, the broader liberalization of the Indian economy. By providing MNCs a moderately business-friendly environment, the Indian government encouraged them to seek new opportunities for using the high-quality, English-speaking Indian labor force. Telecommunications deregulation was critical because it ignited competition that resulted in increased bandwidth, greater quality of service, and lower prices. For ITES, it is telecommunications that provide access to the market, and lower prices improve access. For any nation seeking to follow India's lead, the proper telecommunications policies are absolutely critical for success (Dossani 2002). Adopting IP protection rules may also assist in development.

The ultimate dimensions of the service offshoring phenomenon are difficult to predict. Whereas for the last two decades manufacturing value chains increasingly extended across borders (Gereffi and Korzeniewicz 1994; Kenney and Florida 2004), it appears nearly certain that this will soon be equally true for services. Policy makers in developed nations must begin to prepare for this eventuality by considering what the core advantages of their populations are. We believe that the advantages will come from the sophisticated consumers in developed nations that set the fashion for most of the world's goods and from creative clusters such as Hollywood (Scott 2002), Silicon Valley (Kenney and von Burg 2000), Paris, Boston for mutual funds, northern Italy for a wide variety of goods, Tokyo for consumer electronics, and so on. Increasingly, if routine service activities can be relocated to lower-wage nations, the advanced, developed nations will have to compete in terms of superior creativity (Florida 2002).

For policy makers in the developing world, inexpensive telecommunication is opening a new world of opportunities in the export of services. The opportunities are substantial for Francophone Africa servicing France, Eastern Europe and Turkey serving the German-speaking nations, China serving Japan (because of the similarities in the written languages), and Estonia providing for Finland. Though we concentrated on India in this chapter, the Philippines is already providing services to the United States especially in terms of call centers, video animation, and as a backup for India. All these nations provide opportunities for indigenous entrepreneurs.

For the United States and other developed nations, it is important to note that it is not only low-end "service" work that is being undertaken in the developing nations. There is ample evidence that MNCs are rapidly increasing their R&D activities in developing nations, particularly India and China. This may be especially significant in the case of India, where most of the R&D is undertaken for the global market, suggesting that policy makers in the developed nations should understand that high-end service work may also be at risk and contradicting more sanguine observers, such as Frank Levy and Richard J. Murnane (2004), who assume that the transfer will be almost entirely composed of low-level routinized work. Whether U.S. R&D workers are fully prepared for a globalized economy in which their wages will be compared to those in environments with far lower labor costs is not yet certain.

216   The final concern worthy of mention is that even if most jobs are not relocated, the very fact that they can be may contribute to a global equilibration of wages for work that can be offshored. In other words, workers in developed nations will be required to compete with those in developing nations. Such international wage competition might, in the worst instance, lead to a generalized downward drift of wages that could create a shrinking global capacity to consume. Thus, if service offshoring were to become a far larger phenomenon, the middle-class lifestyle that service jobs provided in the developed nations might be threatened, and it is unlikely that the protectionism often invoked in trade in physical goods would be effective. Such a scenario might prompt a fundamental rethinking of the global income distribution system and the development of what we term a global New Deal.

NOTES

The authors thank Frank Mayadas and Gail Pesyna of the Alfred P. Sloan Foundation for supporting our research and appreciate the comments by the participants of the Manhattan India Investment Roundtable and Harry Rowen on an earlier version. We thank the executives of the forty-six firms that consented to our interviews. The authors also thank, in alphabetical order, the Maharastra Industrial Development Corporation, NASSCOM, and the Science and Technology Parks of India. The authors are solely responsible for the conclusions and opinions expressed in the paper.

1. ITES is a catchall term used for the myriad processes that any bureaucratic entity undertakes in servicing its employees, vendors, and customers. These include human resources, accounting, auditing, customer care, telemarketing, tax preparation, claims processing, document management, and a wide variety of other activities.

2. We define a "business process" as a complete service, such as handling a customer complaint, processing a medical claim, or processing a purchase order. Completing a process requires undertaking a set of activities. For example, in handling a customer complaint it is necessary to understand the complaint, decide on a course of action, undertake the action, and followup to ensure the action solved the complaint. Each of these is an activity that is potentially separable from the others.

3. The decrease in rates was facilitated by technological change, but also by U.S. government pressures on other countries to decrease their fees for connecting international calls (Cowhey 1998; Melody 2000).

4. For the argument that there is or, at least, was a shortage, see Barr and Tessler 1997. For the counterargument, see Matloff 1998.

5. From our interviews, transportation costs per employee averaged $50 per month.

6. Personal interview by the authors, April 2003.

7. According to the Outsourcing Institute, these are high-growth areas in the United States (http://www.outsourcing.com; accessed November 12, 2005).

10

From Linux to Lipitor

Pharma and the Coming Reconfiguration
of Intellectual Property

Steven Weber

The knowledge economy, darling value-creator of the 1990s and 2000s, is in serious trouble. The pain will be anywhere from severe to brutal, depending on how the coming crash in critical sectors plays out. Industries such as entertainment, software, and pharmaceuticals will discover, probably the hard way, that their current uncomfortable state is more than just a temporary rough patch after the late 1990s stock market extravaganza. It is actually a sign of an intellectual property bubble, a more fundamental phenomenon. As that bubble deflates, it will become clear that we are living through a period of deep reorganization for production of knowledge goods.

The irony is that all of this will be good news for the knowledge economy, if companies understand what is happening and manage the transition successfully.

The core problem is that a profound version of the business model puzzle that now faces the software industry is widening into a broader crisis. It is a puzzle that emerges when new ways of organizing people to make things transform complex and high-value-added products and services into something approaching a commodity. The success of the open source software community in creating products that rival and even go beyond proprietary software is reshaping markets for software. And that change is spreading to a range of sectors that depend on controlling and erecting barriers around intellectual property (IP).

A few short years ago open source software was a curiosity that barely made the radar screen of such companies as IBM, Microsoft, and Sun Microsystems. Today, the open source Web server Apache dominates its market segment. Linux is taking away market share from Microsoft's server business and leaping past the proprietary versions of Unix sold by Sun and others. Clusters of cheap Intel processors running Linux have infiltrated the specialty high-end supercomputing market at the same time that thin Linux systems run watches and personal video recorders. If you use Google, you use a cluster of more than 10,000 computers running Linux. Yahoo! runs its directory services on FreeBSD, another open source operating system. If you saw the movie *Titanic* or any of the *Lord of the Rings* trilogy you were watching special effects rendered on Linux machines used at Disney, DreamWorks, and Pixar. Governments around the world are purchasing (in some cases, mandating the procurement of) open source software. What seemed just a short time ago a hobbyists' diversion is now remaking the landscape of a significant piece of the information processing industries.

Open source is not a typical disruptive technology. In fact there is very little about the technology itself that disrupts markets, since open source software (like most software, actually) rarely breaks new technical ground. At the same time, open source software is not just the recognizable problem of a low-cost competitor that takes customers away from higher-cost companies and moves value around the industry.

Open source represents, instead, a new production process that remakes the value chain in software from the ground up. The open source movement has proven that large scale and complex systems of software code can be built, maintained, developed, and extended over time in a nonproprietary setting where many developers work in a highly parallel, relatively unstructured way and without direct monetary compensation. In other words, they work outside company boundaries and build a product whose control is explicitly not protected — not by patents, copyright, nor any other familiar notion of property. In fact open source does exactly the opposite. It rests on a system of property rights that are designed precisely to prevent anyone from controlling the product.

How revolutionary is this move? Push past some of the rhetoric and recognize that open source doesn't stand in opposition to capitalism's core notion of private ownership of the means of production. But it does change what it means precisely to "own" knowledge goods, and thus it changes the loci of advantage and control in software markets. The success of open source in competitive markets is a widely visible demonstration proof of something that some theorists have believed for some time: that the conceptual foundations of our current intellectual property rights regime are visibly failing in the struggle to secure legitimacy and efficacy in the digital age. With this realization there is nothing to stop a massive challenge to the business model logic of proprietary software makers, other than their existing political clout. That may slow the tide, but it will not be enough to stop it.

I am confident that in a few years people will think of a proprietary operating system for computers as a quaint, old-fashioned notion. When that shift becomes embedded in people's minds, no amount of congressional lobbying, legal wrangling, or economic hand-wringing will reverse the transformation of what used to be extremely expensive custom-built tools into standardized commodities that anyone can open up and play with. When business models depend on these "quaint" notions of what can be controlled and protected, they are left hugely vulnerable. Immense amounts of invested capital can be wiped out in short order. The effect can be contagious and devastating.

What the Internet protocol did to telecommunications at the end of the last decade, open source is about to do to the software industry. But the revolution is going to be much more widespread. The telecommunications bubble was ultimately self-contained because it was "just" a specific technology for transporting bits. Open source is a harbinger of a much broader shock because it rests on a different set of property rights. And it is not a sui generis phenomenon, but a more general illustration of how new production processes are emerging in complex knowledge goods, processes that simply bypass standard rules of intellectual property. Businesses that depend on those rules to protect the boundaries of what they can control will fight back tooth and nail, but they are shoring up foundations that are conceptually shaky and becoming politically more insecure every day. Welcome to the political economy of the IP bubble.

CORROSION AT THE CORE

Consider what makes Coca-Cola a proprietary product. The bottles of soda that Coke sells to consumers have a list of ingredients that is surprisingly generic. Coke has a confidential formula that it will not tell you about, on the bottle or anywhere else. The formula is the protected knowledge that makes it possible for Coke to combine sugar, water, and other readily available ingredients with some "secret" flavoring mix and produce something of great value. And when it's done, it's done. You can't reverse-engineer a bubbly liquid into its constituent parts. There is no equivalent of "rip, mix, and burn" for soda. You can buy Coke and you can drink it, but you can't understand it or manipulate its components in a way that would empower you to reproduce the drink, take pieces of it for another purpose, or distribute an improved cola drink to the rest of the world.

Standard intellectual property (IP) economics gives a clear-cut account of why the Coca-Cola production regime is set up this way. Patents, copyrights, trade secrets, licensing schemes, and other means of protecting knowledge assure that economic rents are created and that some part of that money lands in the pocket of the innovator. Without protection, a new and improved formula would be immediately available in full and for free to anyone. The person who invented the formula would have no special and defensible claim on a share of the profits that will be made selling drinks engineered from the innovation. And so the system unravels, because that person no longer has any rational incentive to innovate in the first place. This story says that without exclusive IP, there is no Coke. The digital economy only makes things worse because a free-rider on someone else's IP would no longer even have to buy the sugar, water, and caramel color to mix; he or she would only need a decent computer to make infinite perfect copies of someone else's idea, rendering it instantly and absolutely valueless to the innovator. No exclusive control over IP means no rents, and no rents means no innovation.

Anyone who followed the shooting star that was Napster knows these arguments. But stop for a moment: are they correct in human life, business practice, and real-world economics, as compared to theory? Would composers and musicians have quit composing and performing because their music could be shared across the Internet? Or because anyone, without permission, could rip, mix, and burn pieces of music that someone else created? In fact, we don't actually know the answer to these questions. The courts sentenced Napster to an early death. But a set of peer-to-peer file trading systems like Kazaa (these don't have the central information repository that made it easy to shut Napster down) soon supplanted Napster. By the summer of 2003 at least as many songs were being shipped over these new networks as ever moved through Napster. And it may have made some difference in the economics of the industry. After all, global sales of recorded music have fallen: by 5 percent in 2001, 7 percent in 2002, and another 4 percent in 2003. In September 2003 the world's largest record company, Universal Music Group, cut its CD prices by nearly 30 percent in hopes of recapturing some of the lost buyers.

That hurts. But no one seriously believes this signals the end of music as we know it, or even the end of commerce in music. What it does signal is a corrosive loss of legitimacy and confidence in the IP regimes that support existing business models in music. Napster and Kazaa are not sophisticated technologies. They are mind-sets, pure and simple. And as Peter Drucker said, economic revolutions are made of ideas, not machines.

Forget about where the *money* is for a moment. What Napster really did, by making the business models of record companies transparent to the end user, was to change the way people think about where the *value* lies in the music industry.

Last year in a class of 150 Berkeley undergraduates, I asked: "How many of you think it is appropriate to pay for music?" Notice the bland phrasing. I did not ask how many people thought it was fair, or how many thought it was reasonable, necessary, or rational. About ten hands went up. Don't dismiss the results of this informal poll as youthful indiscretion. These students have taken Economics 101. They are familiar with "the tragedy of the commons." They understand the standard rationale for intellectual property (I taught it to them the week before I asked the question). They don't think any of these arguments are wrong; they simply don't accept them as a fact of nature. And they are going to act on that belief. The RIAA can slow this down, for a while, with high-profile lawsuits against selected "violators." But the damage to existing business models is already done and essentially hard-wired into the mind-sets of this generation.

So what do companies and their public policy allies do? The answer, made up in equal parts of head-in-sand disregard and desperate defense, is part of the peculiar politics of the IP bubble. Consider as an example the "non-circumvention" clause of the Digital Millennium Copyright Act (DMCA). Under the DMCA it is not illegal just to break a copy-protection algorithm or any other kind of electronic lock on a protected digital good. It is also illegal, with only a few exceptions, to build a software tool that can be used to break an electronic lock — regardless of its other potential uses and regardless of what you or anyone else intends to do with it. The justification for this? To preempt the development of technologies that could threaten the copyright regime, under the rationale that once such technologies come into being, it will be too late to deter rule breakers by threatening to punish conduct that violates the rules.

This is quite like outlawing matches because they can be used for arson. There is a huge and increasingly visible imbalance between, for example, the level of protection we provide to the bits that make up an Elvis Costello song and the bits that make up your genetic profile. If I were to take some of the skin cells that fall onto the page you are now touching and sequence your DNA, there is nothing to prevent me from publishing that sequence. If anyone were to use that information to "steal" your identity they could be punished after the fact, but there's no preemptive legal logic for these unique identifiers that prevents me from building or possessing the tools to do that. What makes it worse is that Elvis can always write another song, but you are stuck with your DNA, and once I make it public there is no going back and no other way to protect you. Is the $11 billion music industry really going to be able to sustain this special treatment once people realize what is at stake here?

A little bit of desperation and confusion mixed together makes for strange bedfellows. Even groups that have consistently fought the music industry on behalf of file sharing sometimes are willing to live with an idea like placing a tax on hard drives, blank CDs, or other media that could be used to store the bits that make up a song. Or to collect a general tax on bandwidth, because some of that bandwidth might be used for file trading. This is like taxing glass or trucks to pay the liquor industry for lost revenues from moonshine. You'd need a really good economist to dream up a more inefficient tax or one that would have more negative externalities for unrelated industries.

To criticize is easy; to invent new solutions is harder. But even if the critics have been at times unfair, their attacks have hit home and steadily scraped away at the legitimacy of the music industry's positions. RIAA lawsuits will not shift very many people's perception toward the idea that file sharers are criminals; these heavy-handed tactics are more likely simply to increase resentment. And they seem to have forced more sophisticated file sharers simply to switch technologies, leaving only the most hapless at risk of getting caught. Even within the music industry, the long- term strategists are looking for new answers that are more than just a tweak on a visibly cracking model.

In fact, the software industry has a quite compelling answer. But it is not going to be one that today's music companies like.

THE OPEN SOURCE ALTERNATIVE

When Nicolas Carr wrote that "IT doesn't matter" anymore, he got the story exactly half right. Moore's Law has made raw storage and processing power almost superfluous; there's already more hardware capability than anyone knows what to do with. But there is no Moore's Law for software. Sophisticated software that can make use of all that raw capability has become more complicated, more expensive, less reliable, and more difficult to configure and maintain. Software is now the rate-limiting step in the information economy. What software can do and how it gets built and distributed determines the most important rules for the knowledge economy and competitive advantage within it.

This may be a disconcerting thought, because writing software often seems like some kind of medieval alchemy. But in an abstract sense software is just like any other product, the outcome of a production process that combines human effort with resources (mainly knowledge) in a particular way. In fact the "standard" way of organizing people to write software has been much like the standard way of building a complex industrial good: a formal division of labor that uses proprietary knowledge, guarded by restrictive intellectual property rights, enclosed within a corporate hierarchy, to guide and govern the process. But this is not the only way (see Table 10.1).

By changing what it means to "own" source code, from the right of exclusion to the right of distribution, open source radically transforms the production

TABLE 10.1   Essential features of open source

Open Source rests on an innovative notion of property, where the right to distribute rather than the right to exclude is central.

There are variations in the precise licensing schemes that define open-source software, but at the core are three essential features:

- The source code is distributed freely with the software. Source code represents the software instructions that a human being can read, rather than the binary codes that a computer executes. If you have the source code, you can understand what and how a program is doing what it is doing — and, most important, you can fix, modify, and improve it.
- Anyone can redistribute the software without paying royalties to the author.
- Anyone can modify the source code and distribute the modified source code under the same terms.

The key point of the General Public License (GPL), the most commonly used open-source license, is that you must distribute modified software under the same open-source terms. This is a critical detail, and it lies at the root of what makes open source software a new take on intellectual property. When you "buy" Microsoft Windows, you are actually purchasing only a right to use license, which grants you permission to use the binary instructions on your machine. **And this is not *just* a legal prohibition; it's a practical one.** Since you don't get to see the source code, you can't understand what the Microsoft engineers were trying to do. You can't modify Windows code, and you certainly couldn't give the modified code to someone else.

Open source software simply inverts this logic. You can modify the source code and customize the software in any way you wish. You can then redistribute the software freely as long as you don't prevent other people from doing exactly the same thing.

When people talk about open source as "free software," they mean freedom, not zero price. As Richard Stallman says, think free speech, not free beer. There is nothing in the license that says you can't charge money for open source software or for services around it or anything else that the market will support. What the license says essentially is this: **you can do whatever you want with the software, as long as you don't restrict the freedom of others to do whatever they want with it.**

---

model for software development and debugging. You want to build a new PDA to compete with Palm? Just download open source code and see what you can do, as Sharp did with its Linux-powered Zaurus. You want to build a supercomputer on the cheap? Rustle up some "obsolete" Pentium desktops or Alpha processors, tie them together with Linux in a "Beowulf" architecture, as

Los Alamos National Laboratory did, and you have a 48-gigaflop supercomputer at a sixth of the normal cost.

The popular image of open source as a bunch of eccentric techie hackers writing code for other techie hackers is partly correct, and mostly misleading. Open source developers write code for many different reasons. Some are paid by companies such as IBM and Red Hat. Some are simply fixing bugs in software they use in their daily work. Some are students trying to learn new techniques, and some are aspiring superstar coders wanting to prove to the world just how good they really are. But the place where all this code comes together is not some free-for-all anarchic bazaar where anything goes. Linus Torvalds has final say over what code gets incorporated into Linux and what does not. If you don't like his decision, you are free to take the code and start your own open source project. The freedom to "fork" the code, as software engineers say, is the core right that open source defends.

Of course, if you exercise that right, you had better be able to convince lots of other software developers that your project is better. And that from this day forward they should work with you rather than with Torvalds.

That's the lever that provides Torvalds with control over the process. There is no open source code that he can't use in Linux if he wants to, and there is no code at all that he feels compelled to use if it doesn't do what he wants. Linux doesn't need proprietary IP to keep this innovation ecosystem vibrant and productive.

And Linux doesn't need proprietary code for companies to make money. Yes, the equilibrium price for raw open source code is zero, but raw code does not solve the problems of most customers. Software, even with source code included, is less like an airplane seat and more like a complex recipe for a very sophisticated restaurant dish. Alice Waters publishes her best recipes, but that has not shortened the lines or driven the prices of a meal at Chez Panisse to zero — in fact, it increases demand. So companies such as Red Hat package, customize, and support open source code with a suite of business services keyed to a customer's practical needs. IBM uses open source software to generate demand for its proprietary hardware, software, and consulting services. O'Reilly and Associates sell the gold-standard books and technical manuals that tell people how to make the most of their open source code. Hewlett-Packard uses open source code to enable a business in Web services. And these are just some of the many experiments in business models that sit on the edge of open source.

A principal driver of economic change is technology that transforms what were once expensive and protected service-intensive things into de facto commodities. If you think of open source not as a particular widget but rather as a system of production, then you can see clearly how open source stands to revolutionize information technology markets even if the software itself is not a revolutionary technology. And let's be clear about this: the fact that Linux probably is not running your desktop computer today makes little difference. That is partly because more PCs and computing devices will in fact run Linux and open source programs in the next few years. But having Windows on your desktop is not important for a more fundamental reason, and that is because your PC desktop is becoming much less important. Sun Microsystems said a long time ago that "the network is the computer," and the technology finally is proving that to be correct. Your desktop is like the steering wheel on your car — important, but not nearly as important as the engine. The engine is the Internet, and it is increasingly built on open source software.

Revolutions always have unpredictable consequences, but some of the first-tier implications of this revolution are visible today. When source code is free and open, the market for services around that code has radically fewer barriers to entry. Transparency and competition replace de facto monopoly relationships that were held together by costs sunk into proprietary code. "No secrets" means that anyone can learn to be a packager, customizer, maintainer, or integrator for open source code. And so the days when an IT supplier could lock in a customer through an installed base of code are coming to an end. Yes, IBM and Red Hat would like to be the companies that everyone turns to for building and configuring information services on major open source platforms. If they are good enough at it they might even win. After all, large customers generally prefer working with large and well-established suppliers over working with a network of small players. But if the IBMs and Red Hats falter in responding to their customers' needs or charge too much there is nothing to stop another big player, or equally a mom-and-pop software shop, from instantly offering a competitive service on the customer's existing technology platforms.

THE DRUG DILEMMA

In its essence, the pharma sector faces a scientific and economic problem similar in important ways to the software sector. Put simply, good drugs and good software are really hard to build, and for comparable reasons. Bringing to market a safe and effective human therapeutic compound is an exquisitely difficult task, because the system in which the compound will intervene (much as a software "patch" does) is exquisitely complex, redundant, and idiosyncratic in any particular installation (human being), and in practice (perhaps in some deeper sense) unpredictable in important respects. In its implementation the drug problem is probably harder, since the pathophysiology of even a simple disease is orders of magnitude more complicated than the source code for Microsoft Windows. Yet if drugs were to "crash" the human body even a fraction as frequently as Windows crashes your laptop, we would quickly lose faith in the whole idea of pharmaceuticals.

A decade ago, and mostly drawing on science from a decade before that, these inherent challenges seemed manageable as the sector brought to market a host of new therapeutics (several of which went on to become megablockbusters.) But as everyone within the industry knows so well, the pipeline of new compounds is now thin and is almost certain to stay that way for at least a few years and possibly quite a bit longer. Because of the long lead times from laboratory to pharmacy counter, all the airplanes that this industry could possibly land in the next five years and possibly ten are already visibly in the air. We know that most of them will come up short in clinical trials, and only a very few will reach their intended destination. Of course, the discovery of a single new class of drugs, such as statins or selective serotonin reuptake inhibitors, could generate hundreds of billions of dollars of revenue and rescue the sector from its apparent declining productivity of R&D. There might be a new class of blockbusters in that fleet. Or there might not be. We simply don't know yet.

This is not a result of bad scientific practice, bad business practice, or bad luck. It reflects the essential difficulty of discovering and testing new therapeutics and of the burdens of regulation for safety and efficacy, which together have raised the costs of bringing a drug to market near or perhaps now above the $1 billion mark. That's a pretty significant investment in "white powder," a small molecule that is often surprisingly easy to reproduce once its properties are known (larger molecules and biologics can be more difficult to manufacture but can still be copied). Here, of course, is where the IP challenge for software overlaps that of pharma (even if software is typically protected by copyright and therapeutic molecules by patent). Without the promise of vigorous IP protection for valuable discoveries, who would invest the time, energy, and money in research and development of new drugs? And even if some of today's airplanes turn out to be blockbusters, it is nearly certain that the next generation of drugs will cost even more to develop. This is partly because the low-hanging fruit (the fifty or so best-understood drug targets) may have already been picked. It is certainly because more modern and targeted models of drug discovery (for example, using genomic data for intelligent design) are only at the start of a learning curve where the first manifestations will likely be the most costly. The future of health could be a very promising story, but it will not be a cheap one. Thus if we are going to have a decent shot at reaping some therapeutic benefits of new science, strengthening and lengthening the protection of quasi-monopoly rents to innovators via the patent system is the only bet to make.

Or so goes the modal argument within the industry. If it sounds parallel to the stories that incumbents in music and software tell, that's because it is. In theory the incumbents' arguments make sense. In practice, the Napster experiment and, even more, what has come in the wake of its demise can be seen as interesting partial tests of whether they are the only arguments that make sense. The open source software phenomenon is a compelling data point here. The conceptual foundations of the intellectual property regime that protects proprietary software have been found to be, at best, insecure. Market share is a demonstration proof. If the lawyers who defend the idea of software patents are right, and if the Microsoft executives who call open source a "cancer" of the intellectual property system are right, then they wouldn't actually have to be fighting these battles at all because the markets would score them winners through innovation.

A more modest formulation is simply this: the success of open source puts a question mark over the assertion that exclusive property controls (whether in copyrights or patents) are the only or even the best way to generate and support innovation in complex knowledge goods.

A crack in the conceptual foundation of a legal regime is interesting to academics and lawyers but is unlikely by itself quickly to break the grip of vested interests on policy. The real problem for pharma is that the conceptual fragility of the property rights system that underpins its business model is becoming clear at precisely the same time that the industry's political foundations are being shaken. In fact the politics are cracking in several places. It's hard to pick up a newspaper these days without seeing a nearly schizophrenic assessment of the drug industry. Drugs are too expensive in the United States, too cheap in price-controlled markets in Europe, and still unaffordable for most of the world's population. The productivity of R&D within the industry has been declining steadily for more than a decade, with fewer new chemical entities receiving FDA approval every year (fifty-three in 1996, declining to twenty-one in 2003; in the same period R&D spending roughly doubled to around $32 billion). "Me-too" drugs (that compete with existing therapies and are not significantly better) are rampant; dubious means of extending patent protection for six months to a couple of years (by modest reformulations or testing for safety in children) are the norm. Clinical trial data is "massaged" by sponsors. "Land grabs" of the human genome by companies that file for low-quality patents make it likely that almost any interesting new molecule can be held up in litigation. Putting together a patent portfolio to pursue the development of particular chemical compounds involves a huge amount of legal and business wrangling. The deadweight costs are passed on to consumers. Or so the papers say.

The international politics of pharma are even more fragile. Ask a citizen of South Africa to define a weapon of mass destruction and he or she is likely to answer, "The cost of anti-retroviral treatment for AIDS." The debate over pricing and availability of drugs in poor countries takes place in the shadow of predictions that 100 million people may die in Africa from a disease that can now be treated as a chronic condition in the United States. But if you can sell a pill for a fraction of the "market" price in Africa, why not to poor people in the northeast quadrant of Washington, D.C.? Segmented markets with differential pricing make perfect sense to economists but no sense at all to families of sick citizens. It's not possible to insulate these deals from politics. It's extremely unlikely that reference-pricing discussions in developed economies can be hermetically sealed. The average American knows that Lipitor from a Toronto pharmacy is precisely the same as Lipitor from a Minnesota pharmacy but cheaper, because of price controls in Canada. When the pharma companies claim that reimportation is dangerous, many patients (rightly or wrongly) see shades of the tobacco industry claiming that cigarettes may not cause cancer. It is perfectly plausible that public opinion in the United States and abroad could target big pharma as the principal villain of a future public health crisis.

And would government step in on the side of the industry? For the U.S. government at least, defending the property rights of pharma has seemed to be unconditionally good public health policy and good economic policy at the same time (after all, U.S. pharma is enormously competitive on global markets even if most of the industry's profits are made in the United States). And then there was the anthrax scare of fall 2001. The public wanted massive distribution of Cipro — a third-generation, broad-spectrum, patent-protected, and very expensive antibiotic that was also in relatively short supply. Senators called for overriding Bayer's patent on Cipro under the pretext of a public health emergency — after precisely five deaths. This should sound a clear warning to anyone who thinks that Washington will always be there as the last-ditch defender of pharmaceutical property rights.

And all of this is happening while a scientific revolution in drug discovery and design is probably close to the horizon. But the challenge of early science demands new intra-industry collaborations, public-private partnerships, and an expansion of funding for R&D. This at the exact moment that pharma's injured reputation is making it increasingly difficult to sustain the political legitimacy needed for that kind of multidimensional effort. Land, labor, capital, knowledge, and a public grant of legitimacy to operate are the critical factors of production for this industry—and a collapse of the final factor is at serious risk of diminishing access to the others. Why is it that a U.S. president can set people's hearts beating by announcing a plan to travel to Mars—but not a revival of "the war on cancer" or a massive national or international campaign to create an AIDS vaccine? The troubling answer is that the public has greater trust in NASA—even after the damning reports on the systemic failures that lead to shuttle disasters—than it does in the pharmaceutical industry.

To batten down the hatches and wait for the pipeline to deliver is no longer, in my view, a viable strategy. The inflection point for pharmaceuticals is coming. It might be set off by politics and regulatory change, manifested in some combination of price controls, reimportation, and accelerated generics. It might be set off by industry dynamics, through the creation of open databases and shared precompetitive research, a response to the rapidly increasing costs of discovering and developing new drugs. It might be set off by widespread diffusion of the relatively simple technical tools that make it possible to manipulate cells and, increasingly, the therapeutic compounds that act on cells. We're not yet at the point where "biohackers" working in suburban garages can take advantage of an end-to-end architecture and out-innovate Pfizer and Merck. But we are closer than many people think. And long before biopharma makes the leap to the garage, it will complete the leap it has already begun to make outside the United States and Western Europe, to China, Singapore, India, and beyond.

EXPERIMENTING WITH THE PHARMA FUTURE

Drugs are just too important for pharma to go through a real financial meltdown. But the industry is going to suffer, as the music industry did, from a corrosive loss of financial confidence. Here's one scenario for piecemeal damage that adds up to a breakdown. A company endures the compulsory licensing of one of its most profitable drugs here. Another drug is "repriced" by a government for Medicaid purchases there. More U.S. states allow—even facilitate or mandate—purchases from Canadian pharmacies. At some point politicians name this for what it is—importing someone else's price controls—and legitimate the idea that price controls "work." The major pharma companies respond to all of this with heavy-handed lobbying. The industry group PhRMA (Pharmaceutical Research and Manufacturers of America) overplays its legal and political hand, just as the RIAA did. Suing a grandmother for downloading songs is just a little comical. Suing a grandmother for reimporting a drug that her grandchild needs to stay alive is not funny at all. Given the importance of two things — public trust and a blockbuster product or two — for the stock valuations of the largest companies in the pharma sector, it won't take much of this to create a perceived "bio bomb" where the short sellers move in and stock prices collapse.

It's not hard to see this coming; it's just hard to predict the timing. Which, with a little irony, creates a first-mover *disadvantage* for publicly traded companies. It would be tricky for a major pharma firm to get out in front of the game and aggressively recast its business model for a more sustainable future, because the first company to address explicitly the fragility of its current state and move onto a transition path would probably be pummeled by Wall Street along the way. It might even be the spark that sets off the downward cascade, which could make it very difficult for the renegade firm to cooperate effectively with others after the dust settles. It's an unfortunate but real possibility that even if people in the industry really understood what a more sustainable endpoint for next-generation business models would be, it could be very hard to get to that place smoothly. The more likely outcome is a continued downturn in this critical sector of the knowledge economy, followed by a period of consolidation and underinvestment, followed in the medium term by a persistent lower growth path for the industry.

And then what? Here's the real irony: a much cheaper commodity infrastructure, a large and growing feedstock as more old molecules come off patent, and an approaching scientific threshold in the understanding of cell biology along with the diffusion of tools that let people work with that knowledge together make for an exciting possibility — that we could be on the threshold of a generation of new therapeutics. But that possibility will not be brought to life by companies stuck trying to retain as much as they can of an obsolescing business model.

The economist Brad DeLong correctly points out that the idea of a catalog retail business, which became Sears Roebuck, emerged on top of cheap commodity infrastructure for transport following a bust (one of many) in the railroad sector. That was good news for Sears and good news for consumers. It was not such good news for traditional local retailers. In the open source software economy, IBM has more or less taken Sears's cue and is transitioning to a global services and business hardware-software integration company built on top of the commodity infrastructure called open source code. This is excellent news for anyone who uses information services — in other words, pretty much the whole economy. It is not such good news for proprietary software firms or for companies in information-intensive sectors whose investments in very expensive proprietary software are a key source of their competitive advantage and barriers to entry for others.

The same thing could very well happen in pharmaceuticals. The much-needed boost to productivity in this industry is more likely to come from upstart players who lack any financial interest, or, for that matter, organizational and emotional stakes in knowledge and science that will have become commodities. Think one step beyond shared databases and collaborative research and development initiatives. The stream of information that a molecule leaves in its wake as it interacts with human cells, with doctors' decisions about what to prescribe, with patients' preferences on what they ask of their doctor, and so on, does not need to be locked up as proprietary knowledge. Someone will set up large-scale experiments with open source development processes for therapeutic molecules. Others will become customizers and service providers around those molecules. Perhaps Richard Branson's next venture, Virgin Healthcare, will be explicitly and self-consciously about effective drug marketing (unlike today's large pharma companies, which try to hide very large marketing budgets behind much smaller research and development budgets). Someone will become a real drug-services provider and build a business that is simply about getting the right drugs to the right people at the right time and making sure they use them in an effective way, monitoring their compliance, response, and side-effect profiles with networked wireless sensors. Maybe it will be a cosmetics company, or a next-generation Webvan home grocery delivery service, or an entrepreneurial physician, or, for that matter, a consumer electronics firm like Sony that already has a nearly ubiquitous presence in patients' lives. With an open source environment it doesn't much matter. The "source code" will be available to potential entrants. If Estée Lauder or Virgin Healthcare is good enough, it might become the preferred health services provider for many people; but if it falters in responding to customers' needs or if it charges too much, there is nothing to stop anyone else from offering a competitive service on the same molecular platform.

Of course not all the value creation in the industry will happen within an open source environment (there will still be closed source proprietary research and development in many segments). But a lot of it will. And the contribution to health, which (rather than drugs per se) is the real "product" of the pharmaceuticals industry, could be immense. Some observers will label this a process of consolidation, but it is really quite a lot more than that. We know today that health on a global scale could be improved at least as much through commercial creativity using the science and knowledge we already have as with new molecules. There are enormous market failures in vaccines, which are grossly underprovided and underutilized in poor populations almost everywhere. Childhood diarrhea kills, unnecessarily, in most of the world. The costs of long-term degenerative disease states associated with hypertension, diabetes, and heart failure are breaking the backs of insurance models in the developed world and particularly in the United States. A lot can be done with new business models, new delivery mechanisms, highly customized protocols, and so on that would dramatically increase health and reduce costs.

There is also a lot of pathophysiology that we don't understand about these disease states. But is it certain or even probable that this knowledge will be discovered and developed inside traditional property rights systems, where patents rule? The open source software story suggests that the answer to that question is very likely to be no. There will be coexisting innovation environments, in competition with each other. This too is good news for patients, consumers, governments (which are often the largest drug customers), and probably most insurance companies. It could be bad news for investors in the pharma sector, which will face threats of a persistently lower growth trajectory with downside financial shocks still to come as its IP bubble starts to deflate.

And the politics could get very messy. The offshoring of drug development and clinical trials is likely to become embroiled in protectionism (as has happened in low-level software engineering) just as the cost savings are becoming too important to pass up. Expect to hear as well more national security arguments about the risk of allowing leading-edge basic research in biotechnology to "escape" U.S. companies. Government subsidies could return to center stage and have a disproportionate impact on biotechnology trajectories, particularly when it comes to dual-use issues of interest to homeland security. Governments *will* be involved in the reorganization of these sectors — not because they are "too big to fail" but because ultimately they are too essential not to be rebuilt on firmer foundations.

Knowing this, companies should start right now to look past the anxious present and push for two future political principles. The first is to permit experiments in unlicensed spaces. The Wi-Fi (and, soon, Wi-MAX) example in wireless data communication shows the way. No one knows if there are sustainable business models in Wi-Fi, but at least we are going to have the chance to find out. Wi-Fi has also been a useful demonstration of the fact that there is no state of nature in property rights, particularly for nonexcludable (or difficult-to-exclude) goods such as spectrum. Comcast would like me to think I am stealing when I log on to my neighbor's wireless hub, which is connected to her cable modem. Some of the Bells would like me to think that cities and other municipalities should not, on principle, provide wireless connectivity to their citizens as a public service. I don't see it that way, and neither do the vast majority of people like me. The only firm constraint on experimentation in these kinds of spaces should be the requirements of public safety. The economy does not want my Wi-Fi experiments constrained for the sake of protecting a business model, but we do want constraints ensuring that my Wi-Fi does not interfere with fire department radios. The important question is to define the analogous spaces in pharma. Scare tactics can make it sound like any messing around with body chemistry is far too dangerous to experiment with in an unlicensed space. But we already permit this in adjacent markets — for example, in the huge supplements and natural- or herbal-remedies sector.

The second principle is about finding ways to overcome disincentives to being the first mover in any experiments. This is partly a financial problem, and the government could help by offering subsidy funds, tax write-offs, and other inducements that reduce the market risk that first movers face. The backing of bad bets is inevitable but justifiable on the basis of easing the transition path for the sector as a whole. Public policy should be aligned to treat sunk costs as sunk costs. It makes no sense at this point to try to rescue the investors who chose to pay for the infrastructure that is becoming a public good. The argument that this would block future investment is disingenuous; it has not been true in other analogous situations. More likely it shifts the focus of investment to new players, which is going to happen anyway. As hard as it may be for incumbents to swallow, as a society we should pocket what's been built and move on. Incumbents can move on as well if they wish, but they shouldn't be allowed to stop the tide.

I have a very clear justification for what may sound like a draconian argument. It is simply a question of mindset. Don't let anyone try to convince you that knowledge industries are maturing. The pharma industry — by which I mean the genomic-based, knowledge-intensive, drug-discovery-by-design world into which we are moving — is not a mature industry but exactly the opposite. The software industry is visibly, painfully immature. Let's start taking that immaturity seriously.

In fact drug companies generally think of themselves as science-based organizations. Science matters a lot, and it is the only thing that matters in the long run. But what will matter more in the next few years are commercial and business model creativity and the ability to gently turn the public policy landscape toward facilitating that creativity. On aggregate, there is very little point in buying time for those parts of the system that will not degrade gracefully. That will only make the crash uglier and harder to recover from, because public support for the industry is already so frail. A desperate defense of business models that are visibly cracking will only corrode further what legitimacy is left for the industry. And for an industry such as pharmaceuticals, legitimacy to operate in the public interest is at least as important a factor of production as is labor, capital, or knowledge.

11

The Learning Organization

*A Research Note on "Organisational Change
in Europe: National Models or the Diffusion
of a New 'One Best Way'?" by Edward Lorenz
and Antoine Valeyre*

Tobias Schulze-Cleven

In their paper on the changing pattern of corporate organization in Europe, "Organisational Change in Europe: Models or the Diffusion of a New 'One Best Way'?" Edward Lorenz and Antoine Valeyre address two important questions (Lorenz and Valeyre 2004). First, how can companies leverage organizational learning to drive innovation? And second, how do national institutional maps differentially shape the conditions for this firm capacity? In formulating their answers, the authors reject a dichotomous distinction between Taylorism and lean production. Instead, they conceptualize two distinct post-Taylorist models of organization, a "lean" and a "learning" model, both of which are geared toward the competitive marketplace of the digital economy. The authors find that firms in countries with a high prevalence of the learning model tend to be very innovative. Surprisingly, these countries also tend to have the strongest employment protection legislation. Lorenz and Valeyre build their argument on the recently completed Survey of Working Conditions in fifteen European Union countries, conducted by the European Foundation for the Improvement of Working Conditions (Paoli and Merllié 2001).

THE "LEAN" VERSUS THE "LEARNING" MODEL

The authors' lean model corresponds to the Japanese lean production model originally formulated by Womack, Jones, and Roos (1991). Rather than interpreting the organizational variety in EU countries as evidence of the hybridization of this lean model, the authors conceive of the learning model as a separate phenomenon. In their view, the learning model constitutes a distinct way of delivering flexibility and cooperation within the company. It is socially embedded in a unique way, because it builds on local traditions in work organization.[1] Companies following the lean model display such attributes as the strong use of teamwork, job rotation, quality management, and multiple work-pace constraints. In contrast, the learning model is more decentralized and grants employees a high degree of autonomy. However, both the lean and the learning organizational forms display stronger learning dynamics and higher problem-solving activity on the part of employees than either Taylorist or pre-Fordist traditional organizations.

In their paper, Lorenz and Valeyre ultimately focus on the existence of strong national differences in the prevalence of the learning and lean models independently of occupational category and economic sector. Laying the foundations for this discussion, they acknowledge that the learning model is more prevalent in certain parts of the service sector, such as financial and business services, and utilities. In contrast, the lean model is employed in manufacturing, particularly in the production of transport equipment, electronics, and wood and paper products, as well as printing and publishing. In terms of occupational categories, the learning model characterizes the work of managers, professionals, and technicians. In contrast, lean forms of work organization are more characteristic of blue-collar employees.

DIVERGING INNOVATION PATTERNS

The distinct properties of the lean and the learning organization are evident in the way each promotes innovation. Going beyond Lorenz and Valeyre's paper to bring in recent research by scholars of innovation allows us to illustrate how the lean and learning models tend to follow different approaches to sustaining competitiveness in the digital economy. The two organizational forms' typical innovation profiles correlate with different interpretations of what constitutes knowledge in the digital economy. Embracing knowledge as structured information, lean organizations focus on the gathering and digitization of information. In contrast, learning organizations build on a dynamic conception of knowledge as fluid, contingent on context, dependent on continuous recombination, and ultimately resting in people.

A recent contribution by B. Johnson et al. (2004) stressed that digitization in the digital economy does more than provide tools for thought.[2] Digitization implies that knowledge is created and destroyed at a faster rate, that is, is becoming obsolete more rapidly than before. Giving employees the flexibility and autonomy for experimentation and allowing them to take the initiative in the pursuit of complex tasks, the learning organization tries to leverage the interaction of its skilled employees into individual competence building,

236   organizational learning, and innovation. In contrast, the learning dynamics of the more hierarchical lean model "are embedded in a more formal structure based on codified protocols (e.g., teamwork and job rotation practices) often associated with tight quantitative production norms" (Lorenz and Valeyre 2004). Learning organizations promote what Niels Christian Nielsen and Maj Cecilie Nielsen (this volume) call spoken-about knowledge. Learning in these organizations is a continuous process of accumulating, applying, and combining employees' knowledge. With this strategy, learning organizations should display strengths in the realm of "interpretation" in addition to mere "rational problem solving."[3]

Clearly, the lean and the learning models of organization promote innovation in quite distinct ways. However, according to Lorenz and Valeyre, they both represent the same one of two complementary modes of innovation recently identified by B. Johnson et al. The authors see both the learning and lean models as promoting the DUI (doing, using, interacting) mode of innovation, which focuses on human capital investment, embraces the unity of learning and working, and seeks to stimulate directly applicable know-how. The DUI mode is complemented by the STI (science, technology, innovation) mode of innovation, a more mechanistic, rules-based approach that displays the logic prevalent in the natural sciences and seeks to generate knowledge of the "know-why" type. Firms need to use both STI and DUI modes to be innovative overall. As Lorenz points out, STI needs DUI to overcome bottlenecks in absorbing new technologies and anchor R&D activities in the overall business environment by tacit links to procurement, production, and sales. The lean and learning models are thus two distinct ways of organizing DUI-mode innovation, the main difference being the relatively high levels of employee discretion and autonomy in the learning form. Crucially, autonomy favors the exploration of new knowledge.[4]

The learning model is very successful at generating incremental innovation. Countries that display a high prevalence of the learning model exhibit high rates of patent applications to the European Patent Office. On average, these rates are higher than those of countries in which the lean model dominates as a post-Taylorist organizational form. The question arises as to whether the prevalence of the learning or lean models in a country is determined by the position of national producers in high-tech or high-quality product markets. Admittedly, on average, countries with a high prevalence of businesses using the learning model spend more on R&D than those with a high prevalence using the lean model.[5] However, R&D spending does not correlate with the share of workers in either type of organization at a statistically significant level, if the southern European countries are excluded from the

analysis.[6] Given the learning model's strong record, the potential institutional preconditions for its implementation constitute an important research
question.

THE CONNECTION BETWEEN NATIONAL
CONTEXT AND COMPANY STRATEGY

Lorenz and Valeyre find that learning organizations are prevalent in Sweden,
Denmark, the Netherlands, and, to a lesser extent, Germany and Austria.
Meanwhile, the lean production model is more widespread in the Anglo-
Saxon countries of the United Kingdom and Ireland. These findings contribute important micro-level evidence in support of the continued existence
of varieties of capitalism under the conditions of the digital economy.[7] But
what allows the learning organization to blossom? Can companies adopt the
learning model anywhere, or is there a particular reason British and Irish companies do not adopt the learning model in greater numbers? If learning strategies are dependent on institutional context, can governments facilitate the
adoption of the learning organization as a model?

Lorenz and Valeyre suggest an answer that builds on a recent literature in
political science, highlighting the increasing divergence in labor politics between the Anglo-Saxon liberal market economies (LMEs) and the continental European coordinated market economies (CMEs) under the pressure of
globalization (Thelen 2001).[8] They argue that successful adoption of the learning organization model might presuppose employer coordination on wage
bargaining at a higher level than that of the firm. Accepting the argument of
the political science literature that labor market regulation prevents a collective action problem and thus acts as an enabler of higher-level employer coordination, Lorenz and Valeyre present data on how the relative national
prevalence of the learning and lean models correlates with the amount of employment protection legislation as expressed by an OECD index. These data
show a starkly higher degree of employment protection legislation in continental Germany, Sweden, Austria, the Netherlands, and (with some distance)
Denmark than in the liberal United Kingdom and Ireland.

Lorenz and Valeyre stress two reasons why employer coordination in regional or sectoral bargaining has played an important role in supporting companies' adoption of the learning model. First, employer coordination buffers
the individual company management from distributional conflict, which can
otherwise easily spill over into areas of labor-management cooperation. Employer coordination facilitates local bargaining between employers and workers in pursuit of more flexibility and greater cooperation of labor at the shop

238 level. Second, it provides a solid foundation upon which employers can make extensive collective investments in training and skills, investments that are a precondition for adopting the model of the learning organization.

Lorenz and Valeyre are skeptical about the viability of learning organizations in the deregulated labor markets of the liberal Anglo-Saxon countries. The authors argue that the lack of employment protection legislation translates into low employer capacity for coordinated action around wage and skill provision, making it hard to sustain substantial forms of autonomy in work. In addition to local distributional conflict, which might prevent employers from securing labor's commitment to progressive improvements in product quality, the risk that competitors could poach skilled workers will provide the incentive to underinvest in the provision of training. According to Lorenz and Valeyre, potential substitutes, such as in-house training schemes linked to firm-specific internal labor markets, are likely to prove unstable. Without the broader supporting labor market infrastructure, they are likely to fail to structure careers and provide incentives for skill acquisition. Therefore, Lorenz and Valeyre deduce that employers in a liberal regime would logically choose the lean over the learning model as their preferred post-Taylorist strategy. It would be easier to sustain this relatively hierarchical model of work organization, in which worker autonomy is limited and tight quantitative production norms commonly fix the pace of work.

Lorenz and Valeyre make a much-needed contribution by linking the varieties of labor politics in the different varieties of capitalism with evolving patterns of corporate organization. In doing so, the authors disagree with a simplistic one-dimensional vision of labor market flexibility, which tends to underpin the calls by many economists and editorial writers for more labor market flexibility in Europe. Instead, they argue against a zero-sum trade-off between flexibility at the plant level and higher-level coordination. There are various dimensions to labor market flexibility, some of which are in conflict with each other. Most important, a sharp trade-off exists between certain types of wage flexibility and the functional and temporal flexibility with which companies can employ their labor (Regini 2000).

TAKING THIS RESEARCH FURTHER

Lorenz and Valeyre's analysis provides fertile ground for future research, not only because the authors present hard data and an interesting hypothesis to be tested in further research. However, important questions remain to be answered. Is employer coordination as important as Lorenz and Valeyre suggest? Are there alternative explanations available for this interesting observed

THE LEARNING ORGANIZATION

correlation? How do these potential alternative explanations rate against the one presented here?

239

We need to ask whether the lens chosen by Lorenz and Valeyre is the right one for interpreting the correlation between employment protection legislation and the relative prevalence of learning organizations. Lorenz and Valeyre's explanation follows an approach that was largely elaborated based on empirical research on the German case (Thelen 1991; Soskice 1999). What speaks for or against this approach, given that the highest prevalence of learning organizations is found not in the Germanic countries, but in Sweden, Denmark, and the Netherlands? Future scholarship that builds on Lorenz and Valeyre's research needs to address, first, which institutions are necessary for companies to choose to adopt the learning model, and second, whether these institutions are sufficient for doing so.

In developing answers to these questions, it is important to note that Denmark and the Netherlands trail Germany (with Sweden being about equal) in their degree of employment protection legislation. It is vital to pay attention to the degree to which countries protect against unemployment and how they treat the unemployed. In particular, the distinction between passive (job protection) and active labor market policy (training and placement) as well as national differences in the conditionality, duration, and generosity of high-spending countries' unemployment benefits are critical. Denmark, the Netherlands, and Sweden are important examples of the success of "flexicurity," an approach to labor market regulation that tries to combine the flexibility that companies want and workers should display with the security that workers want and should enjoy.[9] Policy makers in these countries have focused on maintaining workers' employability through well-funded public training programs and the conditioning of generous benefits on individuals' participation in these programs. Such labor market policy sharply contrasts with that of Germany and Austria, countries whose labor laws tend to protect currently employed workers and in which most expenditures flow into consumption by the long-term unemployed rather than investment in skills.

Focusing on the actual policy profiles of the different systems of welfare capitalism, these reflections can give rise to the first of two further alternative hypotheses for why countries with the highest degree of employment protection legislation show the greatest prevalence of learning organizations. The foregoing discussion makes clear how important it is to consider the role of the state in sustaining an environment in which flexible workers with sought-after skills are readily available for employment in learning organizations. In a second line of inquiry, economic sociologists may identify the role of social embeddedness for the learning organization (K. Polanyi 1944; Granovetter 1985).

240   Lorenz and Valeyre stress that the learning model is an alternative to the lean model, deriving its comparative advantage from building on local traditions. They show that Taylorism is comparatively underdeveloped in Sweden, Denmark, and the Netherlands. With older traditions of workmanship only rarely replaced by Taylorism in these countries, it seems likely that learning organizations could use workers' pre-Fordist identities and tap into accumulated social trust.[10] At this stage, both of these rather undeveloped hypotheses point to the learning organization's being supported primarily by the policies and social institutions associated with the social democratic welfare state.

It is too soon to settle on the narrow interpretation offered by Lorenz and Valeyre. Although the prevalence of the learning organization is likely to be highly influenced by the wage-bargaining structures in a particular country, broader policy feedbacks from government labor market policy and the broader welfare state structures might prove very consequential. Scholars increasingly note that welfare state structures play a strong role in sustaining particular production regimes. Social programs provide individuals with incentives and capital for appropriate skill investments as well as nurture collective identities and conceptions of justice into which companies can tap (van Kersbergen 1995; Ebbinghaus and Manow 2001; Estevez-Abe, Iversen, and Soskice 2001). The systems of production and welfare have developed together and continue to condition each other. The success of the learning organization might be the most recent example of this dynamic.

NOTES

1. In particular, Lorenz and Valeyre point to the Swedish sociotechnical principles of the 1970s as sharing many properties of their learning model.

2. On "tools for thought," see Cohen, DeLong, and Zysman 2000.

3. The distinction between interpretation and rational problem solving is developed in another recent contribution to the literature; see Lester and Piore 2004.

4. I want to thank Edward Lorenz for clarifying the relationship between lean and learning and between DUI and STI in personal correspondence.

5. Spending is measured as expenditure as a percentage of GDP.

6. Lorenz and Valeyre write: "The Pearson correlation coefficient recalculated without the four southern European nations, although positive, is not significant at the 10 percent level" (21).

7. The clustering of countries identified by Lorenz and Valeyre corresponds with the patterns established in other recent scholarship. For influential theorizing on the varieties of capitalism, see Hall and Soskice 2001. For the dominant perspective in the comparative scholarship on the welfare state, see Esping-Andersen 1990. Esping-Andersen identifies the Nordic countries as part of a social democratic welfare regime, the continental European ones as part of a conservative corporatist welfare regime, and

the Anglo-Saxon countries as a liberal regime. An important contribution to the literature is also Goodin et al. 1999. Using panel data to demonstrate how different national social program structures matter greatly for populations' welfare over time, Goodin et al. use the Netherlands as a stand-in for the social democratic regime.

8. In contrast to the other chapters in the book, Thelen's contribution stresses the importance of politics over more narrow arguments that are heavily influenced by the new economics of organization.

9. See the contributions by Benner and Vad 2000; Green-Pedersen 2001.

10. For an account that stresses such factors, see Piore and Sabel 1984. A recent treatment of the Danish experience along these lines is Lundvall 2002.

12

SPOKEN-ABOUT KNOWLEDGE

Why It Takes Much More than Knowledge
Management to Manage Knowledge

Niels Christian Nielsen and
Maj Cecilie Nielsen

Unimerco, an industrial tooling group operating at the forefront of industrial development, depends on knowledge for its success. For its CEO, Kenneth Iversen—and for all of his colleagues—leading, managing, and developing knowledge processes internally and with suppliers and customers is a dominant aspect of every workday. Yet, surprisingly, the academic writing, the consulting service offerings, and the software tools labeled "knowledge management" play little or no role in Unimerco's efforts to master knowledge challenges. Unimerco does not buy knowledge management systems or consulting services on knowledge management.

This chapter accounts for the disconnect between the practice of a pioneering knowledge-intensive company and the knowledge-management school of academic theory and consulting firms.[1] We propose a framework for understanding knowledge and knowledge processes that can be truly useful for companies. Building from the Unimerco case, we use this framework to develop a more systematic and comprehensive understanding of the knowledge processes in companies. The fundamental problem of the consulting practice of knowledge management is that it equates codified knowledge with knowledge. As a consequence, the emphasis is on identification, capture, documentation, and dissemination of codified knowledge, not least through the use of IT systems. We argue, however, that codified knowledge in and of itself is not knowledge at all. Similarly, we criticize the school of thought focused on the understanding and deployment of "communities of practice" through which organizations optimize and share tacit knowledge. The central contribution from the communities-of-practice school is the recognition that tacit knowledge is important and, indeed, an irreducible foundation of all knowledge. This research offers companies very real guidelines on how to optimize tacit knowledge processes but focuses on a single dimension of knowledge, tacit knowledge, which is relatively static in nature.

Both these schools of research are flawed because tacit knowledge and codified knowledge are not two separate types of knowledge. We shall discover that companies require a dynamic unity of the two forms of knowledge where iterations from form to form, from a tacit form to a codified form for example, are mediated through other forms of knowledge such as spoken-about knowledge, spoken knowledge, and embedded knowledge. We conceive of a system of knowledge, graphically illustrated in Figure 12.1, within which tacit and explicit knowledge interact and depend on each other, and where the interaction is mediated through the category of spoken-about knowledge, developing these terms and notions as we proceed. This points toward an understanding of knowledge that will allow new ways to describe and capture knowledge processes.

To create a reference for our discussion, let us begin by taking a closer look at Unimerco, a company that is highly dependent for its success on knowledge processes, yet eschews conventional knowledge-management approaches.

UNIMERCO: THE KNOWLEDGE-INTENSIVE COMPANY

Unimerco is a Danish tool distributor and supplier of tool-oriented services. Founded in 1963, it supplied craft-based services to customers that were primarily part of the furniture industry and located nearby. It is now a high-end specialized tool supplier and production optimization service provider for the auto, aerospace, and offshore resource extraction industries and still maintains its presence in the furniture industry. The ever greater sophistication of knowledge about materials and tools has led clients such as Ford and Airbus to turn to sources outside their own organizations for expertise and support. This opportunity has allowed Unimerco to grow into a small, highly profitable global company with revenue of $100 million in 2004 and after-tax profits of $9.5 million, as well as subsidiaries in Norway, Sweden, Germany, the United Kingdom, the United States, and, recently, China, and soon also in Eastern Europe.

Knowledge-driven Opportunities

Unimerco is addressing the market opportunity created by the revolution in materials technology, which — though much less publicized — is as fundamental as the IT and telecoms revolutions. A few universally applied materials that dominated all of industrial production prior to World War II have been replaced by thousands of functionally specific materials — alloys, ceramics, composites, polymers, and more. As each material requires and enables distinct designs, processes, tools, and so on, the ubiquity of thousands of functionally designed materials has revolutionized tools and made highly specialized tooling technologies such as those Unimerco masters necessary.

Because no single company can maintain in-house expertise sufficient to handle all aspects of all the new materials, the materials technology revolution is leading to changes in the global division of labor and strategic outsourcing. Therefore, Unimerco's market niche derives from outsourcing by its customers rather than displacement of competitors and relies on a high level of knowledge sharing between Unimerco and its customers.

Unimerco's value proposition to customers today is the ability to optimize production based on tool optimization. Unimerco has a track record of achieving lower production line total lifetime cost, higher through-put; lower downtime, higher accuracy, and a longer tool life cycle. This is accomplished through a close partnership with the customer; but also through superior tool design, enhanced tool specifications in terms of accuracy, elimination of vibration, fewer separate processes, harder and more durable tool surfaces, and the ability to develop tools with specifications for working in new alloys, light metals, and composites.

In reality, what Unimerco has accomplished is to strengthen the quality of its traditional craft-based tool making workmanship, while responding to new demands on precision and measurement. This is complemented by embedded skills in advanced computerized tooling-machines and the flexibility to follow, identify, master and apply university research on the shop floor. With this in mind, Unimerco has had to codify the core of its knowledge about tool processes so that it can be easily reused. Each application to improve on machine capabilities depends on the individual customer context and a close partnership with Unimerco's machine suppliers. To turn all of this into value for its customers, Unimerco must have sales-engineers who not only understand the new tool specifications and capabilities, but who can also translate this understanding into production optimization opportunities for customers, taking knowledge from one industry and applying it to another. For example, Unimerco has taken early experience with tools for hyper-fast processes in the wood industry (where relatively soft materials allowed such processes early) to pioneer processes in industries working in harder materials. The outcome is that Unimerco has combined the capability to optimize knowledge processes with an enduring tradition of craft and service.

Responding to Knowledge Requirements: Transforming Company Relationships

Unimerco is an instance of how a successful company adapts to the growing intensity of knowledge; its experience helps us see what will be required from companies in the future. As Unimerco has moved to hone and evolve knowledge processes, classical employee, customer, and supplier relationships have been converted into long-term partnerships.

Internal arrangements within the firm from physical architecture through employee ownership to a strong learning culture have been consciously developed to enable knowledge processes. At its core Unimerco has tried to create a responsible and motivated employee community deeply informed and knowledgeable about strategic and tactical issues facing the company. This culture begins with ownership arrangements: Unimerco is 100% owned by its employees, and all employees are offered shares. Though company executives are the biggest individual shareholders, they do not have majority ownership. Anyone leaving the company must sell his or her shares to the internal redemption fund, often paid what amounts to a very comfortable pension. In terms of *personnel policy and leadership style*, the expectation, responsibility, and obligation to share knowledge is complemented by rights to be included in debate and discussion. Unimerco has never used layoffs to deal with market cycles, partly to honor employee commitment to the company, and partly to protect itself from losing its investments in people. *Learning and teaching* are seen as an integral part of the job: employee responsibilities are defined to include their own learning and that of their team and peers. Unimerco pays the tuition for employees attending external educational programs, and although a small company, a wing of the company — the Unimerco University — has been established to support learning activities for employees, customers, and suppliers. *Experimental* hands-on learning is facilitated through reserving a full-scale processing unit for experimental testing of new solutions. Unimerco's values are reinforced by *the architecture* of the facilities which have open space layout integrating administration, sales, R&D, executive functions, and production in the same room. Potential noise problems are addressed with advanced acoustics. A variety of informal meeting places from hubs, to coffee venues, to gyms encourage interchange. The commitment to design and quality is reflected in company investments in facility aesthetics. This architectural design has proven to be so fundamental to the workings of the organization that the company has chosen, at real cost, to replicate it in all subsidiaries.

The internal emphasis on partnerships is extended to *relationships outside the company*. The largest and fastest growing proportion of Unimerco's business is with *customers* who are also *partners*. Unimerco commits to the needs not only of the customer, but also of the customer's customer. Wherever Unimerco sees a possibility for improvement in the customer's production, that knowledge is shared. And increasingly Unimerco is learning from its customers. Customer employees will often take part in internal Unimerco training programs. *Supplier relationships* have undergone a similar transformation.

246   As a rule, Unimerco pursues privileged partner relations with its machine and equipment suppliers. These suppliers are the drivers of significant new knowledge to Unimerco through new generations of their technologies, while Unimerco serves as their primary test user and also as a primary source of new needs and specifications. *Competitors*, increasingly, end up as network partners for Unimerco, which add scope in joint client offerings just as there is a mutually necessary sharing of knowledge. Finally, Unimerco is always involved in a number of intense collaborative projects and extensive knowledge sharing with universities, business schools and industrial research institutes that are *knowledge partners*.

    These organizational changes have increased Unimerco's competitiveness by optimizing knowledge processes. To understand how and why this has worked we need a much more substantial understanding of knowledge.

UNDERSTANDING KNOWLEDGE
AND KNOWLEDGE PROCESSES

The Nature of Knowledge Processes

There are several sources of knowledge about knowledge, including new thinking within the theory of knowledge,[2] a number of different attempts to rethink the theory of science,[3] significant new insights arising from so-called second-generation cognitive science,[4] and some outstanding attempts to understand knowledge processes in companies, particularly those tightly linked to practical experience from knowledge-intensive companies.[5] The sum of these insights is that knowledge is contextual, selective, and segmentation dependent, concept dependent, and metaphorical. Let us briefly discuss each of these key characteristics.

    Knowledge is contextual. No knowledge can be true and no knowledge can be linked to reality independent of context. A statement that appears universal is necessarily an abstraction and can only be true if interpreted *into* a context through a knowledge-adding process. The links between a statement and its context — both in terms of what defined its origin and in terms of its potential application — are multiple and complex but definitely include the fact that knowledge is embodied, that it is personal, social, and historical.[6]

    Knowledge is selective and segmentation dependent. Knowledge reflects a part of reality, the whole of which is infinitely complex and can be perceived in multiple ways. Any perception and any form of knowledge represent a selection of what is relevant and pertinent and what is not. Because of this fundamental relation all knowledge will be abstractive, reductive, and in need of interpretation. Selection determines what is in focus, what is subsidiary, and what is passive background. Selection determines the level of segmentation: are we looking at a physical system of mass and energy, at atomic configurations, at a set of biochemical processes, at biological creatures, or at a social situation? What constitutes the relation between part and whole, not to mention the relation between selected subset and all other possible selections — between that which has been selected and whoever made the selection?[7]

Knowledge is concept dependent. Whenever knowledge is expressed it is dependent on the language in which it is expressed. The content and meaning of a statement vary with its language. At one level this is a matter of the degree of incommensurability between natural languages. At another level we are capable of defining multiple conceptual systems within which we can describe the same phenomena. Each conceptual system is equally consistent internally. However, making the choice to shift, for instance, from Euclidean to non-Euclidean geometry not only results in another physical theory, but also in very different metaphysical assumptions about the nature of the universe and of reality.

Knowledge is metaphorical. This dependency on concept in part follows from the fact that a very large proportion of the concepts we use to talk about our reality are metaphors. Metaphorical concepts are not literal in meaning, and they have several nonidentical meanings and uses, most of which cannot be eliminated even in formal use.[8] This is true also of many of the basic concepts we utilize to formulate and express scientific knowledge such as space, time, cause, and effect.

These four characteristics of knowledge have radical consequences for our understanding of knowledge. Knowledge is not the insight of an autonomous, neutral subject into an independent objective reality. Rather, knowledge is a function of the inalienable unity of an embodied subject and the world in which it is objectively embedded and takes part. The process of codification is reductive, selective, and abstractive. Therefore, any codified form of knowledge is incomplete. To reconnect it *into* context requires an active process, where knowledge that is not included in the codified form is added. The process of application is not one of logical deduction, but of value-adding interpretation. The process is much more than deduction, because the codified form of knowledge cannot include context, criteria of selection and segmentation, idiosyncrasies of concept, or the multiple meanings of metaphor. All of these have to be added to complete codified knowledge as knowledge.

Further, neither the specific selections on which knowledge is based nor the context on which it is dependent can be expressed as part of the codified statement itself. Any codification of knowledge is constituted in conditions

248   that must remain tacit within the given framework of codification. Therefore, knowledge is not and cannot be primarily codified, just as there cannot be a process of increasing codification of knowledge. Growth in codified knowledge is part of the growth in the full body of knowledge, so new codified knowledge will always be just the tip of the iceberg. Codified knowledge in and of itself is not knowledge at all. The argument could be made that tacit knowledge in a rudimentary form would qualify as knowledge — although in that rudimentary form even animals could be argued to be knowledgeable. However, we believe that tacit knowledge becomes an aspect of knowledge as a historical and economic phenomenon only when it is part of a dynamic unity with other forms of knowledge, including codified knowledge. Tacit knowledge in itself is constrained and hardly capable of development.

Any codified knowledge is only knowledge on conditions that cannot be included in its codification and that must remain tacit within this system of codified knowledge. There is no rationally complete system of codified knowledge.

Revisiting the Tradition: Theory
of Knowledge, Theory of Science

This understanding of knowledge is radically different from the standard textbook description, which can be paraphrased: "Science is unified and rational. The development in codified science represents the growth in human wealth and the progress of civilization. Independent and objective knowledge is achievable, and mankind is rising toward an ever more complete insight into reality."

However, the radical difference is less that our understanding is controversial than that textbook ideology ignores the fundamental challenge of theory of knowledge and theory of science. From Plato onward, theories of knowledge are characterized by the attempt to justify knowledge achieved by an autonomous subject about an independent object. Plato argues that no concrete object can be identified and individuated from any number of universals, and no universal can be induced from the specific. His solution is to relegate any specific object to a status of secondary reality.[9]

Descartes finds certain ground for knowledge in the disembodied mind, but again at the cost of what can be known about any reality that is not cognitive.[10] In the empiricist tradition, from Locke to Hume, only the particular independent objects have primary reality. This means that only their primary, nonqualitative properties are real; and since nothing universal can be known, in the end the existence of the world cannot be stated with certainty.[11] Kant attempted a solution through his claim of synthetic a priori knowledge. But even Kant himself recognizes that this still does not solve Plato's problem of the impossibility of bridging the gap between the universal and the particular.[12]

In this sense the tradition of theory of knowledge can be seen as a long line of courageous and very consistent attempts to justify the possibility of knowledge by a subject of a reality that is independent and separate from it. All attempts have turned out to be in vain. The tradition becomes an extended effort to demonstrate that if subject and object are separated and disjunct, then objective knowledge — that is, knowledge about the object that is true of the object and independent of the subject — is not possible.

Obviously, this conclusion pertains to the very possibility of rationally complete systems of codified knowledge, and hence to our very concept of knowledge. Although our statements at the end of the previous section about the nature of knowledge might seem radical compared to the popular textbook ideologies of science, they are not radical at all compared to often painful conclusions within the great tradition of the theory of knowledge.

The same is true when we look at the theory of science. After the breakdown of the Newtonian paradigm and after the failure of conventionalist and logical empiricist attempts to save the idea of objective, independent, and self-contained rationality and truth, all of the twentieth-century theory of science has been in more or less orderly retreat from that concept. From the beginning of the century it was clear that final verification is not possible. Popper's attempt to save the possibility of subject independent knowledge through the imperative of falsifiability proved equally insufficient. After Kuhn, the absolute commensurability of scientific theories had to be abandoned.[13] Finally, it has become clear that one level of segmentation of reality (for example, a molecular level), might be a necessary but never a sufficient condition for phenomena at a higher level (such as biological or social phenomena). This means that any attempt to reduce the explanation of all reality to one unified science is futile and false.[14]

All this has important practical consequences. One reason for the inadequacy of the dominant schools of knowledge management is that they tend to be more aligned with textbook myths about knowledge and science than with either serious theories of knowledge and science or recent findings of cognitive science. The failure to understand the nature of knowledge results in a failure to lead and enhance knowledge processes in companies.

The System of Knowledge: Dynamics and Main Forms

We can now turn back to our project of developing a dynamic concept of knowledge, which will allow us to understand better the knowledge processes observed in Unimerco.

From our four basic characteristics of knowledge, we can define the internal dynamics and the main categories of our knowledge system. While it makes sense to differentiate the concepts of tacit and explicit knowledge, we have seen that the two are not independent of each other, but dynamically linked. Whenever we seek to reflect or work on a dimension of tacit knowledge, the very nature of this operation includes bringing aspects of the tacit into focus and thereby beginning the selection and abstraction process, which in itself is a step toward codification. And whenever we codify or even express knowledge, the codification or the expression is dependent on conditions that are by nature tacit. These dynamics within the system of knowledge prove to be the constitutive feature of the growth, creation, and application of knowledge.

Before investigating these dynamics in more depth, we need to discuss some of the main forms of the system of knowledge and thereby some of the forms in which knowledge appears. The main distinction is between tacit and explicit knowledge. Tacit knowledge is concerned with all those areas of know-how in which we know what is needed to accomplish intended changes without being able to express the knowledge involved. It is clear that by far most of what we do in everyday life is based on the activation of tacit knowledge. It is also clear that much of what in one situation is tacit can be expressed in other situations. For example, the fact that I am right now focusing on what I write and therefore pay no attention to and cannot express the physical process of writing does not mean that it is impossible for me in another situation to focus on the writing process and actually make much of the knowledge involved in that process explicit.[15]

There are, however, many indications and quite a bit of evidence suggesting that some of our tacit knowledge is not only *unspoken*, but also *unspeakable*. This evidence includes observations in brain physiology of the disjunct location of centers of implicit and explicit memory; the fact that whereas the human brain processes a minimum of eleven megabytes per second, we consciously handle a maximum of only forty bytes per second,[16] as well as the logical and empirical observation that the basic knowledge involved in our language capability cannot in itself be expressed.

Another key observation about tacit knowledge is that in its pure tacit form it is not only contextual in a concrete sense, but also absolutely confined to its context. This implies an absence of reflection and a very weak and limited potential for growth and improvement in knowledge. A basketball player can improve his or her layup skills by training and multiple repetitions, but without comparison of success from day 1 to day 10, some comparison with other players, and some reflection on what made the first attempt successful and the second not, improvement is likely to be random and limited. But comparison involves selection, abstraction, and, if not an outright explicit expression of the knowledge involved or aspects thereof, at least the ability to think and talk about some part of the knowledge.

Though there is yet no explicit knowledge, there is a dimension of spoken-about knowledge wherein comparison and reflection are enhanced by anecdotes, by incomplete linguistic expressions, or by metaphors and metonyms. Anyone can get an illustration of the richness and pervasiveness of this spoken-about category by observing a group of children on a basketball court or a group of adults cooking a meal together.

Just as tacit knowledge bifurcates into unspeakable and unspoken, explicit knowledge can be either spoken or codified. Spoken knowledge is verbally expressed knowledge that has not been formalized within a strict formal code.[17] Spoken knowledge plays a significant role both in the process of making knowledge explicit and communicable and in the process of interpreting codified knowledge toward applicability. Very large bodies of knowledge exist in spoken form without ever having been codified. Because attention has been focused overwhelmingly on the codified knowledge of formal science, there has been a tendency to overlook the role of spoken knowledge in the overall system of knowledge. But we must account for it in order to get a full understanding of the richness and complexity of knowledge processes.

We have already discussed the tendency to focus on codified knowledge, to identify codified knowledge with "real" knowledge, and to assume that over time more knowledge will be codified. There is even a tendency to define the concept of the knowledge-based economy through the assumed triumph of codified knowledge. This is a misconception. Codified knowledge can be generated only in a process that starts in a dependence on context, which cannot be expressed within the codification. Essential to this process are selection, abstraction, and reduction, and again the criteria and conditions for such selectivity cannot be expressed within the codification. Finally, despite the apparent formality of the system of codification, any statement will always be language dependent and fundamentally reliant on metaphor and metonyms. We have seen that this implies dependency on nonliteral and multiple meanings constituted outside the framework of codification and even beyond what can be made explicit. In other words, all codified knowledge is incomplete in its codification and complemented only through the addition of knowledge, which within this codification is necessarily tacit. As a consequence, all codified knowledge must be interpreted and added to before it is completed as knowledge, and much of this value adding comes from tacit dimensions of knowledge.

This does not detract from the fact that codified knowledge represents a

252   huge amplification of the reach and scope of our overall system of knowledge. The amplification of tacit knowledge inherent in our basic ability to speak about it is multiplied many times over — and progressively so — as tacit knowledge is exposed to the challenge of codified knowledge. We must not forget, however, that this challenge can be met only by our ability to interpret and add value to the body of codified knowledge in order to synthesize back through the layers of abstraction and selection back to the now reflected context of application. The ability to do this is mostly tacit, and in terms of the framework of codification it is always tacit.

Forms and Processes of the Knowledge System

We are now getting to the core of what characterizes the knowledge system and its dynamics. Tacit knowledge can be enhanced, leveraged, and amplified by being spoken about, spoken, and codified. Any step in this direction enables reflection and amplification. But any step also implies selection, reduction, and abstraction. Therefore, spoken and codified knowledge is incomplete and dependent on the addition of tacit aspects in order for it to be related back to reality and context and thereby to be applied. Growth of knowledge and the key knowledge process in the knowledge-based economy is not the growth of codification, but the intensified exposure of codified knowledge to the tacit and spoken-about interpretation and application, and the reflection of tacit knowledge by the challenge of codified knowledge.

The defining characteristic of the knowledge-based economy is the possibility of a much more intensive and iterative reintegration in processes back and forth between the main categories of the knowledge system.[18] Several points follow from this core insight.

First, whereas some methodological attention has been paid to the process of codification, little work has been done on how to enhance the process of synthesis and interpretation. This also means that the method of codification needs to be revisited, so we need to develop a new methodology of the complete knowledge process.

Second, it follows that no specific form of codification has a given priority over other forms. It is legitimate to recodify a body of knowledge if a different codification promises to be more fruitful in the interaction with tacit and spoken-about knowledge and application. We will see specific examples of this kind of recodification in Unimerco.

And third, in a radical interpretation of the characteristics we have assigned to codified knowledge and its dependence on metaphor, all explicit knowledge is spoken-about knowledge rather than spoken or codified knowledge.

Since reference and meaning are not and cannot be consistently literal and definitive, even our most codified statement of knowledge is less an explicit direct expression of knowledge than it is another way of speaking about knowledge.

Certainly it makes sense at least for conventional and heuristic reasons to stick with the categories of explicit, spoken, and codified knowledge. However, the category of spoken-about knowledge gains further importance. We introduced it as an intermediary step in the process of making knowledge explicit and allowing reflection of tacit knowledge. We see now that spoken-about knowledge is a legitimate step in the process of interpreting and adding value to codified knowledge. To be able to speak about knowledge can be a necessary hybrid that enables us to enrich and thereby apply codified knowledge much more quickly than would otherwise be possible. The nonliteral, multifaceted meanings of the metaphors inherent in codification not only allow for hybrid, spoken-about forms, but make them inevitable.

The system of knowledge unfolds as the dynamic interplay among the categories of tacit, spoken-about, and explicit knowledge. Within this dynamic, tacit knowledge can be unspeakable as well as unspoken, and explicit knowledge can be spoken as well as codified. The dynamic is defined by the interplay of a process of selection, abstraction, reduction, and reflection going in one direction, with a process of interpretation, value adding, and application going in the opposite direction. Spoken-about knowledge emerges as the crucial, enabling form in the interplay that allows the process to intensify and accelerate. The system is illustrated in Figure 12.1.

There are two addenda. The first concerns *embedded knowledge*, which is previously codified knowledge that has been embedded in tools, processes, and technologies. It is a category of enormous and growing scope and impact. A simple example is the way the rules and procedures of calculation are embedded in a pocket calculator. Prior to being embedded, they represented codified knowledge. Once embedded, this knowledge was at the disposal of a large number of users, who could do quicker and more complex calculations than people who did not have access to the tool with its embedded knowledge. A subset of the knowledge compiled in society is available to everyone at the very limited cost of learning to use the tool.

In this sense we are dealing with knowledge in a form that is neither explicit nor tacit in any trivial sense. The knowledge embedded has always been explicit and codified before it is embedded. However, for the user it is very rarely explicit in its full scope and complexity, and the tacit knowledge required of the user is tool- and use-related knowledge rather than the knowledge that was embedded in the tool. This is the reason anyone with mastery of

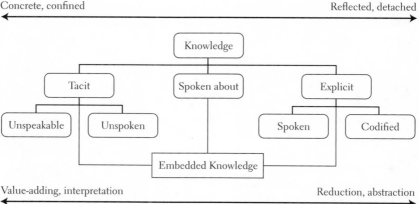

FIGURE 12.1   The system of knowledge and its forms

a pocket calculator can outperform someone using a pen and a notebook for calculations but will most likely underperform if expected to do the same calculations without the tool.

All societies have developed pools of embedded knowledge, some of it embedded in social structures and practices, some of it available in the form of tools and technologies. But the relative dominance and impact of embedded knowledge in industrial and postindustrial societies are mind-boggling no matter how we look at it from the perspective of everyday life, in terms of the knowledge activated in any company, or in terms of the functioning of society. We are far from fully understanding how much of socialization is concerned with building the kind of technology literacy that enables most members of a modern society to activate and leverage a significant scope of embedded knowledge. There is also quite a way to go before we understand the difference between those companies in which embedded knowledge makes employees less competent and more dependent, and those in which embedded knowledge is a real lever for the active knowledge of employees as well as for the company.

The second addendum is the distinction between social or organizational and individual knowledge. Though it is often asserted that tacit knowledge by its very nature is vested in individuals and difficult to share, there are clearly forms of tacit knowledge that belong to an organization as well as to individuals. Just think of the numerous everyday tasks we are competent to perform as part of a team or in a specific social context but could not even start to deal with outside that team or context. Capabilities of that nature seem to represent organizational tacit knowledge. At the other end of the knowledge system, codified knowledge in the form of scientific knowledge appears to be essentially defined by its public and hence inherently social status. But as we have seen, codified knowledge, including scientific knowledge, is only completed as knowledge by the addition of dimensions that are tacit and therefore not obviously social.

Once again we come to the realization of greater complexities in knowledge than expected at first glance. We shall not attempt to map the distinction between social and individual knowledge onto our illustration of the system of knowledge (how can this type of complexity be mapped in two dimensions?), but this does not mean that the distinction can be ignored in further work.

With these additions — of embedded knowledge and of the distinction between social and individual knowledge — we should be ready to take our picture of the knowledge system and its dynamics and use it to describe and understand the actual challenges of knowledge processes in companies.

THE COMPANY IN THE KNOWLEDGE-BASED ECONOMY

Knowledge Processes in Companies

Our analysis up to this point explains why knowledge management has had such limited success in companies. The overemphasis on codified knowledge gives priority to database and communication systems that encourage codification, documentation, and storage of organizational knowledge. The failure to recognize the dependence of codified knowledge on the other forms of knowledge makes the difficulties of this approach inevitable. Codification has meant reduction of context and meaning, and as a result the documented knowledge tends to be underutilized.

Why produce what no one is going to use? Few have been motivated to make the extra effort to codify and document their knowledge. The consequent proliferation of incentive schemes to encourage employees to document and reuse knowledge testifies to the depth of the problems. Without a clear understanding of the specific interconnection of the different forms of knowledge, there is neither a sensible way to select the right knowledge for the right form of codification, nor an obvious way for the organization to apply its codified knowledge.

Another significant school of thought, which is positioned in opposition to the knowledge management tradition, contends that organizations will only know and own what they know if they are able to share tacit knowledge and thereby raise it from an individual to an organizational status.

The roots of this school can be found in studies of apprenticeship learning and in anthropological studies of learning in traditional communities.[19] It documents how communities gradually come to include new members by involving them first in legitimate peripheral positions, from which they can observe

TABLE 12.1   Tacit knowledge and explicit knowledge

	Tacit knowledge	Explicit knowledge
Tacit knowledge	Socialization	Externalization
Explicit knowledge	Internalization	Combination

social practice, and then in more active roles, from supervised simple operations to more independent responsibility for complex and complete practices.
This process not only facilitates the learning of new members, but also en-
hances the sharing of improvements among established members of the community and the ability of the community to speak about the knowledge involved. Finally, it is a way of ensuring that the tacit knowledge of one member
becomes the tacit and spoken-about knowledge of the whole community. Such
analyses of communities of practice have helped in the understanding of the
practices of sharing tacit knowledge and of spoken-about knowledge in companies and have certainly enabled companies to remove constraints from the
sharing of tacit knowledge.

The weakness of this school of thought is that the communities of practice
it describes tend to be relatively static in knowledge, just as tacit knowledge in
itself is confined to its context. To grasp knowledge processes critical for the
knowledge-based economy, we will have to move beyond viewpoints configured exclusively around either tacit or codified knowledge.

In their book *The Knowledge-creating Company*, Ikujiro Nonaka and Hirotaka Takeuchi (1995) show that the crucial knowledge-creating processes in a
company are dependent on the sharing of knowledge: tacit with tacit; tacit
with codified; codified with tacit, and codified with codified. They describe
the process matrix as shown in Table 12.1.

Though we agree that this picture captures some valuable insights, we are
convinced that all the forms of the knowledge system defined above need to
be included in and add value to a full mapping of the knowledge processes.
The real study of knowledge processes becomes a study of the intensity of
the metamorphoses from form to form of knowledge, not once and for all, but
as the actual mode of innovating, enriching, and applying knowledge in a
company.

Mapping Knowledge Processes in Unimerco

Analyzing the knowledge processes of a company starts with mapping specific
processes within a five-by-five matrix of codified, spoken, spoken-about, tacit,
and embedded knowledge. Each element of the typology contains unique and
distinct processes. In the theoretical analysis of the categories and dynamics of

To / From	Codified	Spoken (explicit)	Spoken about (metaphors, anecdotes)	Tacit (unspeakable, unspoken)	Embedded
Codified	UM Cooperation with universities and IRIs	Theory (ion implantation) is concretized into a UM context	System revolution becomes agenda for everyone as "Keep It Simple" (example)	DB data to human process Designs converted into practical solutions in test centre	DB data direct to machine Designs converted into programmed solutions in test centre
Spoken (explicit)	In client interaction the ready spoken to codified gets crucial System building and DB	UM–University Value chain dialogue High-level customer dialogue	Internal communication about customer specification	Training programs and instruction	?
Spoken about (metaphors, anecdotes)	Metaphors function as 'black box' pointers within otherwise codified systems	Metaphors enrich explicit statements	Cross-fertilization of metaphors and anecdotes	Spoken about as catalyst of tacit learning	?
Tacit (unspeakable, unspoken)	Specific operation parameters identified, documented, and entered into DB	Point learners extract tacit knowledge from key persons and translate into training	Anecdotal evidence on specific in-the-field experience (case/war stories)	Apprenticeship Communities of practice	Experience based optimization of layout, etc.
Embedded	New technology specification add to body of codified knowledge	Training programs and manuals add to explicit pool of knowledge	Technology forms frame of reference for nonexplicit dialogue	Knowledge acquired in machines enables and updates work processes Knowledge imbedded in flow facilitates processes	Direct machine to machine interfaces

FIGURE 12.2   Forms of knowledge: The transformation matrix

the knowledge system, we found that spoken-about knowledge plays a crucial role in enabling a tighter and more accelerated integration from codified to tacit knowledge. Based on this we would expect processes evolving around the spoken-about category to be relatively important in any successful knowledge-intensive company.

We identified the distinct types of processes through a preliminary mapping of the knowledge processes in the day-to-day operations of Unimerco. Actual processes were identified for each window in the matrix, with the exception of two borderline cases related to embedded knowledge. Furthermore, as evidence that the full matrix is necessary, none of the identified types of processes were reducible to any of the other types. Results of our first mappings can be seen in Figure 12.2.

In any company there will be a number of processes that fit into each window. In a consulting engagement with one midsized company, close to one

258   hundred processes were identified for one single window. About half of these processes were well functioning, but the rest represented bottlenecks or interrupted the necessary knowledge flows. The following examples from Unimerco highlight important features of the actual processes as well as our understanding of the knowledge system.

Example 1: Knowledge acquisition. Unimerco acquired the company Dandia to get access to Dandia's unique mastery of diamond coating technologies and processes in order to strengthen the service offering to auto industry toolmakers. Initially, the acquisition was a disappointment: it turned out that most of the crucial knowledge in Dandia was vested in a very tacit way in just one person. To deal with this situation, Unimerco selected four point learners to work nearly full time with that person and to learn from him in a classical apprentice type of osmosis (tacit to tacit). Among these four bright and highly skilled persons the learning process evolved day by day into more reflected forms (tacit to spoken about) and eventually reached the point where they were capable of expressing most of the knowledge in spoken form. One significant driver in this process was the fact that Unimerco additionally allocated ten key people, along with the initial four learners, to come to Dandia every other Monday to work with and be trained by the old master. This cohort of ten is now the core team of Unimerco's growing diamond coating business, and their knowledge has spread to the teams working with them (tacit to tacit). The process relies on the language developed to speak about the knowledge (spoken-about to tacit) and on the explicitly spoken form it has taken in the training programs (spoken to tacit). Much of the practical learning takes place in work processes utilizing the advanced diamond coating equipment (embedded to tacit). The knowledge has been progressively codified in Unimerco's norm database (tacit to spoken-about to spoken to codified knowledge), which again is utilized in the performance of any contract (codified to tacit knowledge). During the internal process of codification, the Unimerco teams reached a point where their own codes enabled them to interact directly with and learn from engineering teams from equipment suppliers and from universities (codified-to-codified, often mediated in spoken-to-spoken-about and through embedded knowledge).

Example 2: Recodification of the production database. Although Unimerco has long recognized that critical knowledge is embodied in its people, the company has also emphasized the need for standardization, documentation, and codification of solutions. Substantial investment was made into the Unimerco tools database — a complete collection of all Unimerco tool blueprints, and hence the primary company vehicle for documentation, sharing, and reuse of knowledge. In recent years the database and its use began to be questioned. New tools were not documented in the database, or updates were delayed and made only after active managerial intervention. In reviews with customers there were complaints that Unimerco would try to push one of its existing solutions rather than listen to the real needs of the customer. Internally, the blueprints were sometimes seen as obstacles to continuous improvement of existing solutions. The massive body of codified knowledge was seen as overruling the living knowledge of the Unimerco teams. For these reasons the monumental decision was taken to recodify Unimerco's knowledge database through a fundamental change in segmentation principle. Instead of a database of complete blueprints, it now consists of a much more granular collection of Unimerco norms, each specifying the standards of one tool detail. The codification has gone from a collection of previously made statements to the complete alphabet and grammar utilized by Unimerco in constructing statements. The effects of this change have been very significant: The individual employee or the team has reclaimed mastery of the processes, and the database is the codification of existing organizational knowledge that leverages and can be leveraged in the execution processes. Continuous improvements as well as interaction with customers about solutions have been enhanced, and documentation of new knowledge in the form of elements of new solutions has become an integrated part of the process. Many of the granular components of the norms database have even come to life in the daily language of spoken-about knowledge within the company.

Example 3: Change of organizational systems. Unimerco was one of the pioneers in documenting and even certifying its management systems and procedures with certification of quality, environmental impact, and energy systems. Historically, this had enabled the company to win positions with high-end customers and led to better documentation and greater transparency in processes. But over time the systems began to be seen as counterproductive. They inhibited flexibility and tended to vest control and responsibility in systems rather than in teams. Therefore, in 2002 all the classical management systems were changed from explicit procedural prescriptions into systems of delegation of responsibility and accountability. This resulted in a major realignment of knowledge and responsibility within the organization, which empowered the individual employee and operating group, and enhanced the ability to change and innovate. All in all, it became a major cultural revolution in the company, involving all employees, and enabled through a very intense dialogue in which the change process and all its aspects were identified and spoken about under the evocative heading "keep it simple."

Example 4: The nexus of spoken-about knowledge. As seen in the "keep it simple" example, many of the very intense and complex knowledge processes in Unimerco seem to be anchored in spoken-about elements. Codified knowledge is domesticated into the Unimerco context by substitution of formalized statements with well-known context pointers, metaphors, or anecdotal refer-

260   ences. Learning of tacit knowledge in daily practice or during apprentice arrangements is clearly leveraged by an existing and ever-growing language to talk about key elements, including nonliteral reference to norms from the shared database. Complex changes in systems and codification become a common agenda when included in storylines. Very explicit new customer specifications are interpreted into the Unimerco community of practice in a "spoken-about" shorthand. New practical or field experience is shared in the form of war stories and anecdotes. Large chunks of embedded knowledge are handled through a sophisticated system of contextual reference to practically mastered technology. All in all, there is hardly any core iteration between tacit and explicit knowledge that does not rely on the very rich fabric of Unimerco's system and language for speaking about knowledge. This is true not only of all internal knowledge processes, but also of processes involving external partners or sources of knowledge. In some cases the external partners are in such close and long-term contact that a shared language for speaking about knowledge has had time to evolve. In other cases Unimerco gives high priority to the process of interpreting the codified form in which the external knowledge can be accessed into the internal universe of spoken-about knowledge.

   Though the empirical study of the knowledge processes in Unimerco is preliminary, the evidence suggests that it would not be possible to conceptualize the complexities of the overall processes without at least the framework we have utilized. Much of Unimerco's competitive success is based on its superiority in enhancing the iterative knowledge processes in the complex web between tacit and codified knowledge forms. This superiority is due to conscious as well as intuitive organizational design aimed at optimizing these processes. Spoken-about knowledge is the crucial nexus in the optimization of the iterative processes, and many of the key organizational features of Unimerco have combined to create an unusually rich environment for generating an appropriate language and culture for it.

Optimizing the Knowledge Processes

It is clear that any company needs to optimize its ability to handle knowledge processes in order to sustain success, especially by optimizing the bridges between tacit and codified knowledge. The solutions presently offered by consultants and IT providers under the broad heading of knowledge management fail to provide this link, mistaking codified knowledge for knowledge in total. As a result, these solutions offer little beyond different technological fixes and organizational incentive schemes to promote the capture and dissemination of codified knowledge.

   Neither do the recommendations by the communities-of-practice school enable companies to optimize knowledge processes. As we have seen, Unimerco is a master in this dimension, yet that alone would never be enough to succeed as a knowledge-intensive company in a fierce and ever changing competitive environment. Our recommendation to companies is to take the best from the two schools — knowledge management and communities of practice — and rethink it by forcing them to intersect and by exposing both to the challenges of developing the real internal and external iterative unity of knowledge.

Meeting these challenges involves the design and optimization of *knowledge enablers*; a system of *knowledge receptors*, including *embedded bridges*; and finally, based on this, the careful optimization of *knowledge processors*.

Nonaka and Takeuchi stress knowledge enablers and the need to rethink knowledge management as "enabling knowledge creation."[20] They address the fact that whereas knowledge cannot be commandeered and is hard to control, people do gain knowledge and, if enabled to do so, will share and create new knowledge as part of any work process. Most companies, even those that emphasize the importance of knowledge in their strategies, tend to inhibit rather than enable knowledge, either because they assume knowledge can be managed within the control and command structures and incentives of a traditional hierarchical organization or because they fail to move beyond such an organization.

A knowledge-enabling environment can take many forms. When it is nurtured well, it becomes a source of self-reinforcing knowledge processes. Without a good and vital knowledge-enabling environment, a company will not succeed in competing based on its knowledge processes even if all other conditions for creating a successful knowledge-intensive company are fulfilled.

Knowledge receptors are the sum of all those factors in and about a company that make it capable of assimilating new knowledge without a full, new knowledge acquisition, adaptation, and interpretation process. The factors include established procedures, practices, and relations, and the installed base of already mastered technology, with its embedded knowledge. They also include the language to speak about certain types of knowledge as well as the existing internal codifications and their ability to be expanded to hold new bodies of knowledge.

A simple example with very high real-life impact is the arrival of a next-generation software driver from a supplier for one of Unimerco's tooling centers. The driver might represent ten or more person-years of new knowledge, but at Unimerco it is downloaded on existing machines; it works within both the existing user interfaces and the existing links to the Unimerco norms database. In effect, significant new knowledge is quickly added to Unimerco's capabilities and converted into functional improvements of customer solutions at

262   low cost to Unimerco. Much of the work of mastering this knowledge was done in the past and now is embedded in competencies, procedures, databases, and technology. This embedded knowledge serves as a highly effective receptor of new knowledge. This entire installed base provides a valuable infrastructure for Unimerco and saves the company from having to navigate uncharted territory. Thereby it becomes a powerful barrier to entry by new competitors.

There are two prominent risks for companies with well-developed knowledge receptor systems. The first is a day-to-day risk of lowering the visibility of newly introduced knowledge. When an organization is capable of seamlessly adapting new knowledge, the reception becomes a non-event and is paid less attention. It is well known from office organizations that significant new functionalities in software updates are underutilized when the updates take place overnight and without noticeable changes in the user interface.

The second risk occurs more rarely but has momentous consequences. The more developed and embedded the passive receptor of a company is, the more effectively the company deals with continuous growth in knowledge within its domains. At the same time, however, the company becomes all the more vulnerable to discontinuities. The more the company has relied on the strength of its passive receptors, the less it can deal with new knowledge that falls outside the scope of these receptors. This risk spells a very important management trade-off: short-term optimization will always be at the cost of longer-term flexibility and innovation. This is one of several reasons for the re-codifications undertaken by Unimerco, where the change in segmentation of the knowledge database of the company clearly weakened its ability to absorb new solutions completely within the passive receptor system, but the change in segmentation removed the non-event character of new knowledge and enhanced Unimerco's ability to adapt to discontinuities.

Knowledge processors are fully organizational in nature and are built around all the organizational groupings and practices that are active in each of the many knowledge processes. In Unimerco they include the way the company works with key clients and suppliers, setting up teams of key employees from both organizations and building their ability to interact and share knowledge over time to aid optimization of the client's production process. A further example would be the systematic and real-time communication of client specifications among persons working in sales, design, and production, as well as the knowledge distillation and dissemination process already described with point learners around diamond coating.

There are numerous examples of knowledge processes that are the result of genuine organizational innovation. Direct management intervention or long iterative processes of trial and error have dealt with suboptimal results (misunderstandings by production teams of customer specifications, unrealistic

promises of delivery times by sales engineers) and thereby driven changes in
practice, organization, or roles.

Knowledge enablers, knowledge receptors, and knowledge processors are
clearly all aspects of the Unimerco knowledge organization. They all interact
and they cannot be fully defined independently of each other. Though each
of these aspects will come to be better understood as more research is done,
our hypothesis is that they are needed to grasp the dynamic interplay within
knowledge-intensive companies and open up ways to optimize knowledge
processes more effectively.

In many ways, Unimerco represents a puzzling paradox. On the one hand,
it has mastered the acceleration of knowledge processes in their iterations between tacit and codified forms of knowledge. On the other hand, the cost of
this achievement is an organization, a culture, and a set of competences that
cannot be easily reproduced, which means that Unimerco's most significant
growth constraints are the internal barriers: How quickly can the company
bring new people on board? How quickly can new subsidiaries be developed
and brought up to full Unimerco performance?

CONCLUDING THOUGHTS

What has our analysis of knowledge processes uncovered that will enable us to
reach the next level of understanding, which will be relevant for companies
trying to cope in the knowledge-based economy? We believe that this chapter
contributes three types of lessons: theoretical, practical, and historical.

Theoretically, we know now that the process of discovery involves abstraction and reduction from context, as well as selection and segmentation based
on criteria that are not included within the framework of the codified knowledge. The same is true of the choice of conceptual framework, as well as the
metaphorical nature of some of the key concepts. It will have very significant
theoretical impact to turn this new understanding of the knowledge system
into a positive theory of the components of a real "logic" of application: When
codified scientific knowledge can only be true and applicable through the addition of knowledge that is outside the realm of codification and is based on
the fundamental unity of subject and object, then science is fundamentally
dependent on understanding and meaning—not in the sense that the scientific theory is an expression of arbitrary subjectivity, but because the codified
scientific knowledge can be reconnected only with observation and applied
through the meaningful and interpreting contribution of an acting and per-
ceiving subject.

Practically, this insight into the process of application leads to powerful
new ways of optimizing application of knowledge in organizations. Experi-

264 ence with several companies — a consulting engineering firm, a financial services company, a large-scale consumer product company — suggests that tools and methodologies can be developed that help them identify and optimize their knowledge processes. The key is mapping and understanding knowledge processes in a way that allows the interplay between codified knowledge in its several forms — whether that codified knowledge is held and generated internally or obtained from external sources, including embedded knowledge in technology — with tacit practices through rich interpretive processes.

Companies must over time segment, focus, and build their network relations. They must access codified knowledge and codified substrates of their own knowledge and interpret it into its concrete tacit substance. Seen from the opposite end of the process, this amounts to an ongoing exposure of the company's confined tacit knowledge to the challenge of codified knowledge. There is no way to achieve this without developing a rich spoken-about knowledge culture that can mediate the iterations both internally and with long-term partners.

The historical perspective arises out of the present situation of companies: new competitors, new technologies, new ways of organizing production, and re-formulated business models all place ever greater pressure on firms to generate, obtain, and apply knowledge. Our diagnosis of the new intensity of knowledge processes gives part of the answer to the question, How revolutionary is the revolution? The case of Unimerco suggests some of the historical contours.

The internal effect of the described knowledge pressures and the resulting changes is in part that the traditional form of hierarchy is collapsing and that the separation of different functions is overlaid by processes of direct coordination and sharing of knowledge. The dependent and dispensable employee who acts under the direction of the managerial hierarchy that coordinates the different work functions now becomes the de facto owner of some of the knowledge without which the company could no longer compete. Only the employee can activate such knowledge; problems must be identified and their solutions initiated by those directly involved in the primary work process, and only the employee can master the complexity of knowledge involved in that specific process.

Externally, relations are no longer fully definable in terms of the market transaction. Procurement and relationships with suppliers have changed from a situation where companies dominated dependent subcontractors and put pressure on price to one where companies select suppliers or customers based on their unique knowledge and skills in domains, which the company can no longer attempt to master in house. Because the essence of either relationship includes the sharing of crucial knowledge over time, this new relationship cannot be based on unilateral dependency but must have continuity far be-

yond a transaction between procurement by one company and the sales function of another. Marketing and sales have transformed from mass marketing
and transactions to relationship marketing and the building of loyalty around
a company brand.

In a similar way, we can map changes in relationships with knowledge partners, with shareholders, and with stakeholders. All the relationships of the
company are transformed interdependently. For example, customer brand loyalty is inconceivable without a concurrent alignment of employees around
that same brand and its inherent values. The employees are, after all, involved
in all crucial touch points between the company and the customer.

We have seen how all the constitutive relationships of Unimerco as a company were transformed. We have also come to understand how the intensity of
knowledge processes leads to this transformation of relationships whereby the
very nature of the company is transformed. This further leads to deep changes
in the market institution as well. The scope of these transformations is a key
issue.

The answer to the overarching question about the significance of these
many shifts, however, seems clear: The revolution is as revolutionary in its
own way as the Industrial Revolution was.

NOTES

The article has been produced as part of the EU-funded FP5 project *Tracking the New
Economy Transformation* (IST-2001-37325). The theoretical insights are the result of
joint discussions based on the draft manuscript of N. C. Nielsen's forthcoming book on
the political economy of the knowledge society. The empirical work on Unimerco was
carried out by M. C. Nielsen as part of a Copenhagen Business School team project.
N. C. Nielsen is a non-executive member of the Unimerco Board.

1. Several examples of this can be found in Ruggles 1996; Edvinsson and Malone
1997; Stewart 1997. A majority of the articles on the subject issuing from the consulting world falls within the same mode of thinking.

2. From Maurice Merleau-Ponty (e.g., Merleau-Ponty 1962, 1964) and Michel Foucault (e.g., Foucault 1994) through Samuel Todes (e.g., Todes 2001) to Etienne Wenger
(e.g., Wenger 1998) and modern pragmatism.

3. Examples are Mario Bunge (e.g., Bunge 1959), Paul K. Feyerabend (e.g., Feyerabend 1975), Thomas S. Kuhn (e.g., Kuhn 1962), and Michael Polanyi (e.g., M. Polanyi
1958).

4. Exemplified by Antonio R. Damasio (e.g., Damasio 2003), Mark Johnson, and
George Lakoff (Lakoff and Johnson 1999).

5. For example, John Seely Brown and Paul Duguid (Brown and Duguid 2000),
Ikujiro Nonaka (e.g., Nonaka, von Krogh, and Ichijo 2000), and Lucy Suchman (e.g.,
Suchman 1987).

6. Prominently, George Lakoff (Lakoff and Johnson and 1999), Michael Polanyi
(e.g., M. Polanyi 1958), and Etienne Wenger (e.g., Wenger 1998).

7. Important positions are defined by Mario Bunge (e.g., Bunge 1959), Michel Foucault (e.g., Foucault 1994), and Michael Polanyi (e.g., M. Polanyi 1958).

8. This is, for example, argued by George Lakoff (Lakoff and Johnson 1999), Eugene A. Nida (Nida 1969), W. V. O. Quine (Quine 1953), and Benjamin Lee Whorf (Whorf 1956).

9. See Plato's dialogue *Theaitetos*, not least in the edition with F. M. Cornford's wonderful comments (1935), but also Koyre 1945.

10. Other than going to Descartes's own works, these points come out well in two very different commentaries: Lakoff and Johnson 1999, 391–415, and Rée 1974.

11. An analysis of the main early empiricist philosophers that brings out this aspect very clearly is J. Bennett 1971.

12. Other than going to Kant's *Kritik der reinen Vernunft* and his aesthetics, the two classics that analyze his position in this respect are Habermas 1961 and Lukács 1968.

13. A tour through several of the main positions in this development can be found in Lakatos and Musgrave 1970.

14. This is argued most strongly by M. Polanyi (1958) and by Bunge (2001, e.g., 167ff.).

15. See the brief discussion on focal and subsidiary knowledge below, and for an in-depth analysis see M. Polanyi 1958.

16. Manfred Zimmermann, quoted in Nørretranders 1991, 163–165.

17. We follow the main trend within the newer theory of knowledge by distinguishing between knowledge expressed in spoken, natural language and knowledge expressed in a more formal code as mathematics.

18. This insight can be deepened significantly by looking at the interconnections in a historical perspective. To do that is beyond the scope of the present article, but it is discussed extensively in N. C. Nielsen's upcoming book. Prior to the Industrial Revolution hardly any part of the knowledge involved in production processes was codified. As a matter of fact, all dimensions of knowledge involved were directly integrated in the person performing the work process. The upside of this was a certain type of integrity and dignity for the worker and in the work process; the downside was a very slow rate of innovation and productivity growth. During the Industrial Revolution significant parts of the knowledge employed in production became codified, removed from those performing the physical work processes, and disconnected from the involved tacit knowledge. The knowledge vested in production processes is too comprehensive and intense to be mastered by single workers. Hence planning and control has to be reintegrated with performance of the process. The specific forms of the division of labor, and the institutions developed to organize those forms, that came out of the Industrial Revolution are being changed in quite fundamental ways. This is one of the reasons the knowledge revolution we are discussing might be fundamental in the same sense as the Industrial Revolution was.

19. Prominent examples are Suchman 1987; Lave and Wenger 1991; Chaikin and Lave 1993; Wenger 1998. The school has strong roots in the pragmatist tradition.

20. Their previously quoted book, *Enabling Knowledge Creation*, provides a convincingly thorough mapping of all the many dimensions of knowledge enabling as well as a clear argument why knowledge management does not do the job. Since this article does not leave us room for even a brief overview, we encourage readers and most especially practitioners to seek inspiration and understanding directly from this book.

13

Pooling Knowledge

Trends and Characteristics of R&D Alliances in the ICT Sector

Christopher Palmberg and Olli Martikainen

Inter-firm collaboration is not a new phenomenon. What is new, however, is the rapid growth in such collaboration since the 1980s in parallel with the acceleration of technological change, the internationalisation of firms, and the globalisation of the world economy (Freeman 1991).

New forms of collaboration are also emerging as firms increasingly have to rely on external knowledge to cope with complex technologies on a global scale. Inter-firm collaboration is now often a core strategic activity and frequently extends to research and development (R&D), the most well-guarded, knowledge-generating processes within firms. At the same time, competition is intensifying in both the technological and the market domains. As a consequence, firms must manage knowledge pooling with their collaborative partners in their upstream activities while competing fiercely with these same collaborative partners in downstream markets (Dunning and Boyd 2003). The phrase strategic R&D alliance (hereafter, R&D alliance) is often used to describe this type of inter-firm collaboration (de la Mothe and Link 2002).

R&D alliances are especially pervasive in the ICT sector, and barely a day goes by without press announcements on the formation of R&D alliances between firms. There are several reasons for this. The digitalisation of networks, the application of the Internet protocol (IP), and the general convergence of data communications and telecommunications are blurring technological and industrial boundaries, disintegrating value chains, and reshaping business models in this sector (Bohlin et al. 2000; Li and Whalley 2002). These developments are forcing firms to pool their knowledge, as well as other assets, in order to span multiple technologies, industrial boundaries, and vertical and horizontal value chains. In the ICT sector interoperability of products and

268   services is of special importance owing to the systemic nature of innovation and the presence of network externalities. As a consequence firms also engage in R&D alliances during standardisation and in order to secure access to intellectual property rights to emerging standards.

Previous research unanimously shows that strategic R&D alliances indeed have grown in importance in the ICT sector (Hagedoorn 2002; Caloghirou, Ioannides, and Vonortas 2003). The aim of this chapter is to analyze the broader trends and characteristics of R&D alliances in greater depth as an indicator of knowledge pooling between firms, thereby providing insight into how ICT firms access and manage external knowledge as a part of their innovative activities in the ICT sector (complementing Nielsen and Nielsen, this volume). Towards this end, the chapter employs a resource-based viewpoint of the firm, since this framework offers one viable interpretation for why firms engage in knowledge pooling through R&D alliances. The empirical analysis is based on the world's largest database of R&D alliances (the CATI database) and seeks to update and elaborate on extant research based on the CATI database.

The second section of the chapter introduces a resource-based view on the formation of strategic R&D alliances between firms as a conceptual framework for analysing the trends and characteristics of R&D alliance formation in the ICT sector. The third section introduces the CATI database and presents the empirical analysis. The fourth section concludes the chapter and identifies some future research issues.

STRATEGIC R&D ALLIANCES FROM A RESOURCE-BASED VIEWPOINT

Theoretical interpretations of R&D alliances touch on the core issue of the coordination of activities in the economy. In standard textbook economics, coordination is achieved through market forces as mediated by the price mechanism. Questions related to why firms extend their boundaries and engage in knowledge pooling are largely ignored. The industrial organisation (IO) literature provides some insights, although the main concern is related to R&D alliances as a response to market failures or underinvestment in R&D (Vonortas 1997).

The discussion of the boundaries of the firm can be traced back to 1937 and the seminal article by Ronald Coase. He was concerned with why firms exist and grow, given the theoretical emphasis on the market as an efficient coordinator of economic activities. According to Coase (1937), market coordination is sometimes internalised in the firm owing to costs associated with the price mechanisms. These costs arise during the negotiation and conclusion of contracts related to exchange transactions.

The very influential transaction-costs tradition in economics elaborated
further on these basic Coasian insights. Williamson has argued that different
attributes of transactions give rise to different forms of coordination which
he calls "governance structures" (Williamson 1975, 1985; see also Williamson
1991, 1999). In-house activities are preferred when transactions are frequent
and uncertainties and asset specificities are high, since this raises the costs of
transactions. He refers to this type of governance structure as a hierarchy. In
contrast, outsourcing through markets is preferred under the opposite conditions, that is, when transactions are sporadic and there is less uncertainty and
asset specificity. Assets refer to the subject of the transaction, and they might
include knowledge related to R&D.

Although transaction-cost economics is useful for analysing economic coordination in general and the degree of vertical integration of firms in particular, this framework has limited applicability for understanding why firms increasingly appear to engage in knowledge pooling within a grey zone between
hierarchies and markets. One especially relevant critique of this framework is
that it neglects the complexities and dynamics of innovation. This is because
innovation processes are inherently wasteful: they involve trial and error, experimentation, serendipity, and related uncertainty. The knowledge underlying such processes is typically tacit and complex, making it difficult to articulate, let alone to contract and transact (Antonelli and Quéré 2003).

Teece (1984, 1986, 1992) has provided especially useful insights applicable
to analysing why and how firms pool knowledge, even in situations when transactions are sporadic and uncertainty and asset specificity are high. He complements transaction-cost economics through the concepts of an appropriability regime and complementary assets. An appropriability regime refers to
the environmental factors, excluding firm and market structure, which govern
a firm's ability to capture the profits generated by innovation. These factors
include the system of IPRs, the ability to maintain secrecy, and the degree to
which knowledge is inimitable or tacit (Levin, Alvin, and Sidney 1987). Complementary assets, which often reside outside the boundaries of an innovating
firm, allow an innovator to commercialise technologies. Teece (1986) distinguishes between generic, specialised, and co-specialised complementary assets. Generic assets are general purpose and do not have to be tailored to the
needs of the innovating firm, specialised assets imply unilateral dependence
between the firm and the asset, and co-specialised assets imply bilateral dependency and knowledge pooling.

Teece concludes that markets are a preferable governance structure when
complementary assets are generic and appropriability conditions are weak.
When the level of specialisation of complementary assets increases, however, the integration of such assets into hierarchies becomes more important.

270   This is especially true in consolidating phases of industry evolution, when technological development settles around a specific dominant design, process innovativeness and price competition become more important, and appropriability conditions tighten (Teece 1986). However, if complementary assets are co-specialised and thereby cause bilateral dependency between innovating firms in an industry, knowledge pooling in the grey zone between hierarchies and markets also appears as viable irrespective of appropriability conditions.

The extension incorporating knowledge pooling is discussed in greater detail in Teece 1992. Here the focus explicitly shifts to R&D alliances as a specific form of coordination defined as "a bilateral relationship characterised by the commitment of two or more partner firms to reach a common goal, and which entails the pooling of specialised [knowledge] assets and capabilities" (Teece 1992, 189). Thus, R&D alliances differ from transactions across the markets since, by definition, they can never be unilateral, having only one firm on the receiving side in terms of knowledge assets. They differ from hierarchies since they do not use mergers and acquisitions (M&As) to gain access to other firms' knowledge assets and capabilities, even though they might resemble hierarchies if they are equity-based and durable.

Teece's discussion (1986) of complementary assets also highlights the firm-specific nature of such assets as a rationale for why firms use R&D alliances to facilitate knowledge pooling in specific situations. As a consequence, it can be considered an extension of the resource-based view of the firm (Mowery, Oxley, and Silverman 1998). The resource-based view of the firm seeks to explain performance heterogeneity in firm populations. It assumes that firms vary systematically in the degree to which they control resources necessary for innovation and that these differences are relatively stable over time (Foss 1997). There is a relatively broad agreement within this literature on which types of resources are critical for firms to access and control for the sake of good performance. Critical resources should be idiosyncratically firm-specific, difficult to imitate, and imperfectly mobile across firms, and there should be imperfect competition for their access. This is typically the case for knowledge as a specific type of resource (Peteraf 1993).

The link between the resource-based view of the firm (Foss 1997) and the discussions by Teece (1984, 1986, 1992) is obvious, since the critical resources of one firm easily become complementary assets of other firms. This is especially true of the ICT sector owing to the systemic nature of innovation and the demands for interoperability between different kinds of equipment and networks. As a consequence, R&D alliances might be taken as an indicator of knowledge pooling between firms, based on critical resources or assets which

are co-specialised and complementary to the innovative activities of the firms 271
in question. This insight is the starting point for the ensuing empirical analysis of knowledge pooling in the ICT sector as viewed through trends and characteristics of R&D alliances.

DISTRIBUTION OF R&D ALLIANCES BY TECHNOLOGY FIELDS

The CATI database is a relational database containing data on the year of formation, nature, and technological content of some 15,000 alliances and some basic data on the partners in an alliance. The data is based on systematic reviews of major global business newspapers and journals (Hagedoorn, Link, and Vonortas 2000). The database only covers alliances that contain some arrangements for mutual transfer of technology or joint R&D. The definition used in the CATI database is thereby in line with the one suggested by Teece (1992), as quoted above. Systematic data collection for the CATI database started in 1987 and has continued until the present, complemented with sources from earlier years to extend the scope of the database back to the early 1960s.

The following statistical analysis defines the ICT sector in terms of the technological fields within which firms have formed new R&D alliances, as reported in the CATI database. We will focus here on three fields that can be considered at the core of the ICT sector: microelectronics, software, and telecommunications technologies. The field of telecommunications covers all essential technologies constituting fixed or wireless (cellular) telecommunications systems. The inclusion of microelectronics is motivated by the fact that telecommunications equipment suppliers are major users of components, which provide the brainpower of modern telecommunications systems. Telecommunications equipment suppliers are also major users of software technologies, especially since the digitalisation of wireless networks started in the late 1980s.

Data on the year in which R&D alliances were formed enables an analysis of trends in alliance activity over time. In the context of this chapter, it is taken as an indicator of the degree of knowledge pooling over time and across different technological fields. A general overview of broad trends across the core ICT fields is provided in Figure 13.1.

A relatively steady growth in the number of R&D alliances is evident in Figure 13.1, even though this becomes more erratic after the mid-1980s. By and large, this is consistent with general trends in the growth in the number of R&D alliances across all sectors covered by the CATI database (Hagedoorn 2002). Previous research also suggests that the growing role of the European

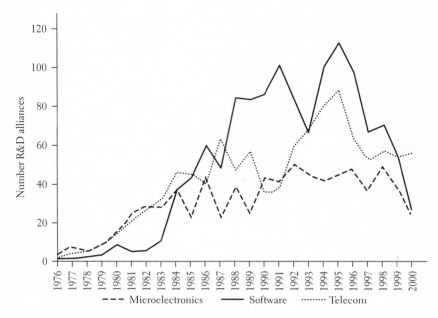

FIGURE 13.1   Growth of R&D alliances in the ICT fields, 1976–2000

Communities (later the European Union) as an initiator, financier, or coordinator of ICT-related R&D alliances through programs such as ESPRIT, RACE, BRITE, and EUREKA played an important role in the European context throughout the 1980s and early 1990s (Hagedoorn and Schakenraad 1993). The late 1990s witnessed a very clear decline in the number of R&D alliances across all ICT fields, with the exception of telecommunications. This decline appears to reflect the general downturn in the ICT markets at the turn of the century.

Apart from these general trends, there are interesting differences across technological fields. The rise of software technologies as an important field of R&D alliance activity during the late 1980s is especially clear, as is their rapid decline in the late 1990s. The peak and subsequent temporary decline in the number of telecommunications alliances in the late 1980s is also interesting. This pattern probably reflects the boom in the standardisation of digital cellular systems, especially in Europe. This boom was spearheaded by the intense development of the GSM (Global System for Mobile Communications) standard in the late 1980s prior to the inauguration of the service in the early 1990s. The so-called basket model applied in GSM standardisation led to the formation of several competing R&D alliances (Bach 2000; Bekkers, Duysters, and Verspagen 2002).

The peak years of telecommunications R&D alliances in the mid-1990s coincide with the development of the Internet protocol (IP protocol) as a new complementary technology to be mastered by the telecommunications operators and equipment suppliers (Organisation for Economic Co-operation and Development [OECD] 2000a). It also coincides with the development of the GPRS (General Packet Radio Service) standard as a platform on the path toward mobile packet data services and, eventually, the UMTS (Universal Mobile Telecommunications System, the European incarnation of the W-CDMA mobile telecommunications standard that defines the European vision for the 3G next-generation standard). After the mid-1990s, the telecommunications field showed less fluctuation and decline when compared to the other fields. One reason might be that increasingly complex standards have led firms to engage in fewer but larger alliances. Moreover, software is increasingly embedded in telecommunications technologies, so software alliances might be replaced by telecommunications alliances over time.

The differences across the technology fields might also be interpreted in light of broader trends toward convergence in the ICT sector. Specifically, the 1980s marked the introduction of commercial microprocessors and digital signalling processors (DSPs), thereby enabling the development of programmable devices such as the personal computer (PC). These developments are probably also captured in the fluctuation and increase in microelectronics related R&D alliances in the mid-1980s, followed by the rapid increase in telecommunications alliances as the major field of application (compare with Duysters, Lemmens, and Hagedoorn 2003). The very rapid increase in software alliances since the late 1980s could also be related to the development of standardised PC operating systems such as MS-DOS, Windows, and Macintosh (Hagedoorn, Carayannis, and Alexander 2001).

The fields in the CATI database can be disaggregated further into subfields that we selected for closer analysis (see the appendix for details of our technological classification scheme). In order to capture changes over time, we divided the data into two periods, 1960–1989 and 1990–2000 (Table 13.1). We sought to contrast the watershed developments related to the introduction of commercial microprocessors and DSPs in the 1980s to subsequent developments related to the IP protocol and the standardisation of digital cellular systems in the 1990s. The division of the data into these periods also divides the number of observations into two equally large datasets and thereby facilitates comparison.

A significant share of all R&D alliances in the ICT fields relates to professional software. Moreover, over time professional software has become an increasingly important field in the alliance activity of firms, accounting for

273

TABLE 13.1   Distribution of R&D alliances across ICT subfields, 1976–2000

	1960–1989		1990–2000	
	Freq.	%	Freq.	%
Microelectronics				
Processors	54	5	110	6
Chips	161	14	229	12
Other generic components	13	1	32	2
Multitech microelectronics	94	8	64	3
Software				
Standard software	64	5	134	7
Professional software	271	23	581	31
Other miscallenous software	0	0	36	2
Multitech software	86	7	84	4
Telecom				
Public networks	10	1	12	1
Private networks and services	120	10	199	10
Cellular systems	40	3	76	4
Other telecom systems	186	16	142	7
Multitech telecom	73	6	196	11
Total	1172	100	1895	100

much of the growth in software alliances illustrated in Figure 13.1 above. Interestingly, the field of cellular systems appears to have been subject to less alliance activity, although a closer inspection of the data reveals a strong upward trend since 1999 in the number of newly formed alliances. This upward trend coincides with the development of third- and fourth-generation mobile standards on top of the GSM.

THE POSITION OF FIRMS AND COUNTRIES IN R&D ALLIANCE NETWORKS

Examining the ranking of firms and countries by the number of alliances they have been involved in illuminates the changing position of firms' and countries' knowledge pooling in the ICT sector, even if a full-fledged network analysis is beyond the scope of this chapter (Hagedoorn and Schakenraad 1992; Duysters and Hagedoorn 1996). The logic here is that the number of firm alliances reflects superiority in terms of firm technological assets. Likewise, the number of alliances at the country level reflects the ability of countries to provide a suitable environment for firms engaged in collaborative R&D.[1]

TABLE 13.2   Ranking of countries by number of R&D alliances
in the ICT sector

1960–1989		1990–2000	
Countries	Freq.	Countries	Freq.
U.S.	1315	U.S.	2883
Japan	320	Japan	321
Holland	254	Germany	222
UK	207	UK	134
France	169	Holland	103
Germany	148	France	86
Italy	114	Canada	76
Sweden	41	Sweden	70
Canada	35	South Korea	33
Belgium	30	Finland	25
Spain	20	Taiwan	24
Switzerland	13	Israel	22
Australia	13	Italy	18
South Korea	12	China	16
Finland	11	Singapore	13
No. of countries	36		40
Mean	76.61		103.28
Std. dev.	226.54		455.40

Obviously, these rankings reflect size differences between firms and countries. Larger countries host more firms and are generally endowed with greater
absolute R&D resources and, hence, should be expected to rank higher than
smaller countries. Large firms tend to be more diversified and also have
greater R&D resources, whereby they should have better prerequisites to form
and manage multiple alliances. It should be noted that we have not been able
to account for these factors in the ensuing ranking lists, and hence they should
be interpreted with due care.

The presentation starts with a broader country ranking to cover the ICT
sector as a whole (see Table 13.2). Thereafter differences in firm and country
rankings (which are omitted from this chapter owing to space constraints)
across the ICT technology fields are discussed in greater detail. As can be seen
in the country ranking, the United States holds an indisputable top position in
the ICT sector as a whole, as a large country that hosts many important firms
in the sector and is richly endowed with R&D resources. This position has
strengthened significantly over time, even though some countries lower down

TABLE 13.3   Ranking of firms by number of R&D alliances
in the field of microelectronics

1960–1989		1990–2000	
Firms	Freq.	Firms	Freq.
Siemens	28	IBM	44
Intel	27	Toshiba	42
Texas Inst.	23	Motorola	37
AMD	20	Intel	35
Philips	19	Texas Inst.	30
Thomson	19	Siemens	22
National Sem.	19	NEC	20
Fujitsu	18	AMD	19
Motorola	18	H-P	18
Toshiba	17	Hitachi	18
Fairchild	13	Fujitsu	15
NEC	13	SGS/Thomson	13
LSI Logic	11	VLSI Tech.	13
AT&T	10	Sun Micros.	12
Olivetti	10	Philips	11
No. of firms	258		354
Mean	2.85		2.62
Std. dev.	4.29		5.19

the list show evidence of catching up in terms of the number of new R&D alliances. By and large, the top of the list has remained relatively stable, with bigger countries such as Japan, Germany, the United Kingdom, and France dominating. Germany has gained a few positions, while Holland, as the only smaller country in the top five, saw its position fall somewhat by the 1990s.

Nonetheless, further down the ranking list there have been greater shifts in the position of countries. A general tendency is the fall in the position of European countries, owing especially to the decline of Italy and Belgium. Finland is an exception owing to its rapid rise in the ranking list during the 1990s. This general European drop appears foremost to relate to the strengthened position of Asian countries, especially of South Korea and Taiwan. China and Singapore also entered this ranking list by the 1990s. The dispersion of R&D alliances across more countries is also reflected in the rising standard deviation.

Table 13.3 presents the top fifteen firms in the field of microelectronics during 1960–1989 and 1990–2000. As a general insight it should be noted that

most firms are traditional suppliers of microprocessors or miscellaneous electronics equipment, while software and telecommunications firms are relatively
absent. The top of the list is dominated by the large U.S. firms such as Intel,
Texas Instruments, and Motorola (which is also in the mobile telephony business), even though Siemens also has a strong presence. Other important firms
are the Japanese electronics conglomerates, most notably NEC, Toshiba, NEC,
Hitachi, and Fujitsu. These firms remained at the top of the list throughout the
1990s, although some shifts are especially noteworthy. Siemens' dominating
position deteriorated, while the positions of IBM and Toshiba strengthened
significantly.

By and large, changes over time are relatively slight, even though new firms
entered these networks. There is only a slight increase over time in the mean
and standard deviation of the number of new alliances formed by the firms.

The ranking of countries by the number of R&D alliances in the field of
microelectronics essentially confirms what was noted above for the ICT sector as a whole. The dominance of the United States is very clear and points to
the fact that smaller U.S. firms have also been frequent partners in knowledge
pooling. Japan ranks second in both time periods, as suggested by the firm
ranking, although it lags far behind the United States in the number of newly
formed alliances. More surprising is the relatively strong position of Holland
in this list, although this is largely explainable by the alliance activity of
Phillips. The positions of South Korea and Taiwan seem to be strengthening,
even though these countries do not host any of the top firms in terms of alliance activity in microelectronics. The presence of Finland from 1960 to 1989
is also notable. This relates to the emergence of Nokia as a significant lead
user of DSPs and chips during the early phases of GSM (Palmberg and
Martikainen 2005).

When turning to the field of software (Table 13.4), a striking feature is the
very dominant position of a few large U.S. computer and software firms,
namely IBM, Microsoft, and Sun Microsystems. These firms appear to have
been partners to a significant share of the increase in newly formed software
alliances in the 1990s, as indicated by a doubling in the number of their R&D
alliances. This is also reflected in the rising mean and standard deviation
shown in the table. The strengthened position of the U.S. computer equipment firms Hewlett-Packard, Oracle, and the microelectronics supplier Intel
is also noteworthy. These developments highlight the superior position of
large U.S. firms in software technologies. They appear to be consistent with
the suggestion above that the standardisation of operating systems restruc-
tured the industry and reshuffled the ICT sector in favour of U.S. firms, with
Microsoft as a case in point.

TABLE 13.4   Ranking of firms by number of R&D
alliances in the field of software

1960–1989		1990–2000	
Firms	Freq.	Firms	Freq.
IBM	40	Microsoft	89
Olivetti	27	IBM	87
Sun Micros.	21	H-P	79
AT&T	21	Sun Micros.	53
Microsoft	20	Oracle	41
DEC	20	Apple	36
Philips	20	Novell	34
H-P	20	Intel	33
Volmac	19	DEC	33
Bull	18	Netscape	20
Siemens	15	Compaq	18
ICL	13	Silicong	18
CDC	11	SAP	15
Fujitsu	11	Texas Inst.	15
Tandem	11	USL	14
No. of firms	444		781
Mean	2.25		2.33
Std. dev.	3.7		6.47

Apart from this concentration at the top of the list, there have also been some interesting internal shifts in the position of firms. The list reflects the fall in ranking of such firms as DEC, Phillips, and Bull, which were major minicomputer manufacturers with their own proprietary hardware and software that was incompatible with the Windows operating system. More significantly, the position of Japanese firms is much weaker in software when compared to microelectronics. Apart from Siemens and Motorola, the list does not include any other major telecommunications equipment producer. AT&T is the exception, and its position as an alliance partner diminished significantly after its breakup in 1984. Major entrants as alliance partners in this field during the 1990s were the U.S. firms Netscape and Cisco (presently a key player in supplying network components to the telecommunications industry).

The country ranking in the field of software also very clearly reveals the dominance of the United States in the ICT sector as a whole. The U.S. position strengthened significantly over time, even without larger firms such as IBM,

TABLE 13.5   Ranking of firms by number of R&D alliances
in the field of telecommunications

1960–1989		1990–2000	
Firms	Freq.	Firms	Freq.
IBM	36	Ericsson	49
AT&T	31	IBM	42
Siemens	25	AT&T	39
Ericsson	25	Siemens	37
Philips	19	Cisco	36
BT	17	Motorola	34
Olivetti	15	H-P	27
Fujitsu	14	Microsoft	26
NT	14	Intel	25
Plessey	14	Sun Micros.	25
DEC	14	DEC	20
CTNE	13	Alcatel	19
Racal	13	NT	15
PTT-Tel.	13	Novell	15
NTT	12	Nokia	14
No. of firms	363		605
Mean	2.82		2.29
Std. dev.	4.1		4.76

Microsoft and Sun-Microsystems. Other countries, including the United Kingdom, Japan, Germany, and France, fell far behind. The relatively strong positions of these countries are surprising since they do not show up in the firm ranking and thereby appear to be based on the R&D alliance activities of many smaller firms. The decline in the position of Italy probably reflects the decline of Olivetti. Interestingly, Sweden ranks among these top fifteen countries in both periods. Likewise, the position of Finland clearly strengthened in the 1990s.

A lesser degree of concentration at the top is evident in the field of telecommunications (Table 13.5). IBM, AT&T, the Swedish telecommunications supplier Ericsson, and Siemens dominate the rankings. Nonetheless, the 1990s witnessed the entry of firms with a core focus outside telecommunications, most noticeably Cisco, Microsoft, and Intel. Accordingly, these firms appear increasingly interested in forming telecommunications-based alliances, even though the alliance activity of telecommunications firms in the fields of microelectronics and software is less evident. The introduction of IP is a potential explanation, because it facilitates convergence between data communica-

280   tions and telecommunications. By and large, it seems that R&D alliances in the field of telecommunications are more or less evenly dispersed across the firms compared with microelectronics and software. This is reflected in the low standard deviation in both time periods.

The position of Motorola and Nokia strengthened significantly in the 1990s compared to the 1980s. From a Finnish viewpoint the rise of Nokia as an alliance partner in the 1990s is particularly interesting. According to Bekkers, Duysters, and Verspagen (2002) the rise of Motorola and Nokia in the rankings during the 1990s was most strongly correlated with the large share of patents they possessed that were regarded as essential to the GSM standard during the late 1980s and early 1990s. Motorola in particular was able to exercise an especially strong influence on the structure of subsequent alliance networks through its aggressive IPR strategy, resulting in the strong position that is also evident in Table 13.5.

The country rankings in the field of telecommunications also point to the strong position of the United States, even though its dominance is somewhat less striking in this field when compared to the ICT sector as a whole. Beyond the United States, the role of European countries seems to be more important in this field as compared to microelectronics and software. This is most likely related to the active stance on pan-European standardisation that many European operators and firms took throughout the 1980s and 1990s. The strong presence of Ericsson in telecommunications-related alliances is also reflected in the strong position of Sweden as a country, especially since 1990. Likewise, the alliance activities of Nokia elevated Finland closer to the top in the country ranking during the 1990s. Also noteworthy is the strengthened positions of China and Taiwan.

R&D ALLIANCES BY ORGANISATIONAL INTERDEPENDENCY BETWEEN PARTNERS

In the CATI database, alliances are classified by organisational interdependency. This classification reveals the particular mode of knowledge pooling that firms use in the ICT sector. A basic distinction is made between R&D alliances involving equity investments and those which are non-equity based. Equity-based alliances are broken down into joint R&D ventures and holdings. Non-equity-based alliances are broken down into joint development agreements or pacts, cross-licensing, or mutual second-sourcing agreements, whereby two or more firms agree to swap component technology specifications on a longer-term basis. Again, this generates insight into broader historical trends across core ICT technologies (Figure 13.2).

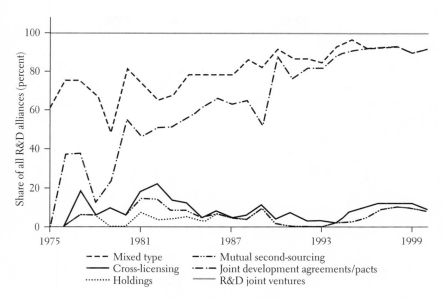

FIGURE 13.2   Organisation of R&D alliances in the ICT sector, 1976–2000

There is a very clear trend toward an increasing share of non-equity-based alliances over equity-based and mixed ones, as illustrated by the rapid increase in the share of joint development agreements or pacts since the early 1980s. Alliances in these core ICT technologies are becoming looser over time as measured by the organisational interdependency between partners and alliances (Table 13.6).

When the same distinction is applied to ICT subfields and changes over time, the rapid increase and prevalence of non-equity-based alliances is further underlined. Prior to the 1990s, around 45 percent of all alliances in microelectronics, software, and telecommunications were either joint development agreements or pacts. In the 1990s, that share rose to 80 percent across all subfields. The share of cross-licensing and mutual second-sourcing agreements was negligible. Nonetheless, the share of equity-based agreements (mainly R&D joint ventures [RJVs]) shows greater variation across the fields.

Apparently, relatively tighter organisational interdependency occurred in the field of telecommunications, as indicated by the higher share of RJVs, even though this share also declined in the 1990s. When disaggregating the data further to the subfield level, the share of RJVs was the highest (around 20 percent) in mobile telephony in both time periods. This could be explained by the paramount importance of standards in this field. These standards might

TABLE 13.6   Organisation of R&D alliances across ICT subfields
(columns sum to 100% by subfield)

	1960–1989		1990–2000	
	Freq.	%	Freq.	%
Microelectronics				
R&D joint ventures	62	19	43	10
Holdings	48	15	6	2
Joint dev. agreements/pacts	150	47	344	80
Cross-licensing	20	6	22	5
Mutual second-sourcing	34	11	2	0
Mixed type	8	2	14	3
Software				
R&D joint ventures	79	19	65	8
Holdings	115	27	32	4
Joint dev. agreements/pacts	222	53	686	82
Cross-licensing	1	0	27	3
Mutual second-sourcing	0	0	0	0
Mixed type	4	1	22	3
Telecom				
R&D joint ventures	144	34	63	10
Holdings	94	22	16	2
Joint dev. agreements/pacts	180	42	513	82
Cross-licensing	2	0	18	3
Mutual second-sourcing	0	0	0	0
Mixed type	9	2	17	3

call for tighter governance structures to facilitate intense knowledge pooling in complex technologies, even though firms subsequently compete in the markets created by the standards. Another interesting observation from a disaggregated viewpoint is that multitech alliances, spanning many subfields, were also characterised by a higher share of equity-based RJVs.

A CONCLUDING DISCUSSION

The pervasiveness of R&D alliances in the ICT sector touches on the core issue of why and how firms coordinate their activities and engage in knowledge pooling despite fierce competition in the downstream markets. Even though transaction-cost economics is a strong device for analysing governance structures in terms of hierarchies (in-house activities) and markets and the

determinants of the vertical integration of firms, both empirical and theoretical analysis of R&D alliances points to several limitations of this approach.
The clearest limitation relates to the complexities and dynamics of innovation, which appears to induce firms to engage in knowledge pooling through
R&D alliance. From the viewpoint of transaction-cost economics, R&D alliances fall into a theoretical grey zone between hierarchies and markets.

Teece (1986, 1992) offers a useful elaboration of transaction-cost economics
by emphasising the importance of complementary assets. Complementary assets are those external technologies and other resources beyond the control of
a firm which are necessary for the commercialisation of technologies. Complementary assets might be generic, specialised, or co-specialised from the
viewpoint of the innovating firm. In cases where complementary assets are co-
specialised, the bilateral dependency between innovating firms in an industry
increases and knowledge pooling through R&D alliances becomes viable, especially since such complementary assets typically are idiosyncratic to specific
firms, difficult to imitate, imperfectly mobile, and/or characterised by incomplete competition for their access (Foss 1997). In this chapter, we take R&D alliances as an indicator of the intensity and nature of knowledge pooling and
analyse trends and characteristics of such knowledge pooling in the core technological fields of the ICT sector.

An immediate and very clear result is that there has been a steady increase
in the number of new R&D alliances since the mid-1980s, followed by an
equally clear decline during the late 1990s (the field of telecommunications is
the only exception). Although this trend is compatible with previous research
on broader trends in alliance activity (Hagedoorn 2002), the drastic, albeit
temporary, downturn of ICT markets in the late 1990s appears to be strongly
reflected in the intensity of knowledge pooling. This trend might also conceal
other types of dynamics beyond the scope of this chapter. It is most probably
a temporary decline, since the R&D alliance activities of firms in the ICT
sector are sure to continue in the future.

Differences across the core ICT fields appear to be related to underlying
technological developments. We identified two such broader developments
that might also explain the clustering of knowledge pooling between firms
over time and across the different fields. The first one concerns the field of
telecommunications, where knowledge pooling appears to accelerate and decline alongside different phases of standardisation. Knowledge pooling was
especially intense during the late 1980s and early 1990s as a result of the development and commercialisation of the GSM standard. After a significant
decline in the number of R&D alliances, a new peak in knowledge pooling
occurred during the mid-1990s with the standardisation of the GPRS and

284 UMTS as well as the development of the Internet protocol. The second development relates to the watershed introduction of digital signalling processors (DSPs) in the 1980s, which coincided with a fluctuating increase in the number of microelectronics-related R&D alliances followed by a surge in R&D alliances in telecommunications and software.

Taken together, these observations confirm the theoretical insights discussed above, especially as they apply to the ICT sector. Standardisation is a necessary consequence of the systematic nature of innovation in the sector, as well as the means by which firms and other players realise interoperability among networks, products, and services. Hence, during intense periods of standardisation, the co-specialisation of complementary R&D assets increases and turns into critical resources for firms which use R&D alliances for accessing such complementary R&D assets in order to innovate and commercialise new technologies. Moreover, broader developments in the ICT sector appear to strengthen this trend and elevate alliances to an increasingly important position in the R&D strategies of firms, excepting the recent downturn in ICT market R&D alliance activity. Examples of these broader developments include the digitalisation of networks, the application of the Internet protocol, and the general convergence of data communications and telecommunications (Bohlin et al. 2000).

Beyond these broader trends, there are interesting differences across the ICT subfields. Significant fields of knowledge pooling include professional software, chips, private networks, and services. Despite the overall decline, the field of cellular systems has recently witnessed an especially rapid increase in the number of newly formed R&D alliances in recent years. Thus, there are reasons to believe that the intensity and nature of knowledge pooling in these fields differs from others in some important dimensions. One possible explanation might be the paramount importance played by cellular telecommunications standards and multiple IPRs to these standards. Previous research points to relationships between the distribution of IPRs, the behaviour and position of firms in R&D alliances networks, and market structures further downstream (Bekkers, Duysters, and Verspagen 2002). This suggests that R&D alliances also can enhance the IPR positions and performance of firms in downstream markets.

Shifting attention to the actual partners in R&D alliances, this analysis reveals a high degree of stability over time amongst a few leading firms and countries in the ICT sector. By and large, knowledge pooling is concentrated in a few leading firms and countries rather than in "second-tier" firms with weaker competitive positions in the downstream markets, although we noted a bias in favour of larger firms and countries in our rankings (compare with

Hagedoorn and Schakenraad 1992). With reference to the resource-based view of the firm that this chapter adopts, it thereby seems that these leading firms control important complementary R&D assets for innovation, and that their control of these assets has remained relatively stable over time. This is the case especially for larger U.S. firms, which appear to dominate in R&D alliances in the ICT sector as a whole. The U.S. dominance is especially clear in the field of software owing to Microsoft, although the United States also hosts a range of smaller firms which are frequent partners to R&D alliances. In contrast to this, it seems that European firms have lost ground while the relative significance of Asian firms is on the rise.

However, closer inspection reveals interesting shifts in the positions of individual firms and countries. A general observation is that the fields of microelectronics and software are subject to greater U.S. and Japanese dominance than the field of telecommunications, where European firms are more prominent. In the field of telecommunications, the significant positions of Ericsson, Siemens, and Nokia probably relate to the GSM breakthrough. This field is also characterised by the entry of Chinese and Taiwanese firms. A noteworthy observation is the rise of Nokia and Finland in software and telecommunications. South Korea is another new outsider alongside U.S., Japanese, and European firms, especially in the field of microelectronics. These shifts do also indicate that new complementary R&D assets emerge in the ICT sector and that the control of these assets is shifting over time at the fringes of R&D alliance networks. This appears to be true especially in the field of telecommunications in which the effects of digitalisation of networks, the application of the Internet protocol, and technological convergence in general are the most profound (compare to Duysters and Hagedoorn 1998).

When attention turns to the organisational modes that firms use for knowledge pooling, a clear trend toward looser types of R&D alliances is observable. In the 1990s the share of equity-based alliances fell to 12 percent, as compared to approximately 20 percent during the 1980s. In the 1990s over 80 percent of all R&D alliances across the ICT fields analyzed here were non-equity-based joint development agreements or R&D pacts. The share of cross-licensing and mutual second sourcing agreements was small. Nonetheless, a noteworthy result is that the field of telecommunications has witnessed a relatively larger share of equity-based R&D joint ventures as compared to microelectronics and software. Thus, firms in telecommunications alliances apply tighter governance structures in their R&D alliances. This might be related to the paramount importance of standards, which might call for tighter governance structures to facilitate more intense knowledge pooling in fields characterised by systemic innovation and demands for interoperability among

286   networks, products, and services. A concrete example is the GSM standard, which led to the formation of several competing alliances between firms, some of which were R&D joint ventures (Bekkers, Duysters, and Verspagen 2002; Palmberg and Martikainen 2005).

The analysis of R&D alliances in the ICT sector in this chapter also raises some interesting research questions for the future. One such question concerns the relationships between technological change and patterns of R&D alliances across technological fields. In this chapter, we suggest that the peaks and bottoms in the number of newly formed R&D alliances, as indicators of knowledge pooling, coincided with major technological developments in the ICT sector. Knowledge pooling in the 1980s appears to have been largely driven by the introduction of DSPs and the GSM. The 1990s appear to have been characterised by the development of standardised PC operating systems and the Internet protocol, strengthening the position of U.S. firms (especially Microsoft). Apart from investigating these relationships in greater detail, it would be interesting to map recent Internet protocol–based developments related to next-generation standards onto patterns of alliances. These recent developments are especially profound with respect to established value chains and business models and should thus also affect the intensity and nature in which firms pool knowledge through R&D alliances (Li and Whalley 2002).

The importance of standardisation in the ICT sector raises additional questions. According to Shapiro (2003), tensions between firms over overlapping patents are especially common in the case of standardisation, since firms thereby bring new technologies to the market. R&D alliances are one means to settle such tensions between firms. Nonetheless, the relationships between standardisation, patenting, and R&D alliances have received relatively little research. It is unclear how the distribution of patents — or IPRs more generally — relates to the structure and power constellations of knowledge pooling. Further research along these lines is important owing to the overall growing importance of IPR issues in the ICT sector.

Finally, and on closer inspection of the data, there is some indication that alliances, especially in the field of cellular telephony, have increased in size over time. The Nokia-led Symbian alliance is one example. Symbian includes multiple significant firms in the industry with a focus on developing a software platform for next-generation mobile phones. Examples such as these raise the broader question of whether competition between firms is being replaced by competition between constellations of alliances, as also argued by Duysters, Lemmens, and Hagedoorn (2003). Further research is needed on how knowledge pooling is managed within such multilateral alliances and how this relates to the other strategic intentions of firms in the ICT sector.

NOTES

This paper draws on Palmberg and Martikainen (2003a). We thank John Hagedoorn and Marc van Ekert at MERIT for providing access to the CATI database, and Mika Pajarinen for assistance with the data.

1. Please note that R&D alliances between two firms from the same country are not excluded in the ranking of countries.

ICT sector	ICT technology fields in CATI	Technical content and definitions	ICT technology fields in this chapter
Micro-electronics	A. Processors, accelerator chips	Computer processors and added ICs	MA. Processors
	B. RISC processors	Reduced Instruction Set Computer processor	
	C. Memory chips, peripheral chips	Memory chips and peripheral driver ICs	MB. Chips
	D. ASICs	Application-specific integrated circuits	
	E. Expansion and other chip boards	Other including audiovisual, TV, car	
	F. Transistors etc.	Generic components	MC. Other generic components
Software	A. Standard software	Operating systems and office software	SA. Standard software
	B. Dedicated software packages	Industrial software including database, CAD and telecom	SB. Standard software
	C. Custom supply dedicated software	System software and software components	
	D. CASE	Computer Aided Software engineering and other tools	SC. Other miscellenous software
	E. Edutainment, games	Education and entertainment	
	F. Internet software	Internet software	
Telecom	A. Public exchanges	Switching systems for fixed networks	TA. Public networks
	B. P(A)BX, LAN, VAN, WAN, ISDN, HIS	PBX, local area, data and value added networks covering speech and data	TR Private networks and services
	C. Cellular telephony	Mobile phones, base stations, switching system and mgmt. tools for mobile networks	TC Cellular systems
	D. Equipment	Transmission and miscellaneous equipment	TD. Other systems
	E. Telematics	Should be telefax, videotex and email systems	
	F. Miscellaneous		
	G. Internet boxes, Net PC	Internet terminals	

III.   MARKET TRANSITIONS: REORGANIZING
        MARKETS, GETTING FROM HERE TO THERE

14

The Peculiar Evolution of 3G Wireless Networks

Institutional Logic, Politics, and Property Rights

Peter F. Cowhey, Jonathan D. Aronson, and John E. Richards

This chapter analyzes how politics shaped the transition between the old world of wireless, 2G and the emerging 3G (high-speed mobile data networks) world. We examine the politics of transition in networked industries using the lens of property rights and political institutions to create a framework that shows how national governments defined and pursued their objectives, how the competing negotiating positions in the world market formed and developed, and how these disputes were ultimately resolved or at least reconciled. The outcome of the transition was the product of a process of global bargaining constrained by those rights and institutions.[1] As a result, contrary to the original plan for 3G, the world ended up with a family of alternative standards for 3G on a variety of different frequency bands.

The legacy of the evolution of wireless through 2G and the emergence of competition in markets that had previously been monopolies, whether private or public, meant that an array of competitors, incumbents and new entrants alike, and a huge variety of technologies needed to be accommodated in transition to 3G. On the one hand, the political impact of the incumbent service and equipment suppliers from the era of monopoly shaped the path between 2G and 3G. There is a politics of transition when dealing with changing market structures involving large incumbents and large degrees of government regulation of the market. On the other hand, the transition took on a particular flavor because of the positions, strategies, and views of newcomers playing a new kind of game dictated by their market resources and by the institutional processes and property rights constraining the market's dynamics.

While some debate the degree of success of 3G, the story here is not about its success or failure or about the particular winners or losers in the new

292   equilibrium, nor is it about which technology provided the best base for 3G.[2] Rather, we provide tools to analyze this and subsequent technological transitions where the stakes in a networked environment are high.

Existing analyses of major transitions in networked markets are in general highly functionalist. They identify various gains and losses from particular outcomes, and analysts then work backward from these outcomes to explain the logic of the transition. These efforts yield important insights about the stakes for the key players. But they miss either how the process of decision making itself alters the mix of outcomes or that the political economy of property rights shapes the preferences of the key bargaining parities.[3] Thus, the excellent literature on various ways that committee systems can facilitate optimal selection of standards misses the political logic for the design of the committees, which depends in turn on how the prior choices about structuring major domestic markets condition these preferences.[4]

As we see it, the potential for a major technological innovation periodically begins to disrupt the status quo, and the resulting question for all stakeholders is how to respond. In networked industries in which governments play a major role, political intervention shapes national and regional positions in global market and regulatory processes by using the terms of licensing to create property rights for companies, thereby shaping the constellation of market interests for the country. The design of regulatory and standard-setting institutions further shapes how these interests and strategies are aggregated and articulated for competition and bargaining in the global arena of markets and regulatory institutions (such as the International Telecommunications Union [ITU]). The powers and roles of these institutions themselves may change as a result of the stresses of the political process.[5] And, just as crucially, these factors alter the form of the technological transformation. Thus, neither the interests of the players nor the logic of the bargaining can be understood without a firm understanding of the politics of technological transitions as implemented through the assignment of property rights and the design of decision-making institutions.

Section I of this chapter lays out the logic of stakeholder positions when there are large incumbents versus new entrants in a regulated networked industry. Section II applies this analysis to the three generations of wireless mobile technologies. And section III concludes with observations about the implications for future transitions.

I. PROPERTY RIGHTS, BALANCING STAKEHOLDER INTERESTS, AND THE POLITICS OF MARKET TRANSITIONS

At first glance, economic theory and game theory seem to lay out a functionalist theory that suggests how considerable gains can be achieved by coordinated government intervention in global wireless markets. There are two potential sources of gains from coordination. First, wireless depends on the use of radio spectrum, which is subject to crowding and interference problems. This suggests government has to either manage the spectrum as a commons or impose some form of property rights regime. Second, there are economies of scale in production of equipment that would benefit from common standards, and there are benefits to consumers that accrue from interoperability of equipment (Besen and Farrell 1991; Shapiro and Varian 1999).

These theories also show that there are competing preferences about how to coordinate. Stephen Krasner has pointed out that the economics of the spectrum or standards problems, a central question here, typifies elements of what game theorists call the battle of the sexes. Take a situation in which a wife and a husband have a choice of going to a ballet or a football game. Everyone would prefer a coordinated approach to action, but there are significant disagreements over which approach (most clearly, in our context, which spectrum should be assigned to what purposes). A similar point obtains, though to a lesser extent, to the setting of standards (Krasner 1991).[6] Resolving these issues is inevitably political as well as technical, and the outcomes will reflect political power and not just economic logic. As this chapter demonstrates, coordination around a single standard may not be achievable if political actors are unwilling to give up their preferred strategy even though it means losing many of the gains obtainable from coordination. For historic as well as logical reasons, governments have a strong regulatory presence in networked industries, so we should anticipate that political intervention is more, rather than less, likely in global telecommunications markets.[7] Certainly, as we shall see, there was political bargaining between the United States and an entente of Europe and Japan over 3G, and the outcomes reflected intense negotiation.[8]

Although emphasizing the exercise of international power — those with power, however measured, are more likely to achieve their preferences — is a step toward acknowledging political factors, it is an inadequate analytic tool. Explaining the ultimate bargain requires more. The analysis must account, first, for the preferences of the national governments that shape their positions. Politics matters mightily for setting these preferences because political leadership selectively empowers different stakeholders in different ways. Political leaders do not just choose between monopoly and competition; they choose particular forms of competition that particularly favor certain types of new entrants. Second, an analysis must account for the role of international institutions in shaping bargaining options and outcomes. In this case, the policy process of the International Telecommunications Union created outcomes that surprised even the most powerful players. Therefore, we argue that the terms under which 3G emerged as a coordinated approach to spectrum allocation, spectrum licensing, and technical standards around the world can best be understood through the literature on property rights and institutions as global bargaining constrained by those rights and institutions.

In a political economy, the logic of domestic and international property rights nicely illustrates the story of wireless networks. Property rights created by governments are essential because they provide incentives for production and responsibilities for costs. However, assigning property rights necessarily entails the transfer of wealth: some receive the ownership rights, while others do not. This creates the incentive to organize politically to win favorable property rights.

In a story of competing interests in an institutionally constrained environment, one might start most simply by imagining a matter of supply and demand. On the demand side, constituents bid for property rights favorable to their interests, and some are more motivated or have more resources to bid for the rights. On the supply side, political actors distribute property rights that achieve political goals. However, matching supply and demand is imperfect. The political creation of property rights regulation gives rise to situations in which market strategies bring about discontinuities and unexpected outcomes. Moreover, the institutions that assign property rights use decision-making rules and procedures that can change the equilibrium outcome in unexpected ways and consequently may be altered in the pursuit of alternative outcomes.

In an international context, national politicians work with international institutions to assign property rights in global markets to maximize both the amount of wealth available for domestic redistribution and the gains from new technology. The result of these often competing goals is that global welfare can be improved, but not improved as much as is theoretically possible. Because domestic interests lobby their national politicians to create property rights favorable to their economic interests, the domestic political fights over property rights become an international power struggle over property rights designed to serve regional interests.

In our case, political actors usually promoted technological innovation by introducing wireless competition to create benefits for particular classes of consumers and new suppliers. At the same time, they also tried to assure significant gains to incumbents from each new generation of wireless technology. This balancing act then shaped choices in the next generation of technology. National institutions gave incumbents a great deal of influence in decision making. International institutions had difficulty overcoming this tendency and ended up in deadlock owing to the need for consensus.

Two market factors made property rights especially important for wireless communications. First, the changing technological foundation promised consumers huge gains, especially if competition spurred price reductions and innovation. Second, the wireless industry is capital intensive and has large economies of scale, strong network externalities, and some path dependency.[9] Consequently, incumbent carriers and their equipment vendors sought favorable technology upgrades on a predictable basis.[10] They favored common planning of new technologies, such as 3G. Externalities and economies of scale meant that stakeholders looked beyond their borders to try to arrange global coordination of technology design and spectrum allocation for new services.

The logic government regulators used in creating a 3G network reflected a classic logic about the politics of market transitions. Regulators tried to create property rights that both ensured incumbents returns sufficient to permit them to abandon existing technology infrastructures and protected the innovation potential of newcomers. Political choices for 3G revolved around policies that allocated and assigned rights for radio spectrum and technical standards that influenced the choice of technologies. Whatever choice was made influenced the distribution of wealth from new property rights necessary to create the 3G infrastructure. In particular, these choices influenced the number of competitors in the marketplace for services and equipment, the terms of competition, and the overall economics of 3G.

The Political Economy of Stakeholders

To begin, consider the nature of parties seeking rents in the market. It is well documented in political economy that groups facing concentrated costs from reform are more motivated to mobilize politically than those firms receiving smaller diffuse benefits from a collective good. This makes such reform difficult (Olson 1965). In theory, commercial side payments among market participants could substantially resolve these issues, irrespective of the initial array of property rights. For wireless telecommunications, however, private property rights are weak because governments license spectrum for a fixed period of time, subject to many constraints. And there are high transaction costs associated with bargaining and enforcement because all parties have significant access to the political process. So politics matters for market reform.

Until the mid-1980s, telecommunications markets were organized by national monopolies. The traditional monopoly carrier, its employees (well paid, national, and unionized), and the equipment suppliers favored by the carrier (which shared in rents earned by the carrier) faced concentrated costs from reform and hence worked together to block any change (Noam 1993). In the United States, this coalition's resistance to reform was eventually broken through a political strategy that employed the court system aggressively.

296   As greater competition became inevitable, the old coalition worked to implement competition in ways that created new sources of market rents. Not only will previous stakeholders put up predictable roadblocks to change, but new stakeholders will attempt to shift policy in ways favorable to their entry. The story of this transition includes four distributional components.

First, as in other markets, those who seek to receive concentrated benefits from reform shape the nature of the new equilibrium by forging counter-coalitions (see Olson 1965; Peltzman 1976). In the case of wireless, the coalition brought together large corporate users (which constituted a large percentage of total long-distance traffic), equipment suppliers outside the traditional vendors to phone companies, and carriers that had identified potentially profitable entry strategies in the market (Cowhey 1990b). This coalition sought particular reforms in a specific sequence as their business needs demanded them.

Second, politicians are not purely passive. Their motives may vary according to the varieties of democracy (European parliamentary coalitions will be driven by concerns different from those of the divided government of the United States), but political leaders can be entrepreneurial. This can include sponsoring changes in market structures. In the United States, we hypothesize that political entrepreneurs advance their individual careers and the welfare of their political parties by reforming markets in ways that win credit from voters.[11] In Europe, the political model would suggest that more insulated bureaucratic elites shaped reform to aid larger political projects, especially the transformation of the European economy. Whatever the particular mix of motives, political leaders may be improving public welfare through entrepreneurship, but they are also managing a contentious political process with strong stakeholders.[12] Thus, political actors choose a particular mix of reforms that provide a solid political foundation for their goals. We should not be surprised when the political leadership frames the choice about a market transition in a way that has a few clear "punch lines" of highly visible benefits to claim political credit and limit the potential for critics to mobilize a successful opposing strategy. For example, reforms in Europe often are justified on the basis of creating "good jobs" through the promotion of press-friendly technologies such as 3G.

Third, the reality of market transitions means that regulators will frequently create competition that is friendly to large competitors rather than simply pushing for higher market performance. The carriers most likely to be assisted with immediate offsetting benefits are the largest firms, which employ the most people throughout the country, and those that provide the most visible services to voters on a daily basis. The former monopolists often also are subject to regulations that confer some continuing rights on other stakeholders such as labor and rural service areas. Even when the political leadership is willing to confront incumbents, it frequently organizes a highly stylized coalition of market interests of particular political significance to the leadership as the "opposition."[13]

Fourth, not all stakeholders with concentrated interests are politically equal. As a rule of thumb, most democratic countries value the interests of domestic consumers and producers more than those of foreign producers and consumers. Foreigners don't vote. Thus, in politically difficult transitions, governments may redistribute rents from foreign consumers and producers to their domestic counterparts. Formal and informal restrictions on foreign investment often transfer rents to domestic competitors or to business partners of foreign investors. Regulatory formulas friendly to creating rents on international services are another option.

Institutional Factors That Shape Property Rights

Regulatory agencies usually make policy decisions about property rights in competitive networked markets. Their expertise is supposed to provide the best combination of increased competition within the constraints of implicit political guidelines about the distribution of gains and losses. They assign three key sets of property rights—the allocation of available spectrum capacity, the assignment of spectrum to specific licensees, and the technical standards for the network—which influence the control of intellectual property.

The form of existing property rights usually reflects the last round of regulatory policy fights, creating continuity for players. These policy fights take place within institutional settings that shape regulatory processes in predictable ways. Though this allows players to have confidence that policy equilibrium will not shift randomly, because of their "stacked" or "sticky" nature, they do not adapt quickly to new markets or technology dynamics (McNollgast 1999).

Originally, during the reign of state telephone monopolies, the dynamics of market change were controlled administratively and closed in most countries. When political leaders began to promote competition to generate more wealth, they created more transparent regulatory procedures to facilitate this goal.[14] But the incumbents facing competition were not stripped of all their institutional advantages. Four regulatory features especially mattered for 3G.

First, the regulatory process is transaction intensive, and large incumbents are prominent in the process. Existing licensees know the process and possess information that regulators prize. Certain prominent companies may wield

298   disproportionate influence because they are viewed as bellwethers of opinion for other stakeholders. This ensures that disgruntled major participants will be listened to in a timely manner when they complain to regulators.[15]

Second, if regulatory processes become more transparent, it is more difficult to give remedial financial assistance to particular market stakeholders. Governments become more indirect in their aid, by propping up incumbents through proceeding slowly or by acting weakly with respect to reforms that favor competition.[16] Or, regulators may package a market reform that could hurt incumbents with another action that visibly benefits them. In most countries 3G spectrum and services were supposed to benefit incumbents by expanding the total size of markets even as governments licensed more entrants at the end of the 1990s.

Third, institutions that set regulations or standards are designed to induce voluntary agreements among participants as much as possible. (The regulatory process tries to at least narrow the range of disagreement.) To do so they have to produce credible information. This poses a serious challenge if the pace of innovation, and thus standard setting, accelerates. Specialized streamlined forums may arise that challenge traditional institutions such as the ITU, and thus the credible information yielded by the standard-setting process in a traditional standards organization may decline. This in itself makes it harder to achieve cooperation.[17]

Fourth, institutions vary in their ability to make binding decisions. Two dimensions are particularly important for this analysis. To some extent, institutions may be able to facilitate bargaining based on credible ex ante commitments. But global institutions, particularly the ITU, could not produce rules regarding the conditions for being part of the standard-setting process that required meaningful commitments to limit IPR ex ante. Just as critically, when faced with conflicts among key stakeholders, institutions vary in their ability to resolve them. As the number of decision points or veto points in a policy process increases, it becomes more likely that the process will maintain the status quo or produce a decision skewed to serve the needs of players with the strongest veto power (Austin and Milner 2001; Tsebelis 2002). Most national regulators use majority decision making to resolve deadlocks more credibly even if their preferred procedures usually try to induce consensus-oriented outcomes. In contrast, many international institutions, such as the ITU, contain more stringent unanimity rules in decision making, further increasing veto power, although political and economic pressure may induce reluctant parties to acquiesce (Greenstein 1994).

Concern for existing stakeholders balanced with introducing more competition and new technology explains why there was an ambitious effort to create a global blueprint for 3G. The approach to 3G was designed to provide incumbents with rewards in return for greater competition. The ambitious level of global coordination envisioned by 3G planners reflected the assumption that policy decisions would be largely an insiders' game, a reasonable view in 1985, when the undertaking began, during a largely monopolistic era. But the consensus-driven process in the ITU can break down. For one thing, the pace and complexity of standards setting grew greater, so the traditional standard-setting bodies became less credible sources of authoritative information on strategies as firms (and governments) used bodies other than the ITU (e.g., the Internet Engineering Task Force in the 1990s) (Hjelm 2000). Even more importantly, consensus is in trouble as the range of stakeholders expands and their interests diverge. The institution could not deliver credible ex ante commitments by companies to cooperate on standards (especially licensing IPR), and (for reasons to be explained shortly) the decision system was becoming less able to resolve deadlock. Ultimately, the ITU is an intergovernmental organization where only governments have decision-making power and other stakeholders are mere observers. As we shall explain, the U.S. government and a small group of other governments intervened forcefully to shape the ultimate outcome of the ITU process. The result of all of these forces was a larger incidence of lost influence by the ITU to other standard setters, stalemate, or unexpected compromises, as happened with 3G.

Defining Property Rights for 3G

National and international regulatory institutions assigned three sets of property rights for 3G. As seen above, features of the regulatory process meant that these rights did not always foster maximum competition within the markets. In the case of 3G, property rights disputes also resulted in multiple global standards for 3G.

The first set of property rights revolved around the process for defining and sharing intellectual property (IP) rights and the selection of standards for global wireless networks. A new generation of wireless services emerged from a global collaborative planning process between carriers and equipment suppliers coordinated through the ITU and regional and national standard-setting processes. Participation in these processes, the terms of operation, and the conditions imposed on the use of IP in the standards process all have shaped global technology.

The second set of property rights revolved around rules governing the allocation of radio spectrum for specific uses, including the rules of service governing the use of licensed spectrum. Spectrum allocation refers to the decision regarding how much spectrum to allot to particular services or groups of services, and on which frequency ranges.[18] All governments treated the spectrum as a commons that required careful licensing to avoid interference problems among rival uses. Even if there were private ownership alternatives, political leadership had few incentives to explore them.[19] Revisiting spectrum allocations opened the way for politicians to earn credit by micromanaging a valuable resource. In addition, government control made it easier to satisfy the large demands for spectrum by military and police services, which few political leaders wanted to oppose.[20]

Licenses were granted in predictable and restrictive ways. In the United States, for example, government spectrum licenses traditionally limited the ability of spectrum owners to change the type of service (e.g., fixed versus mobile wireless), limited the ability of single providers to own more than a limited amount of spectrum in a given market (e.g., spectrum caps), and restricted ownership transfer. Licenses were in effect for a set period of time (e.g., fifteen years). (This mix began to change slowly in the late 1990s.) Asian and European governments often went further, even dictating the type of technology platform that spectrum users could employ to offer services. These types of process typically favor incumbents with operational or informational advantages.[21] As a result private property rights for spectrum were weak (Owen and Rosston 2001).

The third set of property rights is associated with assigning licenses. The number of licenses, the method for selecting licensees, and the sequence of assignment of licenses shape market efficiency. Over the past two decades the number of licenses slowly increased, thereby creating more marketlike systems for providing services. But the sequence of licensing decisions since the early 1970s provided hefty market rents for the original incumbents and then for their initial challengers.

Private bargaining among property rights holders breaks down because the regulatory process has the discretion to frequently revisit issues vital to interpreting the value of property rights (e.g., how new decisions on spectrum might influence the value of existing licenses). Therefore, property rights are not secure, and private bargaining among companies often is ineffectual.[22] Consequently, the private sector encourages government to micromanage spectrum problems.

In short, competition in telecommunications markets prompted major changes in market structure and improved market performance in many countries. Yet political leaders try to ease the risk for large competitors from the transition to greater competition. These same politics of transition, explored in the next section, raised the costs of the transition to 3G and prompted

governments to help some competitors at the expense of others. The regulatory
process could not handle the reconciliation of these trade-offs, thereby delaying
3G rollouts.

II. THE POLITICAL ECONOMY OF THREE
GENERATIONS OF WIRELESS

The argument so far is that there is a politics of transition in which property
rights matter for resolving global coordination issues because they dictate the
rights and roles of different marketplace participants. Property rights define
the rules of the game under which economic actors pursue their interests and
thus influence the economics of markets and the behavior of governments and
firms within these markets (Richards 1999). In the context of 3G, property
rights mattered because they set the economic incentives, costs, and benefits
associated with both the supply of and the demand for 3G technologies for
both equipment vendors and service providers. This also meant that regional
interests had incentives to back certain types of property rights regulation.
Within this preference structure, institutional structures matter to how bargaining occurs over competing structures. This section demonstrates how property
rights and institutions shaped 3G technologies and examines why this mattered for marketplace growth, evolution, and dynamics. The political economy of 3G begins in technology and the policies chosen for the first two generations of wireless services.

As 1G and 2G market growth soared, mobile wireless became the darling of
the financial community and a focus of strategy for even the most traditional
carriers. The introduction of competition in telecommunications services,
done largely in the hope of bolstering national competitiveness in information
technology and what is often called the knowledge economy, posed many
challenges for the former monopolists. Political considerations caused them
to be shackled by high costs and inefficient workforces in their traditional businesses. Fortuitously, mobile services allowed them to create new subsidiaries
that earned far more revenue per employee. Table 14.1 illustrates this point by
showing the revenues for wired and wireless divisions of several major carriers.

The expansion of former monopolists into wireless services eased many political problems about transitions to competition. However, by the late 1990s
there was increasing competition in wireless and the industry faced the
prospect of slowing market growth for voice services. This would reduce
the profitability of major carriers. Companies and regulators faced a fundamental political dilemma: how to keep competitiveness while restoring
growth (Jagannathan, Kura, and Wilshire 2003). For both sides, 3G seemed to

TABLE 14.1. Revenue per employee
of major wireline and wireless
carriers, 2002

Sprint PCS	$1,024,522
Sprint FON	239,368
NTT DoCoMo	2,211,281
NTT	429,045
Telefonica Movile	714,285
Telefonica	200,336
Vodafone AG	185,386
Vodafone Group	691,467
Verizon	285,193
SBC Communications	227,598
Deutsches Telekom	214,819
AT&T Wireless	457,939
AT&T	414,440
France Telecom	206,794
Bell South Corp.	261,292
Industry Average	315,629

Source: Data from Multex fundamentals. http://
www.multexinvestor.com/mgi (accessed January 6, 2003)

be a solution to the problem. The 3G networks could reenergize market growth as the number of voice-only cell phones declined and those with data connections rose rapidly.[23] Equally important was revenue, with attractive margins coming from increased roaming by customers across national borders. Hence a global network was very attractive for business.

As the serious final decision making over the transition to 3G began in the late 1990s, consistent regional and global roaming across national borders still was rare except within Western Europe and parts of Asia. The temptation to create viable global footprints was huge, despite gigantic global investment costs. The gains from global branding were enormous, and large carriers strove to achieve sufficient size and scale to drive seamless international networking. At least until 2001, financial markets rewarded such multimarket strategies of large companies for three reasons. First, global branding would matter for upper-end business customers, and controlling those customers would get global carriers better terms from content providers. Secondly, global scale would increase carriers' bargaining power with equipment suppliers.[24] And finally, global operations required deep pockets, and incumbents were the only players with that kind of financial capacity.

However, unlike most 2G arrangements, which favored only incumbents, the key challenge in 3G was to provide new market returns for incumbents even as regulators introduced more competitors. Countries with the largest markets intended 3G to be contested by several competitors. However, this required the incumbents to believe that there was a larger market to share. The 3G technology was a political gamble for Europe in particular. New policy reforms meant there would be more competitors, but technology innovation was supposed to boost the total size of the market and keep margins high. Many other countries adopted a similar logic.

The property rights fight took three forms.

The Role of Standard Setting and IP

In most countries (the United States is a notable exception), governments owned and operated the telephone carriers until the 1980s. Inevitably, governments were heavily involved in the standard-setting process for telecommunications. Traditionally, most carriers worked with a small set of preferred nationally or regionally based suppliers in a closed standard-setting process.

Global standards processes reflected this legacy of limited competition in global markets. The setting of wireless network standards is globally synchronized through the ITU, which operates by consensus. The global ITU standards process is formally organized around and fed by leadership out of the major regional standards bodies. (Created in 1865, the ITU had 189 member states and over 650 sector members at the end of 2002.) Significant variations in national standards were common, and efforts to coordinate new global services and standards had to plan on these variations because ITU decision making was consensual. Thus, in second-generation services, there was huge variation in rollouts.

A major change in property rights occurred in the standard-setting processes before the end of service monopolies in most countries. As the economies of scale in the telecommunications equipment industry became larger, major suppliers hungered for more global markets. In a series of tough trade negotiations starting in the 1970s, the United States insisted that the cost of opening its equipment market was contingent on the reciprocal opening of other national markets around an open procurement process guided by "open, industry-led, and voluntary" standard-setting processes (Cowhey 1990c; Drake and Nicolaides 1992). These reforms gave large firms easier access to each national market, though a century of monopoly-oriented behavior still had to be overcome. Because the industry still treated telecom technology development as a long-term technology planning process involving global coordination of standards and industrial policy planning, during the 1980s and early 1990s the game remained tilted toward traditional suppliers and carriers.

Second-generation technology emerged in the late 1980s, a period of limited competition in Europe and Japan, and the standard-setting and licensing policies of the time reflected this fact. These early digital wireless services involved technologies that promised better quality, lower costs, and more user capacity. Naturally, the technologies were set forward within the international standards process.

Regional Features of 2G

The earliest major plan for coordinated 2G emerged in Europe, where political leaders saw 2G as a chance to dramatize the benefits of integrating European markets and policy. In 1982 European elites decided to design a single common standard, GSM, a variant of time division multiplexing access (TDMA).[25] The process designed to create standards for GSM, which took place within the European Telecommunications Standards Institute (ETSI), used a weighted voting process to assure a prominent role for incumbents.[26] The one non-European firm with a prominent position in the market and a wide array of GSM patents, Motorola, became locked in a dispute over the terms for licensing its intellectual property. Motorola was a large equipment manufacturer with a major global position but lacked switching systems and had a smaller share of the EU market than the European leaders. The ultimate compromise had Motorola cross-license its patents to the major incumbent suppliers in Europe, a deal that allowed Motorola to thrive as a supplier of selective radio equipment in Europe. Predictably, second-tier (and Japanese) equipment suppliers complained that the terms for patent pooling for GSM favored the largest European companies.[27]

Though ETSI standards are voluntary, the European Union was allowed to adopt an ETSI standard as a European norm and did so by creating policies that de facto required all carriers to use GSM.[28] This built economies of scale around GSM service, allowing it to evolve into the dominant global technology for 2G. The EU considers GSM to be its greatest recent success in industrial policy.

The Japanese followed a system designed to produce standards that were just different enough from other nations to impede supply by foreign firms and favor a few select Japanese suppliers.[29] The Japanese standard (PDC) made some headway in penetrating the Asian market but did not flourish outside Japan. Still, the large, closed Japanese market provided large-scale economies and high profit margins that financed Japanese suppliers as they adapted their equipment to foreign markets.

In the 1980s, when Japanese equipment exports to America surged and U.S. importers had little success in Japan, noteworthy trade disputes proliferated. Initially the U.S. government focused on forcing Japan to reform its standard-setting and procurement systems. Next, America insisted that Japan license a wireless carrier that used Motorola technology. Then, Japan was pushed to reallocate spectrum to make the new competitor viable in the Tokyo market (Mastanduno 1990; Schoppa 1997). Nonetheless, the market was still restricted when Japan ventured into 3G, its dominant market share tied to standards incompatible with those of Europe and the United States.

The United States has long followed a strategy focusing on market competition over coordination. Its continent-size national market allowed it to create large economies of scale without single standard setting. Additionally, America's political leaders were against ceding too much power to any company. Even the AT&T monopoly rested on a weak, loophole-infested legal foundation, later broken by the courts (Cowhey 1990b). By the 1970s a few industry associations, rather than any individual carrier, dominated the standards process. The Telecommunications Industry Association and Cellular and Telecommunications Industry Association, the key groups, featured open membership and voluntary standards. The FCC, for its part, adopted a technology-neutral strategy that resulted in two dominant technology camps, CDMA (code division multiple access) and TDMA (time division multiple access and some of its variants).[30]

Summing to this point: In response to the growing importance of telecom services to corporate users, the rumblings of competition, and the accelerating pace of technological innovation, the major carriers embraced more ambitious, integrated technology plans for new services. Thus, all the major telecom players became involved in a multiyear effort to roll out 2G. The legacy of the political economy of the global market was a huge variety of 1G and 2G technologies that needed to be accommodated in the process, but GSM had the leading global position.

The Challenge of 3G

As if diverse patterns of standard setting and property rights weren't enough to delay widespread 3G rollouts, the technology of a newcomer added another wrinkle to the process. First, European and Japanese companies decided to create the 3G successor to GSM based on CDMA rather than TDMA. This decision was based on the idea that CDMA (or a variant such as W-CDMA) had the greatest potential for allowing sharply improved efficiency by using limited spectrum allocations to transmit large amounts of data (Lembke 2001). This created a huge problem because, to a degree not fully appreciated at the time, a single U.S. company, Qualcomm, controlled the key intellectual property rights of CDMA. A series of patent suits did nothing to weaken their supremacy.[31]

Qualcomm's entry into the intellectual property rights (IP) platform immediately undercut the typical arrangements for telecom networks in global standards bodies. Traditionally, it was standard for the major suppliers to cross-license their intellectual property rights on a cost-free basis while developing major new standards. It was logical. Everybody needed the IP, so, rather than deadlock about the precise distribution of payments, the top tier of suppliers gained by using low- or zero-cost licensing to grow the market. These arrangements grew to the point that in recent years big regional bodies would not embrace a standard unless there was agreement to license the relevant IP to every IP holder under the standard.

With 3G, the ITU faced a problem unlike any it had encountered in the past. On one hand, the formal ITU rules about licensing are artfully ambiguous about expected terms for licensing, but no standard can emerge without the consent of all significant holders of IP.[32] Thus, the licensing requirements do not produce an ex ante commitment to cooperate. In this case Qualcomm controlled a decisive piece of IP. Moreover, this IP was its main competitive asset. Qualcomm could not give it away and survive because the company was too new and too small to fight it out in a competition hinging on advantages in economies of scale in manufacturing, distribution, and marketing. Therefore, Qualcomm insisted on collecting royalties and playing a significant role in designing the 3G architecture, even though it was not a traditional player in standard-setting processes. It had virtually no profile in Europe, where ETSI dominated. In short, because a newcomer, Qualcomm, was relatively new to the inner corridors of standard setting globally, the strategic information available to all key players was weaker than is normal. Miscalculation was easy.

Key players slowly realized the implications of Qualcomm's claims. European and Japanese incumbent suppliers wanted business as usual and therefore wanted to weaken Qualcomm's licensing position. Europe and Japan also sought to incorporate a series of design features from GSM that they saw as improving CDMA's performance in 3G, called W-CDMA. These features might also create new intellectual property that could weaken Qualcomm's control and provide Europeans with IP bargaining chips to press Qualcomm into giving them better licensing terms.[33]

Qualcomm considered these design features arbitrary and worried that the design changes simply made the transition to 3G more complex and time consuming. Even more worrying, Qualcomm believed that the numerous changes incorporating features of GSM architectures hurt a principal

TABLE 14.2   Worldwide mobile communications subscribers by technology (in millions)

Technology	1998	2003	2005 (est.)
GSM/GPRS[a]	138	864	1019
CDMA 2G	22	46	17
CDMA 2.5/3G	0	135	275
W-CDMA	0	2	45
PDC (Japan)	38	62	60[b]
iDEN	3	13	15[c]
TDMA/other 2G	16	149	142[d]
Analog	80	37	22[e]

[a]GPRS is a 2.5G technology that was not available until 2003

[b]Japan is the primary developer and user of PDC

[c]IDEN is a Motorola technology primarily used by Nextel in the United States. The purchase of Nextel by Sprint PCS means that it will convert to the CDMA family.

[d]TDMA is found especially in the Americas. Lehman Bros. predicts almost no new spending on TDMA technologies by 2006. The majority will move into the GSM/W-CDMA family of technologies due to conversion by American carriers such as AT&T/Cingular Wireless.

[e]1st generation technology that is being phased out.

Sources: December 1998 numbers are from GSM Association statistics posted in the table for Quarter Four, 2004 (www.gsmworld.com) except for analog numbers derived from confidential market research for an equipment supplier supplied to the authors. Figures for 2003 and 2005 are from Global Equity Research, Lehman Brothers, "Infrastructure," presented to Lehman Brothers' 2004 Global Wireless Conference, May 24–25, New York City.

advantage of its 2G CDMA systems, the promise that they could be upgraded more or less cheaply and quickly to 3G. This was vital to Qualcomm because 2G systems were going to remain a large part of the total world market for wireless equipment for many years to come. Table 14.2 shows the large role still played by 2G in 2003 and the estimate for 2005. If the transition to 3G was likely to be complex regardless of the choice of 2G technology, then there was less of a downside in selecting the second-generation market leader, GSM.[34]

Given the high stakes, it was not surprising that the major carriers soon divided up on 3G depending on their 2G architectures. Second-generation carriers with a base in TDMA or GSM (primarily Europe and Japan) supported W-CDMA, and those with a base in CDMA supported extending CDMA to 3G (especially North America).[35]

The European Union recognized that replicating its rules for GSM by dictating a single mandatory standard for 3G had potential liabilities under new WTO rules. So, it crafted a position that required each member country to ensure that at least one carrier in its market would employ W-CDMA (dubbed

308   UMTS in Europe). By doing so it allowed for the possibility of more than one technology standard for 3G, but it was commonly understood in Europe that the rule was likely to "tip" the market toward W-CDMA because of network externalities. The certainty that there would be comprehensive European coverage for one standard gave an incentive to all carriers to deploy that standard so that their customers had European coverage in their travels (Gandal, Salant, and Waverman 2003).

Carriers in markets with multiple technology standards for 2G had to resolve conflicting interests. In some countries, such as Canada, CDMA was the choice of the dominant incumbent. In those cases they championed Qualcomm in the ITU process. More often CDMA was the choice of one of the newer entrants. This led the dominant incumbent to favor W-CDMA, especially for reasons of business competition. NTT DoCoMo, for example, had a strong interest in urging the ITU to choose W-CDMA as the only 3G option because its technical specifications would make the second-generation network of its rival, DDI (now KDDI), much less valuable for the third generation.[36] Similar stories, each with their own national nuances, appeared in Korea and China as they introduced greater competition.[37] Hence, not surprisingly, the potential for gain in 3G defined player positions more than any sense of technological superiority.

At the ITU, the European Union and Japan favored a single standard for 3G on the basis that this would yield the largest economies of scale and simplest interoperability of systems around the world. They favored the version of 3G, W-CDMA, backed by their largest carriers and equipment vendors. Qualcomm responded by refusing to license its IPR to this proposed ITU standard.[38] Under ITU rules, without this agreement it was nearly impossible to set a global standard. For decision making the ITU uses a one-country one-vote system. Deadlock is usually avoided because government and commercial interests seek some measure of certainty about standards and spectrum allocation. Although informal polls are used to gauge relative standings of positions on some spectrum allocation debates, in practice consensus is needed in order for there to be progress. In addition, member governments have committed to work within ITU allocations on spectrum. National bargaining positions must take these ITU dynamics into account.

Qualcomm's position was strengthened by the support of a few key governments, particularly the United States. Qualcomm worked intensively with Lucent (which had virtually no sales in Europe) and U.S. carriers committed to CDMA to rally support in Washington and triumphed, despite objections from GSM carriers. One key was that Qualcomm and CDMA had become a showcase of how spectrum auctions could induce new technological successes and American exports. The Clinton administration worried that a global standards process might undermine the success of this "showcase" of the reform process. Moreover, a Democratic administration dedicated to free trade had to be especially sensitive to charges that it was not being tough enough on manipulation of the global market by rival technology powers. In many ways the Clinton administration's trade policy was justified politically as "tough love"—the United States would open its markets more extensively to the world but would be very tough on the conduct of its trading partners.

The divisions among U.S. companies over the policy meant that the White House had to pick its approach carefully. It justified its intervention on the basis of the established U.S. position that standard setting and licensing for 3G should be technologically neutral (i.e., no country should dictate that its 3G licensees must embrace a specific technology). This meant that the U.S. government pushed the ITU to adopt either a single standard acceptable to CDMA operators or multiple standards.[39] On the European side, Johan Lembke's (2002a) work makes it clear that the European Commission understood how important standards bodies could be as a battleground for competing interests. Lembke quotes the commission discussing the growing number of standards covered in part by IPR.

In a situation where companies increasingly act on a global basis and make their decisions based on a global strategic perspective, the pursuit of technological leadership by increasing the value of IPR portfolios could hamper efforts to ensure backward compatibility and global harmonization. The ongoing transformation of formal standards organizations worldwide from groupings of experts seeking consensus on technical matters into battlegrounds for the assertion of competing commercial interests does not facilitate the smooth introduction of new technologies, for example, future mobile communications.

In the end, the strong regional component of the ITU decision-making process meant that the existence of regional blocs supporting both technologies amounted to a veto of any plan centered on a single standard. Most of Europe and Africa, large parts of Asia, and a few South American countries relied on GSM and supported its successor, W-CDMA.[40] GSM was therefore the dominant system. The question is how the CDMA camp blocked GSM's domination of 3G. As Table 14.3 shows, the answer rests in a solid commitment on the part of important CDMA operators throughout Latin America and Asia, which meant that the W-CDMA camp could not paint the bargaining on standards as an issue of North America versus the world. Even large operators such as Telefonica and Bell South, which did not use CDMA in their home markets, embraced CDMA in many of their South American properties, where

TABLE 14.3   Countries with CDMA
operators, 2002

Asia	Americas
Australia[a]	Argentina[a]
Bangladesh	Bermuda
Cambodia	Brazil[a]
China[b]	Canada[a]
Hong Kong[b]	Chile[b]
India	Colombia[b]
Indonesia	Dominican Rep.
Japan[a]	Ecuador
Japan[a]	El Salvador
Malaysia	Guatemala
New Zealand[a]	Haiti
Philippines[b]	Jamaica
South Korea[a]	Mexico[a]
Taiwan, Peru[a]	Puerto Rico
Thailand	United States[a]
Singapore[b]	Venezuela[a]

[a] Countries with major commercial operations in place or about to be launched in 1998 when the 3G battle flared to its peak.

[b] Countries that had smaller commercial CDMA ventures in 1998 or larger planned ventures with greater uncertainties about their launch.

they were market leaders. GSM's relative strength therefore meant less, given this regional opposition.

Despite these blocs, ultimately a compromise was reached. The major suppliers recognized Qualcomm's IPR. Ericsson, the last major company to license from Qualcomm, purchased Qualcomm's network supply business to shore up its CDMA position. On the other end, Qualcomm compromised on its 3G design, and the GSM camp was able to build its own version of 3G, W-CDMA, which Qualcomm had previously rejected. This horse trading meant that, contrary to the ITU's original 3G plan, three versions of 3G were sanctioned.[41] The first, CDMA2000, was a direct descendent of Qualcomm's 2G cdmaOne technology. The second, W-CDMA (wideband-code division multiple access, also called Universal Mobile Telecommunications System, or UMTS), is descended from GSM and incorporates some of the features that Qualcomm had resisted. The third, TD-SCDMA (time division–synchronous code division multiple access), is an idiosyncratic blend of CDMA and TDMA that will most likely drop out of the marketplace unless it is widely adopted by China.[42]

Delays caused by the variety of standards of 3G buildout plans had important consequences for the economics of the market. A new system, 2.5G, emerged as a transition offering. The attraction of 2.5G is that it can be deployed on 2G networks as an upgrade. In this contentious arena a dispute over what is a 2.5G system was bound to arise. The specific merits of the dispute matter less for this analysis than the fact that investing in 2.5G capabilities (around 50 kilobits per second) often moved back rollout of 3G.[43]

In sum, regional patterns of standard setting and IP regulation developed that enhanced certain patterns of market advantages. Though this variation posed few problems for carriers in 2G, their legacy influenced a huge portion of the story in the transition to 3G. The diverse technology base and market structures would create large market advantages by making the technical transition easier or more difficult for different players depending on the form of 3G standards and IP regulation. This was a major political economy problem. Spectrum allocation only accentuated the challenges.

The Allocation of Spectrum

Spectrum is the second form of property rights critical to our story. The objective of the radio regulations of the ITU is "an interference-free operation of the maximum number of radio stations in those parts of the radio frequency spectrum where harmful interference may occur." Regulations that supplement the treaty governing the ITU have the "force of an international treaty" (Hudson 1997). Every three years there is a World Radiocommunication Conference (WRC) that makes decisions on new spectrum allocations and other policies to avoid interference among spectrum uses. As in standard setting, the WRC requires consensus decision making.

The global end game at the ITU shapes national responses, but the roots of spectrum allocation remain national. First-generation service relied on analog technology. During 1G, efforts to use spectrum plans to bolster regional suppliers over out-of-region suppliers led to a variety of idiosyncratic national spectrum plans. Countries saw commercial services as local, a view that served as a self-fulfilling prophecy. This did not lead to much coordination, and it was nearly impossible to use a phone outside of its country of origin because of the different spectrum bands in different countries.

The legacy of 1G was an embedded base of varying spectrum allocations that is difficult to modify owing to the property rights of incumbents. It takes high levels of political commitment and, therefore, political reward to significantly rewrite spectrum plans. Second-generation technology revisited the issue of spectrum allocation because all agreed that it would require larger

312 allocations in a different band than the previous generation. As in standard setting, the EU and the United States moved in different directions.

The EU's choice to standardize around GSM technology was half the battle. The EU also bridged differences in plans for national spectrums. For GSM the Council of Ministers issued an EU directive requiring the use of a single band.[44]

Three factors permitted this outcome. First, European operators believed that spectrum harmonization would grow the service market, especially for lucrative business users, more quickly on a single band than if the EU adopted a variety of technologies and band plans. This provided a benefit to operators to offset the loss of market protection afforded by idiosyncratic national band plans. Second, European equipment makers recognized that if they did not create a major new European market for GSM, they would have to lay off large numbers of their workers. Third, as noted earlier, GSM became a flagship project for political leadership to show the benefit of market reforms (Sandholtz and Zysman 1989; Cowhey, in Hufbauer 1990; Pelkmans 2001).

Significantly, the EU member states retained their control over spectrum planning and licensing. Although all players saw the advantages of unifying the internal market to seize network externalities and scale economies, they still wanted their friendly home governments to control the details of spectrum allocation and licensing. This gap in the powers of the EU ultimately had major consequences for 3G licensing.

On one level the European experiment was a great success. The GSM technology worked. Consumers responded enthusiastically to a truly continental service. During the 1980s the market-oriented features of wireless also were appealing when compared to the moribund marketing for traditional phone service. The European success fueled interest in the growth of global digital services and thus emphasized international harmonization of band plans. African administrations, long tied to European suppliers, agreed to follow Europe yet again. Asia adopted a mixture of band plans, but the European consumer success in selling GSM led national governments to tilt toward the European band plan.[45]

The United States took a different direction with regard to spectrum management to complement its policy of technology neutrality. Unlike the EU, America already enjoyed unified spectrum band allocations. A single 1G analog network covered the United States, and the large continental market could generate large economies of scale in equipment supply even without global harmonization. In addition, powerful players already occupied bands in use in Europe and did not wish to abandon them to help create transatlantic harmonization. Further complicating the situation, the U.S. satellite industry had ambitious plans for mobile satellite services using low earth-orbit systems. These systems needed spectrum that overlapped with possible 2G and 3G systems. Objections by incumbents and conflicting desires by new entrants produced paralysis that made political leaders in the administrations of the first Bush and Clinton reluctant to alter existing spectrum plans (U.S. Congress, Office of Technology Assessment 1993). So the United States selected more flexible bands for 2G. Canada followed the U.S. plan because Canada's chief industrial and financial centers are tightly tied to the United States and its flagship equipment firm relied on sales in the United States. To get along, other countries in the hemisphere agreed to follow the U.S. allocation decision, at least in modified fashion.

Regional dynamics determined the bargaining positions of actors during 3G spectrum planning in the ITU. European suppliers and carriers began the 3G process with the goal of creating a uniform global band and a homogenous network environment (W-CDMA).[46] Because of the dominance of GSM in Asia, Asian spectrum band allocations for 3G approximated those of the EU. Therefore many European and Asian carriers systematically considered building a global footprint from the start. In contrast, at the 1992 World Radiocommunication Conference the U.S. position favored a commitment to facilitating mobile services, without giving special priority to 3G over 2G or mobile satellite services. To the displeasure of other countries, the United States did not clear the spectrum designated elsewhere for 3G in the United States until late 2002 because of infighting among carriers and resistance by the military and public safety agencies holding the desired spectrum.[47] Even then, the United States declared that 2G spectrum could be used for 3G, thereby creating diversity in the global spectrum band. As a result, critics of the United States argue, global economies of scale in equipment suffer, thereby raising costs for consumers because even phones on the same standard often must contain chips designed to work on two sets of frequencies to allow global roaming.[48]

Assignment of Licenses

The third, and most evident, form of property rights is the matter of licenses. As in standard setting and spectrum allocation, regional patterns of market behavior held steady in the assignment of licenses.

Not surprisingly, the United States led the charge for more competition in license assignment. There each of the original seven regional Bell operating companies was awarded one of two wireless licenses in its home territory. Like the few other early creators of duopoly, the United States embraced non-market-based criteria for awarding the second wireless license. Methods for selecting licensees varied, but "beauty contests" (administrative selection of a sound company promising good performance) and lotteries were popular. This practice benefited equipment suppliers that were clamoring for an increase in the number of competitive operators so they would have more customers to buy their wares.[49] The small pool of new entrants then acquired some of the property rights of incumbents because they became prominent players in the regulatory process determining future spectrum allocation and assignment policies.

The spectrum licenses for 2G systems in the United States, obtained by auction, allowed freedom in choosing the services to be offered and were neutral on the technology to be used. As a result, by the mid-1990s the United States had at least five competitors in every region of the country and rival technology camps. However, domestic carriers ruled the scene because of U.S. restrictions on foreign investment in wireless carriers. Until the WTO agreement on telecom services in 1997 liberalized foreign investment rights in major industrial countries, no foreign carrier contemplated controlling a U.S. wireless carrier. Even then, Deutsche Telekom (now called T-Mobile as a wireless provider) did not venture to purchase VoiceStream until 2000. Thus, only one global megacarrier controlled a wireless carrier in the United States.[50]

One consequence of the technology and service neutral licenses in the United States was that 3G could be deployed on 2G spectrum if one wished. Thus, 3G could be deployed on a band other than the one recommended by the ITU. When the additional spectrum conforming to ITU band plans for 3G was eventually made available, it, too, was assigned by auction with technology-neutral licenses. But incumbents dominated the bidding.

When wireless mobile phones became possible around 1983, most European governments responded by extending the property rights to spectrum for services provision to the incumbent wired network carriers; many governments did not even bother to separate the setting of policy from the operation of the national phone company. Competitors were gradually introduced through the assignment of a second license by beauty contests.

The EU hoped to recreate the success of GSM through quick deployment (the goal set in 1998 was 2002) using uniform spectrum and standards. As in 2G, the policy tool was a measure to assure regional roaming on a single standard for 3G. It did so first by requiring separate licenses for 3G services on a single designated band (this meant that a 2G carrier could not just upgrade to 3G on its 2G spectrum). Second, the EU insisted that there must be one carrier in each member state that adopted a standard that ensures EU wide roaming (e.g., W-CDMA). This is what led to de facto convergence of all carriers on W-CDMA in 3G (Gandal, Salant, and Waverman 2003). The net effect of these efforts on the equipment side was to shore up the continued dominance of European suppliers for the GSM family of mobile network equipment. Lehman Brothers calculated in 2004 that Ericsson, Nokia, Siemens, and Alcatel had an 81 percent share of the market for 2G and 2.5G in the GSM family. The combined share for W-CDMA was 84 percent (although Siemens had NEC as a partner and Alcatel teamed with Fujitsu [Randall 2004]).

Since 3G was to take place on "virgin" spectrum, incumbents could play a role in 3G only by winning licenses in key markets. Most countries auctioned and some used a version of beauty contests. In either case only companies with large resources could play except in niche markets, as illustrated by the over $100 billion spent in 3G auctions.[51]

On the auction side Britain licensed five and Germany licensed six 3G competitors. Italy had only five final bidders for five licenses, later reduced to four.[52] Spain and France allowed in fewer new competitors initially. A small pool of supercarriers dominated the new 3G market, mainly because of the high cost of auction licenses and the potential advantages of having coverage in multiple national markets for building a network. In Europe, for example, a small number of traditional incumbents (BT, DT, FT, and Telefonica, primarily) along with two newer supercarriers (Vodafone and Hutchinson) commanded a large share of the key licenses.

In 2001 3G temporarily imploded, especially in Europe, under the weight of the collapse of the Internet and telecommunications bubble. It became evident that the collapse of stock market valuations of carriers and the weight of heavy debt burdens on European carriers, new and old, might require deep job cuts. Bankruptcies became a real possibility. This dramatically increased pressure on many European countries to revisit their licensing strategies. European governments began to seek ways to ease the financial burdens on carriers to deploy 3G. In 2001 the Dutch government assisted KPN, the traditional carrier, in a new financial offering to allow it to refinance debt.[53] France extended the licenses of the two 3G licenses from fifteen to twenty years and reduced the upfront 5 billion euro fee to 619 million euro plus an annual royalty payment to be based on earnings.[54] Bouyugue Telecom, which had dropped out of the auction because of the high price, was quietly promised a license on the same terms. The French government subsequently had to fashion a direct financial subsidy to France Telecom.[55] Germany granted relief by allowing carriers to share the building out of certain network infrastructure, thus reducing significantly the financial burden on all bidders but especially on the smaller, new entrants. The United Kingdom took this same course, allowing some back-scratching among larger carriers.[56] Net savings may reach 30 percent on capital expenditures.[57] Still, as competition authorities conceded,

316   network coordination could lead to other forms of parallel price and service offerings that reduce market performance. Another option was to change the licensing terms to provide financial relief. Rather than deal with the awkward problem of direct subsidies for existing licensees, the EU agreed that 3G licenses may trade spectrum and licenses starting in July 2003 as a way of obtaining financial relief.[58]

In general, Asia relied less on auctions and allowed fewer competitors while frequently dictating the choice of technology in the licenses for service.[59] Fewer competitors generally led to less pressure on carriers in the telecom slump of 2001. For example, in the second generation the Japanese government promoted the deployment of its favored technologies on the networks. When Japan expanded entry in the mid-1980s the government organized licensing on the basis of a beauty contest (Noll and Rosenbluth 1995). This made it easier to indirectly steer both the equipment and services markets. For 3G, the government once again used a beauty contest to favor the three incumbent wireless carriers.[60] This created a dual market by selecting companies on both sides of the 3G technology debate. The KDDI group, a descendant of the carrier involved in the Motorola trade war, adopted the cdmaOne and cdma2000 standards. DoCoMo, NTT's mobile wireless group, built around the W-CDMA effort, as did the group affiliated with Vodafone. Korea allowed only three competitors and set the standards to be followed by each licensee. (As of early 2004, the licensee authorized to deploy W-CDMA had been slow to do so.) In like manner, Hong Kong and then China carefully split their operators' licenses for 2G so that the largest went to the GSM camp while CDMA was assigned to a newer entrant.

The legacy of 2G networks created a natural tendency toward fragmentation of 3G. The roots of the major wireless carriers in the traditional wireline network carriers necessitated protecting existing regional patterns of licenses, IPR, and spectrum allocation. National and regional institutions reinforced the existing market bargains, ranging from the diversity of U.S. standards through the careful tilting of standard setting in Europe. Given these roots, the balance between benefiting existing licensees and stakeholders even as governments took credit for introducing more competition proved to be too difficult to coordinate under a single global standard and spectrum plan. The key to understanding the political economy of 3G was the relationship between the small pool of incumbent carriers and equipment suppliers and a newcomer (Qualcomm) with strong market position owing to its IPR leading to different incentives. The decision-making system at the ITU could not manage the division of interests that had grown out of the divergent regional politics of spectrum allocation, property rights practices, and licensing.

III. THE IMPLICATIONS FOR THE NEXT TRANSITION

Except in countries with strong CDMA carriers for 2G, hopes for an early roll-out of 3G were usually disappointed.[61] But by 2005 it appeared that 3G still remained the most likely backbone for a general medium-speed, wireless data network, especially for those demanding mobility. Nonetheless, the delay also made it easier for more diversity to emerge in wireless. The world will not have a single neat technology or market model. Wi-Fi has already made a major mark for "campus" (university or office) environments, for example. Even as 3G plays out there is discussion of a vigorous push toward 4G that would introduce a variety of wireless technologies, especially on unlicensed bands, to permit much higher speeds and other capabilities. While 3G offered the opportunity to integrate multiple standards, 4G may create the possibility of integrating multiple technologies. As chipsets gain in power and complexity, the idea of devices with integrative capacity through technology becomes a reality.

At the moment, the push toward 4G is too often an example of not learning from experience. For many of the advocates of 4G, 3G was the right idea but failed either because of bad timing (prematurely pushing for high-speed wireless before better technologies were available) or poor execution (including the corporate battles over rollouts). But this misses the point. Third-generation technology assumed that massive global coordination of spectrum, standards, and licensing policies was possible in a timely way. But the stakeholders in wireless communications, even in the insiders' community, have diversified significantly, while the coordination mechanisms remain relatively weak. Just as significantly, the goal of 4G assumes that we know what the future should look like. This severely taxes our ability to forecast in any competitive market with significant technology innovation.

Forecasting is difficult, but this article emphasizes that understanding the transition from 3G to whatever is one day called 4G requires appreciating the rules of the political game.[62] However strong government institutions' informal mandate to cushion the largest incumbents from the most brutal forms of change, competition is a much larger part of the landscape than in 1985, and some of the new entrants of the 1990s are now part of the established roster of players. Moreover, digital technology itself has made it easier to get cross-entry across segments of the communications and information technology market. Cisco, Intel, Microsoft, Samsung, LG, and Qualcomm are all major players on the world scene—a much different roster of players than in 1990. Technological advances also allow increased interaction of products without standardization (e.g., smart phones promise to switch standards and spectrum as needed; Aronson and Cowhey 2004).

318   One possible model for the future is closer to the modal type of the information industry. Collective efforts on standardization of technologies and supporting business processes have a pluralistic view of the future. Given the speed of innovation and the diversity of players, it is impossible to have a single authoritative source of information or decisions for all related technologies. There are competing models of the future and many different collective efforts to advance those visions (Lehr and McKnight 2003). Though markets, technology communities (e.g., the Internet Society), or (occasionally) governments may evolve a single standard for particular key parts of the landscape, the goal is not to have a general consensus model of the future. Moreover, different decision rules for setting standards and sharing IPR exist in different fora, and the ability of a forum to craft a rule that is ex ante acceptable to all key participants is part of its appeal (or failure). Thus, the capabilities associated with 4G can be nurtured through much more vigorous test bed processes and narrow, specialized standards setting. The IPR process broke down in the standards process for 3G because a monolithic design raised the costs for the players. Indeed, the solution to the 3G was to back away from unified planning and deployment.

NOTES

1. Various authors from both the U.S. and European perspective (see Lembke 2002b) have highlighted the need to build theory about the ways in which governments use regulation and standard setting for regional advantage. While previous attempts outline this general purpose, this chapter seeks to make the intellectual jump between description and theory.

2. We agree with Lembke that social scientists cannot easily provide definitive assessments on the technical merits of competing technology proposals, and in any case the debate is not central to this argument (Lembke 2002b, 56).

3. For an admirable effort to examine decision making in standards bodies from the perspective of antitrust policy, see Lemley 2002.

4. For an excellent example of the literature on committees, see Farrell and Saloner 1985.

5. Walter Mattli rightly points out that rigidities in global intergovernmental standards organizations may propel leaders to shift authority incrementally to private and regional organizations. The EU created a new telecommunications standards organization, ETSI, partly out of frustration with the ITU process. See Büthe and Mattli 2003; Kahler and Lake 2003.

6. Our approach is closer to that of Austin and Milner 2001.

7. In short, governments are already in the business of influencing market equilibriums in telecommunications. So a proponent of government intervention does not have to win a first political battle over getting a major level of political intervention in the market.

8. The Euro-Japanese alliance was primarily coordinated by industries with similar business interests. Once these ties were forged, they had implications for the options of governments and regulatory bodies for reasons that follow from our discussion of the politics of transition in this chapter (Lembke 2002b).

9. Where network externalities exist, networks grow more valuable to individual users as more people use or are connected to the network. See Owen and Rosston 2001.

10. Once a carrier has installed a supplier's network equipment, the supplier is locked in. Equipment vendors calculate that once their equipment is installed, they are unlikely to be displaced. Since global carriers prefer suppliers with global support capabilities, this limits entry for network and, to a lesser extent, handset equipment. (Based on interviews with European and Asian suppliers, November 2002, December 2002.)

11. In short, politicians supply new property rights to constituents in exchange for campaign contributions and votes. On political entrepreneurship see Cox and McCubbins 1993.

12. The structure of government institutions, the nature of electoral systems, or the form of executive power (e.g., parliamentary or presidential) influences how these strategies play out in a particular country. Our analysis of global markets handles these factors on an ad hoc basis. See Tsebelis 1995.

13. See the discussion of how U.S. political leaders organized smaller high-technology firms in Cowhey 1990a.

14. International institutional processes also change. For example, the introduction of competition in major industrial markets led to an agreement that separated national regulators from national telephone operators (reversing traditional practice). The WTO basic telecommunications services agreement set forward this regulatory principle in 1997. See Cowhey and Klimenko 2001.

15. See McNollgast 1999 on "fire alarm" mechanisms that are built into administrative procedures. Political leaders can give a higher profile to outsiders in the regulatory process. Normally they do not. By making it costly for parties to participate in the regulatory process, politicians save valuable time.

16. For example, an incumbent may bargain for "transition rules" that allow more market protection during as markets reform. The winners from competition will accept slower transition because otherwise the dissidents might tie up the decision process in less predictable and/or lengthier ways. Many countries offered a multiyear period of selective competition as a transitional device in their WTO telecommunications commitments to open their markets. See Sherman 1998.

17. See the discussion of ETSI in Kahler and Lake 2003.

18. The laws of physics make bands differ in their radio propagation characteristics, so spectrum is not equally tractable for all tasks. For example, spectrum bands over 100 MHz permit straight-line transmissions that can be power efficient.

19. The absence of private property rights partly reflects the high transaction costs of assigning and monitoring individual property rights in the early days of radio technology. It emerged from a tradition of state building that reserved commons for government ownership. See Hazlett 2001.

20. In most industrial countries the military controls about 30 percent of the spectrum.

21. The political process is arcane and fiercely contested. Advocates debate what would constitute a threat of interference and the plans for reallocating different pieces of spectrum to different uses. These proceedings raise enormous informational problems for government decision makers. The glacial process cumulatively favors incumbents.

22. Given weak property rights, compromises among companies may not emerge without a credible enforceable guarantee. Political decision-making processes shape possible trade-offs, as has happened in many government programs for technology transfer. See Cowhey and McCubbins 1995.

23. Ovum data as reported in "Wireless Briefing," *Red Herring* (March 2002), 68–69.

24. The size of the market of your "flavor" of 3G influences the total cost structure for the technology. Within that cost envelope any individual carrier's buying power depends on factors such as the size of its potential purchases.

25. The key consensus on the outline of the standard was reached by 1987. The principal player initially was the European Conference of Posts and Telecommunications (CEPT) when they created the GSM MOU (Memorandum of Understanding). This later evolved into a global organization for promoting GSM. See Tan 2001.

26. National and regional standard-setting processes varied. Effective participation required both a significant commercial presence and the ability to fund staffers who could dedicate extensive time to the standard-setting process. Voting, if used, often was weighted according to market revenues and required super-majorities (e.g., two-thirds majorities). In contrast to the one company–one vote principle of the U.S. Telecommunications Industry Association, the ETSI used weighted voting. Market revenues mattered significantly in the weighting. Manufacturers dominated the voting. In 1997, 49.5 percent of the members were manufacturers, 15.8 percent were public network operators, 9.18 percent were national government authorities, and 12.4 percent were research bodies. Also see Hudson 1997.

27. It took the Japanese suppliers several years to acquire the IPR licensing agreements, a delay that gave the European firms a major lead (Bekkers, Duysters, and Verspagen 2002).

28. On how the EU used a combination of spectrum and standards policy to assure a common approach to 2G, and for an excellent discussion of government intervention to make standards setting credible, see Pelkmans 2001.

29. Japan's procurement policy was opened to international scrutiny when Japan agreed to extend the GATT procurement code to the NTT. On Japan's procurement system and political economy see Noll and Rosenbluth 1995.

30. A bipartisan consensus was made possible politically by the diversity of U.S. industry. The FCC declared technology neutrality, agreeing that in general government could not select the right mandatory technology even if there were cases where it might be hypothetically advantageous to do so. See Farrell and Saloner 1985.

31. The key dispute involved Qualcomm and Ericsson in litigation that began in 1995. This was resolved in an agreement announced on March 25, 1999, that included cross-licensing of patents and Ericsson's purchase of Qualcomm's terrestrial infrastructure business. Vitally, from the viewpoint of Qualcomm, the agreement included a

stipulation that licensing would be done for all three proposed versions of 3G (Hjelm
2000).

32. Traditionally some standard-setting organizations, including the ITU, demanded "royalty-free licensing." Many others now require "reasonable and nondiscriminatory" licensing. This discussion relies on Patterson 2002. In 2000 the ITU Telecommunication Standardization Bureau stated: "The patent holder is not prepared to waive his rights but would be willing to negotiate licenses with parties on a nondiscriminatory basis on reasonable terms and conditions." The bureau does not set precise criteria for these conditions and leaves it to negotiations among the parties. But the relevant factors for setting royalties include costs for development and manufacturing plus profits (Patterson 2002, 1053–1054 and n. 40).

33. In 2002 another group of European vendors announced that it would set an absolute cap, at a relatively low level, on royalties charged for W-CDMA technology use. Qualcomm quietly rejected the cap and observed that it held about 50 percent of the IP on W-CDMA, thus making any royalty offer that it did not agree to meaningless. Other companies rejected Qualcomm's estimate of its holdings, but the effort to create the cap floundered nonetheless.

34. Concern over 2G sales was precisely why neither side followed the economic logic of compromise to grow the market size set out in Shapiro and Varian 1999.

35. The collective approach of numerous industrial actors sought to ensure compatibility across markets for preferred methods of technology. See Lembke 2002a for more on how the objectives of both firms and national level actors were driven by regional interests.

36. The key event producing the W-CDMA initiative was a successful negotiation on common interests among the largest expected winners in Europe—DoCoMo, Nokia, and Ericsson. Alex Lightman points out that if the ITU had standardized only around W-CDMA specifications, the chip rate in the system would have been incompatible with seamless upgrading from second generation CDMA systems. See Lightman 2002.

37. The United States had no comparably dominant wireless incumbent. AT&T was a TDMA carrier as were the wireless groups of several large Bell operating companies. Verizon and Sprint ran the flagship CDMA networks. So the carriers quarreled bitterly over the U.S. position in the ITU on standardization.

38. Qualcomm notified the standards bodies involved in 3G that it held patents that were essential to all proposed versions of 3G. It offered to license, on reasonable and nondiscriminatory terms, a single converged ITU standard for 3G or its own proposed standard. It declared that it would not license other versions of 3G, such as the EU's W-CDMA standard. "Qualcomm Supports Converged Standard for IMT-2000," press release of June 2, 1998 (CDMA Development Group, http://www.cdg.org/news/ jun_98 .asp; accessed December 17, 2002).

39. For example, on October 13, 1999, Secretary of Commerce William Daley, U.S. Trade Representative Charlene Barshefsky, and FCC Chairman William E. Kennard released a letter to EU Commissioner Erkki Liikanen protesting EU policy.

40. Most low-income developing markets rely more on European suppliers of network equipment than they do on North American suppliers. European companies,

seeking larger markets, attempted to enter these markets far earlier than their American counterparts.

41. Qualcomm collected IPR royalties on all versions of 3G.

42. In November 2002 China appeared to tilt in this direction by setting aside spectrum reserved for this technology. However, as of 2005, the Chinese policy on 3G was still under debate. Both the United States and the European Union raised issues about mandatory standards with China. Siemens was the primary foreign partner for TD-SCDMA efforts.

43. See, for example, the discussion of EDGE in Morgan Stanley, "Telecom Services and Equipment: Cross-industry Insights," February 1, 2005.

44. Even after additional bands in a higher frequency opened for 2G, the EU still required use of GSM.

45. Developing nations benefited enormously from 2G. In most of them the wireline network was severely underbuilt compared to demand. Wireline monopolies were inefficient for many reasons, including overstaffing, inflated procurement costs, and corruption. But a major problem—pricing—was not related to costs. Governments charged too little for local phone service and too much for long distance. The high profits on long-distance services never sufficed to build out the local network, but served as a political barrier to realistic pricing of local services. Wireless services provided a political escape from this trap because national governments treated 2G as a premium service and therefore could charge premium rates. See Cowhey and Klimenko 2001.

46. CEC, Commission of the European Communities (1998k) Amended Proposal for a European Parliament and Council Decision on the Coordinated Introduction of Mobile and Wireless Communications in the Community, COM (98) 496, Brussels, July 27. Council of the European Union (1999) Council Decision on the Coordinated Introduction of Third Generation Mobile and Wireless Communications System (UMTS) in the Community, 128/1999/EEC.

47. Officially cdma2000 carriers endorsed reallocation, but their real preferences were unclear because they could launch 3G without new spectrum. Opponents included the politically powerful UHF television broadcasters. See ITU 2001.

48. Some phones will be able to handle both 3G modes, to be both dual band and dual mode. This increases costs for production in a market where consumers demand low prices.

49. Governments subsidized carriers by not charging them for using valuable resources. The rents created by this choice were shared with labor and equipment suppliers. See Shy 2001.

50. CDMA carriers (Alltel, Sprint, and Verizon) focused on the large North American market, while GSM and TDMA carriers (AT&T Wireless, Cingular) looked at alliances with non-U.S. carriers. Vodafone owned a minority share of Verizon.

51. The licensing system required large resources and was usually designed to produce only one or two new entrants. See Klemperer 2002.

52. Klemperer (2002) argues that the auction designs of countries varied considerably in their merits. High price in itself was not proof of a poorly designed auction.

53. Debt levels for major European carriers included 65 billion euro for Deutsche Telekom and 64 billion euro for France Telecom. See Edmund L. Andrews, "Lower Goals for Telecom in Europe," *New York Times*, March 20, 2002.

54. *Telecommunications Reports International*, October 26, 2001, 1–2.

55. John Tagliabue, "France Struggles to Ease Debt of Phone Company," *New York Times*, September 14, 2002.

56. *Telecommunications Reports International*, September 28, 2001, 12, 20.

57. For example, E-Plus and Group 3G collaborated in Germany, and British Telecom and Deutsche Telecom agreed to do so reciprocally in Germany and Britain. Skeptics suggested that real savings would amount to 5 to 15 percent. *Telecommunications Reports International*, April 27, 2001, p. 4.

58. Kevin Delaney, "France Telecom Approves Plan to Reduce Debt," *Wall Street Journal*, December 5, 2002, A3.

59. Most of the former British colonies of Asia eventually tilted toward auctions — Hong Kong, Singapore, Australia, and New Zealand — with varying degrees of success. Hong Kong had four carriers (Ure 2003).

60. Japan did not consider increasing the number of competitors until mid-2005. On Japanese policies, see http://www.itu.in/itunews/issue/2001/08/licensing3g.html (accessed March 10, 2002).

61. A CDMA carrier for 2G had two consequences. That carrier had a more seamless technical path for upgrading to 3G, and this upgrade path, as in Japan, could spur non-CDMA carriers to make quicker transitions to 3G.

62. For example, 4G may simply be a highly upgraded version of 3G plus some combination of upgraded Wi-Fi, Bluetooth, digital television for mobile devices, and broadband satellite systems.

15

Factors for Success in Mobile Telephony

*Why Diffusion in the United States
and Europe Differs*

Heli Koski

Production and usage rates for information and communications technology (ICT) differ considerably among countries and regions (see, e.g., Koski, Rouvinen, and Ylä-Anttila 2002; European Information Technology Observatory [EITO] 2004). One of the most visible differences in ICT use between the United States and the European Union lies in mobile telephony. In 2003, the share of mobile telephony subscribers was about 88 percent in the EU area, whereas it was only 54 percent in the United States. Given that the United States has generally been ahead of European countries in ICT technology adoption, this wireless telecommunications technology gap seems puzzling.[1] An investigation into the underlying reasons for this technology gap is important, because the diffusion of mobile telephony may contribute to economic development and welfare (through such mechanisms as improvements in the quality and timeliness of decision making).

U.S.-European differences in the development of mobile telephony are striking and concern both analog first-generation (1G) and digital second-generation (2G) mobile phones.[2] At the time the United States mandated Advanced Mobile Phone Service (AMPS) as the technical standard for 1G wireless telephony, European analog markets remained fragmented and featured multiple incompatible standards. Consequently, the diffusion curve for analog mobile phones in the United States slopes up higher and declines later than it did in EU countries. In 2000, when the last two new analog mobile handset models were launched,[3] only about 1 percent of EU citizens were still analog wireless service subscribers, compared to about 11 percent in the United States.

For its 2G mobile telephony markets, the EU chose GSM (global system for mobile communications) as a single standard. In contrast, the U.S. government decided to allow multiple digital mobile telephony standards to compete in the market. In the United States, the diffusion of digital telephony has been sluggish, whereas in Europe the number of GSM subscribers has grown rapidly.[4] In the North American market, CDMA (code division multiple access) has won a dominant position, with more than 40 percent of mobile service subscribers in early 2004, whereas GSM's share of wireless users was only about 13 percent (http://www.cellular.co.za, January 2004 [accessed November 13, 2005]). Recent growth in the number of GSM users in North America, however, suggests that the United States might eventually follow the global trend in 2G mobile diffusion.

The U.S.-European comparison demonstrates that regional mobile telephony markets differ not only in terms of their technology compositions but also in the speed of aggregate mobile telephone diffusion. How can we explain these cross-country differences in technology mix and diffusion speed? To what extent are the global markets for mobile telephony influenced or even constituted by regional technology and competition policy? This chapter sheds light on the role of telecommunications policy and market responses for the success of wireless telephony standards.[5] In particular, it tries to explain the observed differences in mobile phone diffusion between the United States and the countries of the European Union.

The analysis in this chapter suggests that differences in EU and U.S. regulatory policies have strongly contributed to the differences in the diffusion speed of 2G wireless telephony in Europe and the United States. In the United States, standardization policy relying on user-driven innovation in 2G mobile communications combined with a roaming policy that supports 1G technology by mandated compatibility had a detrimental effect on 2G wireless diffusion. U.S. regulators employed a liberal 2G market approach while simultaneously adopting the principle of technological neutrality. It was a mistake for the technologically neutral policy to concern itself only with competition between 2G standards and not across wireless telephony generations. In contrast, among EU countries, competing 1G technologies were not supported, and the combination of standardization and market liberalization provided favourable conditions for 2G mobile telephone diffusion.

DIFFUSION OF MOBILE TELEPHONY:
THE UNITED STATES VERSUS EUROPE

During the first half of the 1980s, the European Union and the United States did not differ much in their diffusion rates of analog mobile telephony (see Figure 15.1). Early analog telephone use was led by the Nordic countries, for which, however, even at the peak of diffusion subscription rates, never reached much more than 10 percent of the population. As Figure 15.1 shows, the

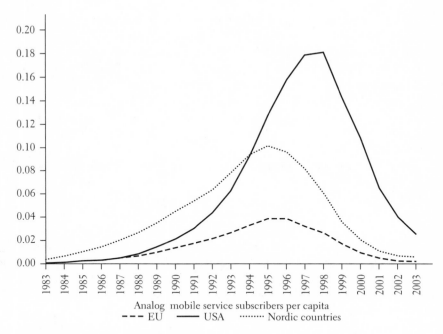

FIGURE 15.1   Analog mobile telephones per capita (%), 1983–2003

United States reached a larger installed user base of analog mobile telephones than the EU countries. In Europe, the collapse of the installed analog user base began in the mid-1990s, whereas in the United States, the number of analog mobile telephone subscribers continued to grow until the late 1990s.

Commercial digital (or 2G) mobile phone services were introduced in 1993 in the United States and between 1992 and 1995 in the EU countries. By the mid-1990s there was a salient difference in the penetration rates of 2G mobile phones between the United States and EU countries (see Figure 15.2). This gap has only widened during the past years. The EU average rapidly caught up with the Nordic countries in the late 1990s and slightly passed them in 2001. In 2003, the share of digital mobile telephony subscribers of the population among EU countries was about the same as in the Nordic countries at about 88 percent, whereas the corresponding share in the United States was only about 52 percent.

The growth in the total number of mobile telephony subscribers per capita also reveals some notable differences in diffusion dynamics between the United States and the EU. During the second half of the 1990s, the EU witnessed fast growth. The growth in the mobile telephone subscriber base peaked at over 70 percent between the years 1997 and 1998 (Figure 15.3). The

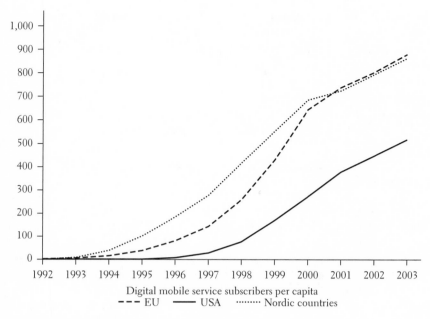

FIGURE 15.2   Digital mobile phone subscribers per capita, 1992–2003

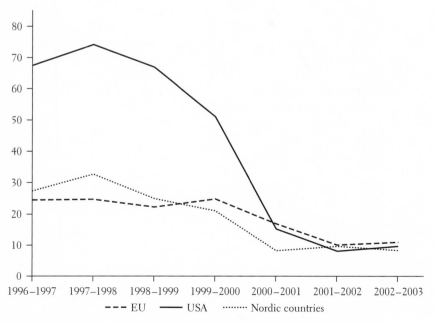

FIGURE 15.3   Growth in mobile phone subscribers per capita, 1996–2003

328   annual growth rates were less dramatic among the Nordic countries, because their markets for mobile telephony had matured earlier than those of most EU countries. The relatively low growth in the U.S. mobile subscriber base did not accelerate after the mid-1990s as it did in the EU.

Interestingly, although the EU fell behind the United States in 1G wireless telephone diffusion, the regions' respective positions reversed after the introduction of 2G mobile telephony. What explains these differences in the mobile subscribers per capita between the United States and European countries?

EXPLANATIONS FOR THE EUROPEAN LEAD IN MOBILE TELEPHONY

The primary focus in this chapter is on the role of technology policy in the observed differences in mobile telephony diffusion between the United States and the EU countries.[6] Competition and standardization policy are argued to affect technology adoption and diffusion both directly, through consumer demand (e.g., policy affects the expected user values of technologies), and indirectly, through behavioural incentives created for producers and service providers (e.g., policies can strengthen the incentives for product compatibility). Research on the effect of various technology policy on diffusion patterns has assumed that the observed relationship is independent of the changes in regulatory strategies over time. In stark contrast, I argue here that the interaction between regulatory strategies over time has been an important contributing factor to the observed differences in diffusion of mobile phones between the United States and Europe.

The following discussion reviews the reasons suggested in explanations of cross-country differences in technology diffusion and evaluates their relevance in the context of mobile phone diffusion.

Market Entry and Competition

Competition policy and market structure are among the first factors typically mentioned in policy-oriented economic studies of technology adoption. Various empirical studies unanimously suggest that competition has facilitated diffusion of mobile telephony (Gruber and Verboven 2001). The empirical investigation of Koski and Kretschmer (2005) further finds that competition has primarily facilitated mobile phone diffusion via its impact on service pricing and that non-price competition has had only a minor role in promoting diffusion.

Among the EU countries, until 1996 there existed typically only one mobile telephony licensee per country. During the years 1996–1997 a duopoly was the most common competitive setup, and it was only after 1998 — the year that EU countries had a legal obligation to fully open up their telecommunication markets to competition — that the mode number of mobile telephony licenses reached three. Figures 15.2 and 15.3 illustrate that between 1996 and 1998, when the number of cellular license holders increased in the EU area, there was concurrently strong growth in mobile telephony diffusion among European Union countries.

While there generally seems to be a clear positive relationship between competition and technology diffusion, competitive conditions can hardly provide an explanation for the observed slower growth of mobile telephony in the United States compared to the EU. In fact, the United States has been more advanced than European countries in facilitating market entry and competition through mobile license allocation. By 1996, there were already six competing mobile telephone service providers in certain geographically segmented U.S. mobile telephony markets (Organisation for Economic Co-operation and Development [OECD] 2000a). In 2000, 91 percent of the U.S. population had access to three or more mobile service operators (Federal Communications Commission 2001a).

Clearly, factors other than market structure in mobile telephony service provision account for the underlying differences in the speed of diffusion of mobile phones between the two continents.

Standardization

The pros and cons of standardization are disputed in various forums; the question whether technologies generally and network technologies particularly should be standardized or left to the market has kept both academic researchers and policy makers occupied (Katz and Shapiro 1985; Dranove and Gandal 2003). The advocates of standardization emphasize the benefits from compatibility. Standardization reduces both user and producer uncertainty and enables the maximization of benefits from compatibility, thus resulting in higher expected benefits for users and faster technology diffusion. There are various pieces of empirical evidence supporting this view both from mobile telephony markets (Koski and Kretschmer 2005) and from other network technologies markets such as digital video (see Dranove and Gandal 2003).

Since the early days of cellular telephony, the countries of the European Union have followed a different standardization approach in their telecommunications market than the United States. No uniform analog mobile telephony standard existed among European countries. The NMT (Nordic Mobile Telephone) standard that was developed in the Nordic countries was adopted in many EU countries (e.g. Austria, Belgium, Netherlands), but the large countries pursued other strategies. For instance, France complemented NMT with its own country-specific analog standard, RC2000. The analog cellular networks in Germany were based on the country's own C-450 analog technology, and those in the UK on the widespread TACS technology. In contrast, the Federal Communications Commission (FCC) adopted the mandated technical standard, AMPS, for analog cellular services. The AMPS standard did not only survive the phase of 1G mobile telephony. In addition, all U.S. cellular carriers were required to provide analog services to their analog customers and to roamers visiting the coverage area of their network.[7] In practice, as long as this rule holds and there are analog mobile phone subscribers in the United States, each carrier is obliged to build and retain its cellular networks' AMPS compatibility.

When planning and introducing digital (i.e., 2G) mobile telephony services, U.S. and European policymakers again adopted opposite standardization policy approaches. This time around, EU countries chose GSM as the single standard for their 2G mobile telephony markets, whereas the U.S. government decided to allow multiple digital mobile telephony standards to compete in the market.[8]

The diffusion of analog mobile telephony was notably greater in the standardized United States than among nonstandardized 1G markets in Europe (see Figure 15.1). Similarly, the diffusion of digital mobile telephony has clearly been faster in the standardized European markets than under the laissez-faire U.S. approach to the 2G market (see Figure 15.2). These findings are in line with both the predictions of theory and previous empirical findings on the impact of standardization on diffusion. The obligation of all U.S. cellular carriers to provide AMPS-compatible services has strengthened the position of analog cellular technology in the United States and possibly hindered carrier investments in digital cellular networks. Because the U.S. market for digital wireless services has involved plenty of uncertainty owing to various competing standards, it has been safer for wireless carriers to delay investments in digital wireless networks.

Technological incompatibility combined with a geographically fragmented market structure in digital wireless telephony has led to problems with roaming in the United States. Mobile phone users have not always been able to use their phone outside the coverage area of their own service provider. The first digital handsets did not allow roaming between different radio frequencies and/or standards, i.e. they were neither multi-band nor multi-mode. The world's first dual-band, dual-mode CDMA phones were only shipped in the fall of 1997,[9] two years later than the first single-mode CDMA phones.

Even though U.S. regulatory authorities imposed strict roaming rules for analog telephony, telecommunications regulation has not included similar re-

quirements for digital telephony. Wireless carriers have to provide manual
roaming for other carriers' digital customers, but only for those that use technologically compatible handsets with their networks. Thus, when a customer is
using a handset that is technically capable of accessing a licensee's system and
there is no roaming agreement between the wireless carriers, he has to pay for a
roaming service on the spot by dealing with the customer-service department
of the local carrier—a rather cumbersome and costly procedure for the user.

In Europe, domestic wireless roaming has involved fewer problems than in
the United States. First, the European domestic cellular markets have been
less fragmented, because the cellular licensees typically offer nationwide coverage for their networks. Second, thanks to the common 2G standard in the
EU countries, multimode roaming has not been necessary. It seems quite possible that the slower take-off of the U.S. digital phone diffusion is also related
to the timing of introduction of multiband, multimode telephones. In the
United States, though the FCC has not imposed automatic roaming obligation for wireless carriers, the markets seem to be gradually taking care of the
roaming problem. Over the past few years U.S. wireless carriers have come to
realize the importance of nationwide network coverage and signed numerous
roaming agreements extending domestic roaming.[10]

In summary, the U.S. standardization and roaming policy which long privileged the analog cellular telephony standard, AMPS, created uncertainty regarding the future value of competing 2G technologies. Promoting analog
mobile phone use hindered intergenerational transition from analog to digital
technology. The U.S. market environment has given cellular telephone carriers incentives to wait instead of to invest in new digital networks. My closer inspection has shown that the American approach to 2G wireless policy has been
less technologically neutral than is usually assumed.

Investments in Communications Infrastructure

Investments in the wireless telecommunications infrastructure contribute
to its quality and coverage and thus influence the demand for mobile telephone services. Figure 15.4 compares cumulative investments in the mobile
telecommunications infrastructure (per capita) between Europe and the
United States. We can only conduct a rough comparison of U.S.-EU differences, because the available data covers only the years 1992–1999 and is patchy
among EU countries (e.g., only one observation, for 1998, exists for Finland).
After the introduction of digital mobile telephony in the early 1990s, investments in mobile telecommunications infrastructure increased in both the
United States and Europe; however, the mobile infrastructure investments of

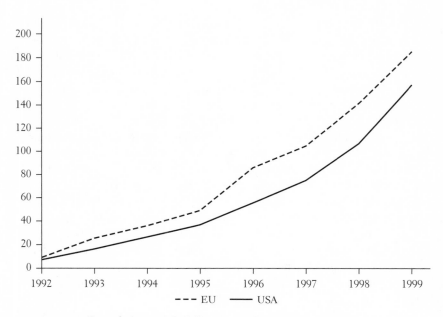

FIGURE 15.4   Cumulative mobile infrastructure investments per capita

the European telecommunications operators, particularly after the mid-1990s, were larger than those of the U.S. cellular carriers.

Figure 15.5 illustrates total public telecommunications operator (PTO) investments per capita in Europe and in the United States. During the early 1990s, the (relative) level of investments of U.S. and European telecommunications operators did not deviate much from one another. Starting in the mid-1990s, telephone operators' investments grew sharply in the United States, and in 2000 they were more than fourfold compared to those in the early 1990s. The European Union did not witness a similar growth in total PTO investments per capita even though investments in mobile infrastructure among the EU countries increased substantially.

Figures 15.4 and 15.5 demonstrate that European and U.S. telecommunications operators have focused on developing different areas in their service provision. Investments have been more strongly directed toward wireless communications in Europe than in the United States. It seems plausible that greater mobile infrastructure investments have meant better mobile service quality in Europe than in the United States and have further contributed to the European lead in mobile phone use.

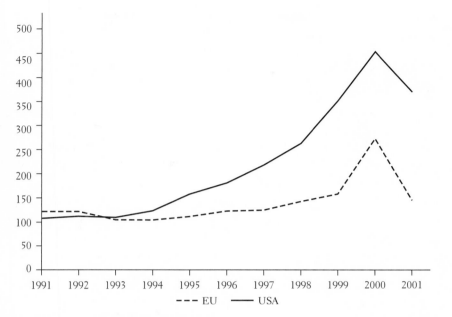

FIGURE 15.5   Total PTO investments per capita

Intergenerational Effects

Intergenerational effects from old to new generations of network technologies may be either positive or negative (Liikanen, Stoneman, and Toivanen 2002). On the one hand, a larger installed user base of a technology may facilitate adoption of its next generation because of factors such as learning and the widespread availability of information about the technology among potential adopters. On the other hand, a relatively large installed user base of an old technology may hinder the adoption of a new technology, because consumers expect to derive greater benefits from compatibility when using the old one. Liikanen, Stoneman, and Toivanen (2002) find positive intergenerational effects between the diffusion of 1G and 2G mobile telephony, that is, a larger installed user base of 1G has facilitated diffusion of 2G mobile telephones. This seems consistent with our aggregate figures of diffusion among the EU and Nordic countries. The Nordic countries have been ahead in the adoption of analog mobile telephones (see Figure 15.1) and also witnessed earlier diffusion of 2G mobile phones than other countries (see Figure 15.2).

The United States does not fit into this picture; even if the statement on intergenerational effects applied to averages in countries, it seems to be an exception to the rule. In 1994, the relative size of the U.S.-installed user base of analog mobile phone users was about the same as in the Nordic countries and clearly higher than among the EU countries. However, during subsequent years, in the United States growth of digital mobile telephony was slower than in the Nordic countries and in the EU countries. It seems that in the United States the larger installed user base and greater coverage of analog mobile networks may have hindered digital telephony diffusion. To what extent carriers' (investment) behaviour versus consumer behaviour account for this so-called installed base effect is impossible to evaluate. As the discussion above suggests, the U.S. cellular standardization and roaming policies are likely to have played an important role in this respect.

Other Explanations

In addition to the U.S. standardization policy, there are other factors that may at least partly explain why the United States has fallen behind the EU in mobile telephony diffusion.

Pricing. An important difference between wireless telephone services in the United States and Europe is pricing. In Europe, the caller is the one who is responsible for the costs of a telephone call he makes. The U.S. system follows what is called the receiving party pays (RPP) principle, which means that the receiver gets the bill for the mobile phone calls he receives. In addition, there is a cheaper alternative to using mobile phones for local calls in the United States. Residential wire-line customers have access to unmeasured local call services. By paying a monthly fee, customers can make local calls free of charge. In contrast, in Europe, local calls are a measured service, as part of which the cost of a call is determined by its duration. Thus, the pricing structure for local calls in European countries is more favourable to substitution through mobile phone usage than is the case in the United States. These factors, costly incoming calls and costless alternative to local area calls, probably hindered mobile phone diffusion in the United States.

Markets for pagers in the United States. Analog telephones do not allow transmission of data such as text messages. Paging offers a substitute portable technology for digital mobile telephony in transmitting data, and it has been more popular in the United States than in Europe. In 1997, the installed user base of pagers in the United States accounted for more than 35 percent of the world's paging market, whereas the installed user base of pagers in the EU covered only about 6 percent (European Union 1998). In 1999, there were still 45 million paging units in service in the United States.[11]

Roaming conditions are one of the reasons why paging sustained its popularity in the United States even after the introduction of digital mobile te- lephony. In 1997 there was nationwide network coverage for paging, whereas digital cellular phone roaming was very limited. In addition, there is no cost to receiving a paged message, whereas incoming cellular phone calls are costly.

CONCLUSIONS

The U.S. standardization and roaming policies have combined to support the adoption and use of analog mobile phone technology at the expense of digital telephony and aggregate mobile phone diffusion. First, while the FCC adopted a standard for analog mobile telephony (AMPS), for digital technology the resolution of standard(s) competition was left to the markets. Second, AMPS was granted a long-term privileged position; the U.S. wireless carriers are still required to provide roaming for all analog customers, whereas they are under no obligation to provide services for other carriers' digital customers. In the geographically fragmented U.S. mobile telephony markets, these policy actions supported the installed user base of analog telephony and, through the uncertainty they created, gave carriers incentives to hold back on investing in new digital networks.

Furthermore, U.S. consumers had two portable substitutes for digital mobile services, analog telephony and paging services, both of which had nationwide roaming coverage. Digital roaming was not technologically possible during the first few years of digital mobile telephony and remained limited even after the introduction of multiband, multimode handsets. The popularity of paging services in the United States further hindered the transition from analog to digital mobile telephony and increased the gap in mobile phone use between Europe and the United States during the 1990s.

The pricing structure of wireless services and its substitutes has been more favourable to mobile phone use in Europe than in the United States. The receiving-party-pays principle (i.e., costly incoming calls) and unmeasured local calls (i.e., costless alternatives to making local area calls) have probably also played a part in slowing mobile phone diffusion in the United States.

In summary, both the U.S. regulatory and market conditions have been less favourable for the adoption of digital mobile telephony than those in the EU, thus leading to a slower diffusion speed of mobile telephony. The reasons for these developments cannot be found in the liberal market regulatory policy of the United States as such, as is often claimed, but rather in how this policy was implemented. In fact, the U.S. roaming regulations supporting the 1G wireless standard and further warping market incentives for 2G investments neither represent a laissez-faire approach nor remain technologically neutral.

336   The lesson that can be learned here is that interactions between regulatory strategies over time matter. An intergenerational transition from coordinated standardization to a liberal market approach may be detrimental for the adoption and diffusion of the next generation of technology. If policymakers decide that an intergenerational switch from standardization to a laissez-faire approach is necessary, they should pay close attention to minimizing the adverse support for the old generation of technology.

An intriguing question is how the transition from 2G mobile telephony to 3G wireless communications will take place, whether and how countries diverge in diffusion of 3G and its substitutes. The success of the next generation of mobile phones offering improved data services depends partly on the history of wireless markets (i.e., the development of 1G and 2G mobile telephony markets) as well as on the development of the markets for 3G substitutes. The majority of European consumers have become accustomed to mobile phone use, a state of affairs that supports a shift to 3G telephony. In contrast, in the United States, investments in wireless local area networks (WLAN) have been greater, and its use more widespread than in Europe. The ability of the 3G service suppliers and mobile phone manufacturers to develop interesting service concepts to attract different customer bases in the United States and Europe will determine, by and large, how 3G wireless telephony succeeds compared to its potential competitors such as WLAN services. The importance of the location independence of services (e.g., how much users value 3G mobile phone services such as taking and transmitting photos "anywhere") is likely to be one of the key determinants of 3G success.

The final factor affecting 3G diffusion is technology policy directed to promote the launch and adoption of the next generation of wireless communications technologies.[12] In Europe, policy has promoted a single candidate, universal mobile telecommunications system or UMTS, for the next-generation wireless communications — quite contrary to the general technological-neutrality principle of EU technology policy. However, the unrealistically high 3G bids of the European telecommunications operators have been a major hindrance to 3G network investments and development in Europe. Ironically, during the past few years the United States has been rapidly catching up in wireless communications use and may well be leading Europe in the adoption and diffusion of the next generation of wireless communications technologies.

NOTES

1. In 2003, ICT expenditures per capita were about 1,517 euro in Western Europe, whereas they were over 50 percent higher, about 2,328 euro, in the United States (EITO 2004).

2. The analog wireless telephony networks are able to transmit only voice. The de- velopment of digital cellular networks means not only improvement in service quality and more security for the customers but also enables the transmission of data and, for instance, access to multimedia services via mobile phones.

3. Source: EMC World Cellular Database.

4. In early 2004, GSM technology was also the global standard or, using the terminology of the management literature, dominant design in mobile telephony, reaching close to 1 billion subscribers or over 70 percent of the world's market share. It seems that the majority of the world's 3G wireless networks will also be based on GSM technology: "around 85 percent of the world's network operators chose the 3GSM standard to develop 3G services." Source: http://www.cellular.co.za/news_2004/Jan/122303-1_billion_gsm_users_in_2004_09.htm (accessed November 13, 2005).

5. See also Leiponen (this volume) for a closely related discussion of firm-level standardization strategies on mobile communications markets.

6. We employ here Mowery's (1995) definition of technology policy: "policies that are intended to influence the decisions of firms to develop, commercialize or adopt new technologies."

7. This requirement is scheduled to expire in 2008 (http://wireless.fcc.gov/services/cellular/operations/findingaserviceprovider.html#differences). See Federal Communications Commission 2001b for further discussion on analog cellular compatibility standards in the United States.

8. See Bach 2000 for an in-depth discussion of the GSM standardization procedure and the role various institutions play in it.

9. Source: http://www.qualcomm.com/press/releases/1997/press753.html (accessed November 13, 2005).

10. See, e.g., "AT&T Wireless Extends Domestic Roaming" (*InfoSync World*, April 24, 2003, http://www.infosyncworld.com/news/n/3467.html [accessed November 13, 2005]); "Cingular Continues Hunt for Nationwide Presence" (RCR Wireless News, April 7, 2003, vol. 22, no. 14: 8.

11. Source: "Delegates" 2000.

12. See Glimstedt 2001 for a detailed discussion of 3G standardization in mobile telephony.

16

National Styles in the Setting of Global Standards

The Relationship between Firms' Standardization Strategies and National Origin

Aija Leiponen

Technical standards are ubiquitous in industries with network technologies. Standards enable interoperability — without widely accepted standards, trains, telephones, televisions, and many other modern technologies would not work. Standards tend to lock markets into technological trajectories for a significant period of time because of substantial fixed investments in network equipment and technological processes, not to mention the sunk costs of developing standards in the first place. As a result, technical standards may substantially shape markets: the distribution or concentration of intellectual property rights and technological capabilities may determine the long-term nature and intensity of competition in an industry.

The emergence of technical standards is a complex process where social, political, and commercial interactions among interested parties influence the unique outcome. Economic research has identified two polar standardization processes: market-driven standardization (de facto) and politically driven standardization (de jure). However, in recent years, "hybrid" or committee-based standardization processes that emphasize cooperation among companies as opposed to government agencies have been dominant in many industries (Farrell and Saloner 1988). Wireless telecommunications is a recent example of this mode of standardization (Funk and Methe 2001; Bekkers, Duysters, and Verspagen 2002). However, standardization processes and their effects on markets are not yet well understood. The chapter by Cowhey, Aronson, and Richards in this volume illustrates the complexities involved in international standardization cooperation and the possible implications for market emergence. The current chapter focuses on firm behavior. The goal is to shed light on firms' standardization strategies, where private actors play a central role but need to cooperate with their rivals in developing technical standards.

Because technical standards in wireless telecommunication no longer are set by government agencies or left for markets to decide, standardization has become a central part of firm strategy. One anecdotal indicator attesting to this is that many leading technology companies have hired specialized vice president–level standardization managers in recent years. In contrast, strategic management research has not investigated standardization as an integral part of innovation and technology strategies. Shapiro and Varian 1999 remains one of the very few references in this area. Their analysis builds on theoretical industrial economics and focuses on the choices related to compatibility between technologies and control of intellectual property. These choices turn on the characteristics of the competitive field (whether there are many or few suppliers) and the nature of the innovation (whether it is an incremental or a radical improvement). However, a lot of room remains to enhance our understanding of firms' standardization strategies — in particular, regarding how firms organize standards development activities.

Despite the global reach of wireless telecommunication technologies and standards, firms from three continents, North America, Europe, and Asia, appear to behave differently in their standardization activities. As discussed by Cowhey, Aronson, and Richards in chapter 14 and by Zysman and Newman in the introduction, path dependencies created by national policies, traditions, and cooperative linkages continue to matter in this age of global competition. One particular source of national differences may be institutional legacies and the degree to which they complement international standardization institutions (Mattli and Büthe 2003). Whereas continental European companies have a longer tradition of international (within-Europe) coordination and cross-border cooperation, telecommunications authorities in the United States have followed a much more liberal market-oriented and technology-neutral approach emphasizing competition among potential technologies (Tate 2001). Asian countries, Japan in particular, have tended to choose technical standards through a more centralized process whereby the government — in cooperation with major companies — imposes standards on the industry (Funk 1998; Tate 2001). These national systems of standardization have generated incentives for firms to develop certain types of capabilities and cooperative arrangements, which tend to persist over time. National origins may thus continue to have a bearing on global firms' strategies toward standard setting.

This paper examines whether American, European, and Asian firms' standardization strategies differ with respect to cooperative orientation and activity in international organizations. These differences are of interest to the

340   companies operating in this area, as they enable firms to see how others are positioned in the standardization networks and how this might matter for their performance. Even though most of the companies analyzed in this study are large and operate internationally, their cooperative networks are path dependent in that partners and modes of cooperation persist over time. Patterns of cooperation may become institutionalized in the sense that their evolution lags behind economic and competitive changes in the environment (Cook 1977; Gulati 1995). The chapter also aims to contribute to management and economic research on standard setting by developing a typology of standardization strategies illustrated through the descriptive data analysis of firms engaging in these different strategies. Finally, policymakers interested in standardization processes and outcomes should find the analysis illuminating because it presents new evidence of the kinds of firms that are central, and thus powerful, in standard-setting networks.

Organization of firms' standardization activities is studied here using newly compiled data from firms participating in the Third Generation Partnership Project (3GPP), an international standards development organization formed to coordinate standard setting related to the universal mobile telecommunications system (UMTS) standard. The 3GPP is the successor of the Special Mobile Group of the European Telecommunications Standards Institute (ETSI). Though the origins of this organization are in Europe, it currently has worldwide presence in terms of both membership and adoption of the technical specifications created.

In the next section I discuss conceptual issues related to firms' standardization behavior. The third and fourth sections present descriptive analyses using a dataset on firms' standardization activities within the 3GPP, their memberships in other standards development organizations, and participation in international technology alliances. The final section summarizes the results and discusses their implications for managers, policy makers, and future research in this area.

WHY DO FIRMS COOPERATE IN TECHNICAL STANDARDIZATION?

Industrial Dynamics

The economic industrial organization theory suggests that it is very important for firms in network technology markets to gain a sufficient market share for products and technology platforms. This is because of demand-side network externalities (Katz and Shapiro 1985) and supply-side scale economies (Klepper 1996; Simons 2003), both of which generate first-mover advantages. Network externalities speed up the convergence of markets toward one technology, because once a technology has obtained a critical mass of users, it is increasingly advantageous for subsequent adopters to buy products based on this platform: the benefits of the product itself are compounded by the size of the network that users can access by subscribing to this technology. As a result, the market share of the winning technology can grow initially at an exponential rate (Arthur 1989). Strategically, then, it is essential for firms to be among the first to enter new network markets and to obtain a critical market share. Otherwise they may be forced to switch to the competing technology or exit the markets altogether.

In practice, however, network externalities rarely operate quite as strongly as suggested by this simplified model. Other factors, such as political differences between regional or national markets, may support multiple technological standards. Similarly, if users have very strong preferences for the different features or qualities of the competing technologies, the market may remain fragmented. For example, Apple Macintosh computers have retained their niche market share around 2 percent despite the network externalities associated with IBM-compatible personal computers. Macintosh computers have been seen as particularly suitable for certain applications such as graphic design, and they also have benefited from a very loyal user base. Nevertheless, network dynamics make it very important for firms to enter the markets early and quickly rally support for their technological solution from suppliers and clients. Various cooperative strategies discussed in this chapter can facilitate these goals.

The telecommunications sector is characterized by even stronger network effects than the computer industry because communication technologies have absolutely no value without the capacity to exchange messages. Reliability of telecommunications is also an important consideration in governments' control and security functions, and consequently, public authorities have traditionally played a central role in telecommunication standardization (Schmidt and Werle 1998). Telecommunication standards have usually been developed in international organizations such as ITU (International Telecommunications Union) and regional or national organizations such as ETSI (European Telecommunication Standards Institute) and ARIB (Association for Radio Industries and Businesses, Japan). Gradually during the late 1980s and through the 1990s the role of government authorities in these organizations has been reduced while commercial firms have gained in importance.

Currently wireless telecommunications companies are relatively free to develop technologies and promote them as standards. They can remain outside of any standardization cooperation and try to employ market forces to gain market acceptance. This was the approach of the equipment firm Qualcomm and the operator Nextel in second-generation networks during the 1990s. Alternatively, companies can participate in formal standardization processes,

342   where successful equipment providers' technologies will become incorporated into the jointly developed standard.

Governance of Standards Development
in Wireless Telecommunications

The evolution of wireless telecommunication technologies and markets continues to depend heavily on successful development of technical standards. The first generation of analog systems (1G) was developed on national and subregional bases. As a result, multiple incompatible technologies were adopted within Europe, North America, and Asia. In most countries, these technological choices were made by government authorities together with domestic telecommunications operators and equipment suppliers. The second-generation technologies (2G) introduced digital transmission of information. Now, governments and companies both realized the potential benefits from transnational cooperation. After much competitive jockeying on the part of the commercial parties, European governments adopted the GSM technology, which was jointly standardized through the European Telecommunications Standards Institute (ETSI). North America and Asia continued to have multiple competing technologies. In addition to GSM and TDMA, which are closely related technologies, a newcomer to the industry called Qualcomm introduced a radically different technology based on code division multiple access (CDMA) air interface. However, a late entry and subsequent smaller economies of scale prevented Qualcomm and its allies from becoming a more significant contender in the global race for market share in the 2G wireless telecommunications systems.

As information and communication technologies continue to converge, the next step in the technological competition is to enable better data exchange rates over wireless communication systems. This is the goal of third-generation (3G) systems. Again, despite the initial negotiations, the world markets ended up divided into multiple competing technologies. European telecom operators adopted the Universal Mobile Telecommunications System (UMTS), which builds on existing GSM networks but uses the wideband CDMA air interface (W-CDMA), while North American and Asian markets have adopted either UMTS- or CDMA2000-based systems. CDMA2000 builds on the existing CDMA networks.

The race to launch networks and sell these systems to yet undecided operators around the world created extremely high stakes in the standardization game. Indeed, a striking organizational phenomenon took place in wireless telecommunications in the late 1990s: equipment providers appeared to be founding new industry fora, associations, or alliances almost weekly. Organization theories did not seem to have an explanation for such behavior. Why were so many different cooperative organizations needed to coordinate one technical system, albeit a complex one? Moreover, these cooperative organizations were often founded by a core group of the same leading companies in the industry, and thus it was not a question of different subgroups within the industry organizing to compete (Axelrod et al. 1995).

In addition to public and semi-public industry associations, firms formed private strategic alliances and cooperated in open international standards bodies. Rosenkopf and Tushman (1998) suggest that during the early stages of the technology life cycle, when the technology is in ferment, companies manage technological and market uncertainties by engaging in interfirm networks and alliances. As a result, cooperation constellations are reconfigured under these turbulent conditions. Empirical observation suggests that cooperative organizations may also be used to "market" new solutions and applications to other industry players. Also, these kinds of ad hoc cooperative organizations may enable faster development of technical specifications than is possible in public and bureaucratic standards development organizations. Finally, a more cynical interpretation of industry association activity is that leading firms use associations to hide a tacit collusion that is intended to solidify the technological trajectory to their advantage, without including smaller firms or firms with different technological positions in the decision making. Industry fora can thus be used to lock in the cooperation agenda and key technical solutions to the founding firms' technologies before other companies are allowed to participate in public, open standardization processes.

Whichever the underlying reason for industry associations, coordination among firms in an industry based on network technologies is critical to achieving interoperability. This coordination can assume many different forms of governance and operate through several distinct cooperative organizations simultaneously. Table 16.1 displays an attempt to list the possible organizational forms for standard setting. The extremes of letting the markets or the government select technical solutions have all but disappeared from wireless telecommunications, but almost all of the middle ground forms of coordination — associations, alliances, forums, and negotiation — have been used in the standardization of third-generation systems. Indeed, these intermediate forms of committee-based standardization have largely taken over previous de jure and de facto processes. This perhaps vindicates Farrell and Saloner's (1988) analysis that committees are more efficient than pure markets in standardizing complex technologies, particularly if market processes are allowed to interact with more political committee processes.

TABLE 16.1   Forms of governance for coordinating technical standardization

Mode of coordination	Explanation	Example
De jure	Public bodies (ministries, departments, agencies) select the solution and impose it on the industry	○ Choice of NMT in the Nordic countries (early 1980s)
Open SDOs	Firms and governments meet in open and public Standards Development Organizations to coordinate standards development	○ Development of GSM in Europe (ETSI, mid-1980s) ○ Development of WCDMA (3GPP, since mid-1990s)
Semi-open forums	Firms coordinate specification development through gradually opening forums	○ WAP Forum ○ UMTS Forum
Private alliances	Firms cooperate through private alliances	○ Symbian to develop operating systems for wireless devices ○ Other joint ventures, R&D alliances
Negotiation	Leading firms negotiate directly with each other	○ Development of NMT in the Nordic countries (early 1980s)
De facto	No cooperation: markets select the winning technology or technologies	○ PC operating systems (1980s) ○ 1G digital communication in the US

It is an interesting question whether these different cooperative arrangements are complements or substitutes. Transaction-cost economic logic would suggest that firms match organization form to the characteristics of the transaction, in this case, standard-setting activity. In this scenario, there is no reason to observe firms engaging in multiple different types of organizations to achieve one goal. Alternatively, standard setting can be viewed as a political game of influence rather than an economic problem of maximizing efficiency. In this perspective, firms might use the different channels of influence to promote their own preferred technical solutions toward industry adoption. Firms might then use many different forms of cooperation to achieve one single goal, and we would observe strong positive correlations among the arrangements.

The Process of Cooperative Standards Development

Standards development includes both aspects that align rival firms' interests and those that foster conflicts. For instance, market growth and formation of cooperative relationships are generally beneficial to all agents in the market.

Interests are in conflict when firms compete to incorporate their own technology in the standard and to obtain information about rivals' strategies while attempting to protect information about their own strategies and technologies from leaking to their rivals. Balancing mutual and conflicting interests — cooperation and competition — is therefore one of the key strategic issues in standardization.

Depending on the firm's business strategy, either cooperation or competition may be emphasized in standardization strategy. The American technology supplier Qualcomm is an extreme case again. Its strategy has been based on leveraging a high-quality and high-potential radical innovation (the CDMA air interface), and as a result, it has focused on technological competition in the market rather than seeking consensus and support through cooperation. In contrast, Motorola and Nokia have had a more balanced approach to competition and cooperation. Their technological capabilities have been based on incremental innovations, production efficiency, and management of complex technical systems. Yet another approach is followed by many telecommunications operators who emphasize cooperation in their strategy, because for them, the size and growth of markets is much more important than any technological choice.

There have also been a few imitators of Qualcomm's success in "performance play" strategy (see Shapiro and Varian 1999, 204). Small technology companies such as Phone.com (currently Openwave) have attempted to leverage their intellectual property in standards negotiations. Phone.com managed to convince leading equipment suppliers with its software protocols related to the WAP standard, although this innovation has not fared well in the market. Although occasionally opportunities may emerge for smaller companies to control lucrative technological niche markets with their IP (e.g., PrairieComm in baseband processors), it is generally difficult to achieve broad success without an exceptionally significant innovation.

Participating in standards negotiations necessitates special skills on the part of firm representatives. Both technical and social competencies are required of the delegates in order to become influential players in the social network (Dokko and Rosenkopf 2003). Individuals need a strong technical background in order to obtain working group chairmanships and to be able to influence committee decisions more generally. Additionally, these technical experts need to be able to coordinate and negotiate among the different participants with conflicting views and interests. Negotiation and communication skills are thus necessary. Influential company representatives cannot push their own company's solutions too openly or indiscreetly, for most committee work is based on consensus decision making whereby broad coalitions are needed in choosing among solutions. Engineers participating in standards negotiations and

346   standards managers who were interviewed in this research project emphasized
        social and communication skills as key factors behind committee influence.
        For example, working group chairmen are elected based on their technical
        competence and their skill in finding consensus solutions that are acceptable
        to many participants (Dokko and Rosenkopf 2004).
            In addition to individual technical experts participating in standards nego-
        tiations, companies rely on formal and informal coalitions in trying to influ-
        ence specification outcomes. Company representatives often communicate
        informally one on one outside of the official agenda in order to gauge and mo-
        bilize support for their technical solutions. More formal coalitions behind work
        item proposals and change requests are formed to promote official acceptance
        in the committee work. Coalitions are typically formed around dominating
        technology companies with large market shares. Smaller players may have a dif-
        ficult time promoting their proposals except in special cases, where possessing
        very central intellectual property in a technology subfield is necessary.
            Though requirements for skills and other resources in cooperative stan-
        dardization are substantial, attempting to promote a technology to become an
        industry standard through sheer market forces may be an even more daunting
        challenge. In particular, extraordinary technological capabilities and patent
        portfolios are necessary for success. A large market share may also help a com-
        pany get the attention and consumer awareness it needs. However, even here,
        cooperation and alliances have their place, because complementary products
        and services need to be launched simultaneously. For example, launching
        new kinds of smart phones requires, at a minimum, cooperation between
        the supplier of terminals and a service provider, to make sure networks and
        telecom services support the new technology. Additional alliances with com-
        plementary equipment, component, or application suppliers may be advanta-
        geous as well. However, these types of cooperative arrangements take place
        privately outside of open standards development organizations.

COOPERATIVE STANDARDIZATION ARRANGEMENTS
BY 3GPP MEMBER FIRMS

This section illustrates firms' standardization strategies through a descriptive
analysis of a dataset that contains information about 3GPP member firms' cooperative activities. In particular, the focus is on firm memberships in other
standards development organizations (SDOs) and relevant industry associations, their private alliances, and their activities in the 3GPP working
group coalitions. Data sources include public Internet websites of 3GPP and
other SDOs and associations, and the Dutch MERIT-CATI (Cooperative

Agreements and Technology Indicators) database of joint ventures and strategic alliances.

The basic sample of firms consists of 3GPP members as of the end of 2003 (about 380 organizations). However, only a subset of member firms has been active in making work item proposals and change requests to technical specifications. Moreover, the chapter focuses on commercial business organizations, and therefore governmental, research, and other agencies are excluded from the analyses. Thus the analyses here rely on a sample of about 140 companies.

Industry associations and organizations provide public records of their membership, and they are usually open for new members after the initial charter has been set by the founders. These associations provide settings for discussing technological solutions among equipment suppliers and promoting them to operators and suppliers of complementary technologies. Some associations have very formal rules and procedures for jointly developing technical specifications; others focus on informal discussions and promotional events. The analysis uses the number of memberships as an indicator of firms' industry association activity.

Organizations considered here to be relevant to the development of third-generation wireless networks are listed in Table 16.2. Generally, larger firms and equipment suppliers tend to be most actively engaged in a number of different cooperative organizations — they have sufficient financial means and human resources. Interestingly, many of the 3GPP members are also active in organizations that develop and promote the main competing standard, CDMA2000 (3GPP2 and CDMA Development Group). Other organizations on the list promote the UMTS/W-CDMA standard (e.g., UMTS Forum, GSA) or develop complementary technologies or applications (e.g., Bluetooth, IPv6, Hiperlan2). The table demonstrates that many of the 3GPP members participate in different standards development and technology promotion organizations.

Within 3GPP, the focus is on the work item process. When participants want to introduce a new item for the agenda of a working group, the organization's rules require that they get support from at least three other participants in the group. Thus, some cooperation is a prerequisite for companies trying to make any proposals about the direction of work within the standardization organization. Observations of participation in these work item coalitions yield indicators of network centrality for the descriptive analysis. By treating coalitions as cooperative linkages, centrality indicators can describe each firm's position in coalition networks, and they are derived using standard network analytical tools. Firms' network positions are measured here with two different but correlated centrality indices (Wasserman and Faust 1994). *Degree centrality* simply

TABLE 16.2   Membership of 3GPP firms in a sample of other related
SDOs and industry associations, 2003

Association	Purpose of association	#3GPP members
3GPP	Develop technical specifications for 3G mobile system based on evolved GSM core networks	383
UMTS Forum (merged with OMA in 2003)	Promote the global uptake of UMTS 3G mobile systems and services	92
Open Mobile Alliance (OMA)	Facilitate global user adoption of mobile data services	79
Cellular Telecommunications & Internet Association (CTIA)	"The voice of the wireless industry": Represent its members in a constant dialogue with policy makers	62
3GPP2	Develop global specifications for ANSI-41 network evolution to 3G	55
CDMA Development Group (CDG)	Promote the adoption and evolution of CDMA wireless systems	42
Bluetooth Special Interest Group	Develop, publish and promote the preferred short-range wireless specification for connecting mobile products	41
Wireless Ethernet Compatibility Alliance (WECA; currently Wi-Fi Alliance)	Certify interoperability of wireless Local Area Network products based on IEEE 802.11 specification	29
HiperLAN2 Global Forum	Provide common connectivity for mobile communications in corporate, public environments and home to ensure Interoperable products	25
3G.IP (disbanded in 2004)	Promote a common IP based wireless system for third generation mobile communications (operator driven)	24
IPv6	The next generation protocol designed by the IETF	24
Mobile Wireless Internet Forum (MWIF) (folded into OMA 2003)	Drive acceptance and adoption of a single mobile wireless and internet architecture that is independent of the access technology	23
Global Mobile Suppliers Association (GSA)	Promote GSM/EDGE/WCDMA worldwide	18
Voice XML Forum (Sponsor & Promoter members)	Promote the Voice Extensible Markup Language	18

TABLE 16.3   Fifteen most central actors in 3GPP work item
coalition networks, 2000–2003

Firm	Degree centrality	Firm	Betweenness centrality
1. Ericsson	1118	1. Ericsson	149.029
2. Siemens	975	2. Nokia	123.929
3. Nokia	940	3. Siemens	89.065
4. Vodafone	700	4. Vodafone	77.065
5. Motorola	693	5. Motorola	71.826
6. Nortel Networks	683	6. British Telecom	68.808
7. Lucent Technologies	475	7. France Telecom	63.723
8. Alcatel	460	8. Orange	58.705
9. AT&T Wireless (AWS)	436	9. Lucent	58.319
10. T-Mobil	387	10. Nortel Networks	56.012
11. Telia	277	11. Alcatel	49.713
12. Orange	274	12. T-Mobil	37.846
13. British Telecom (BT)	230	13. NTT DoCoMo	30.717
14. France Telecom (France)	245	14. Telia	27.33
15. NTT DoCoMo	193	15. AT&T Wireless	25.693

measures the number of ties (coalition partner encounters) of each node of the network. The more ties an actor has, the more likely it is to receive the information that flows through the network. *Betweenness centrality* computes how often node k is along the shortest path between nodes i and j. Betweenness thus measures the extent to which k is able to control information flow within the network.

Table 16.3 lists the most central firms according to the two different centrality indices. In terms of their network positions, large European and North American equipment companies such as Ericsson, Siemens, Nokia, Alcatel, Nortel, and Motorola are the most central. Telecom operators (apart from Vodafone, which is central in both measures) are in the second tier of central players. In contrast, very few Japanese companies occupy central positions in the 3GPP work item process. Additionally, smaller technology or R&D suppliers are found at the fringes of the network. The results based on the betweenness centrality measure differ slightly from those based on the degree centrality measure in that telecom operators such as British Telecom, France Telecom, and Orange are clearly more central in terms of betweenness. These operators appear thus to have been successful in forging strategically important coalitions as opposed to gaining influence through a sheer volume of coalition activity.

TABLE 16.4   Centrality of 3GPP member firms in private
alliance networks, 1995–2001

Company	Degree centrality	Company	Betweenness centrality
1 IBM	75	1 IBM	1049.4
2 Toshiba	73	2 Motorola	1025.5
3 Motorola	66	3 Siemens	672.7
4 Hitachi	60	4 Ericsson	652.9
5 NEC	58	5 Intel	560.2
6 Mitsubishi	51	6 Microsoft	546.7
7 Intel	47	7 AT&T	540.9
8 Fujitsu	46	8 Hitachi	472.2
9 Siemens	44	9 NEC	446.7
10 Microsoft	42	10 Lucent	346.5
11 Sony	42	11 Compaq	327.1
12 H-P	38	12 H-P	287.5
13 Sharp	32	13 Toshiba	247.4
14 Matsushita	30	14 Alcatel	230.1
15 Oki	26	15 Fujitsu	228.1

Finally, the CATI database provides information released by firms about their cooperative arrangements since 1987. This dataset concerns alliances of 3GPP member firms between 1995 and 2001. There is a slight bias in the CATI database toward large companies, so the alliance activity of small technology companies may be underrepresented. According to the CATI database, many 3GPP member companies engage privately in strategic alliances with other members even outside the public industry fora. However, many of these alliances are not necessarily aimed at technologies related to wireless communications. Information technology companies have many other active technology areas where cooperation is a useful strategy. Nevertheless, even unrelated interactions may prove important when an opportunity arises to cooperate in technologies that are indeed related to wireless communication. According to Table 16.4, large North American and Japanese information technology companies dominate the global alliance activity among 3GPP members. Further, equipment manufacturers clearly dominate this type of cooperative activity. Few telecom operators appear in the alliance dataset, not to mention the top 15 most active alliance partners. It is also interesting to note that for most firms, more than half of their alliance partners are also members of the 3GPP (not reported in the table). 3GPP thus represents a meaningful innovation network for these firms.

The structure of the private alliance network looks very different from that of the 3GPP work item coalition network. Computer manufacturers and more generalist information technology companies are at the center of this network, while the wireless or general telecommunications equipment companies are in relatively more peripheral positions (Table 16.4). Interestingly, the raw data suggest that Japanese companies are tightly clustered together, while Toshiba appears to take on a gatekeeper role between other Japanese firms and the rest of the world. As a result, Japanese companies have high degree centrality scores but are substantially lower in terms of betweenness centrality. Nokia and Ericsson are in very similar and not particularly central positions in respect of the alliance network. However, Ericsson has a very high betweenness centrality score, suggesting that with relatively few alliances this company has been able to get into a position of effective information acquisition and influence.

STANDARDIZATION STRATEGY: CHOICE OF COOPERATIVE ARRANGEMENTS FOR STANDARD SETTING

This section applies simple statistical analyses to determine how firms' characteristics are related to their approaches toward cooperative standardization. Particular focus is on how nationality influences the choice of standardization strategy, controlling for intellectual property (IP) portfolio, size, and industry segment. Based on the earlier discussion, North American firms, with their market-oriented standardization tradition, are expected to be more active in private alliances than European firms. European firms have a long tradition in SDO participation and can thus be expected to have stronger network positions in 3GPP coalitions and overall be more active in SDOs.[1] Asian companies have been able to rely on their governments' leadership in standards development, and therefore they are expected to lag behind American and European companies' SDO behavior. Additionally, the relationships between the different forms of standards cooperation are explored: do companies rely on one single form of standards cooperation, or do they engage in a multitude of cooperative arrangements? In the latter case activity in one form of cooperation would positively predict activity in other forms.

These questions are examined through partial correlation analyses of 3GPP firms' alliance network centralities, SDO membership activities, and centralities in the 3GPP Work Item coalitions. Alliances include all joint ventures, joint development agreements, and other strategic contractual arrangements from 1995 to 2002, as captured by the CATI database. As far as network centrality in 3GPP work item coalitions, the focus here is on degree and betweenness centrality, which yield slightly different results.

In addition to the main variables of interest described above, the analysis includes indicators of firms' patent portfolios, size, and industry segment. Technological capabilities, measured by patent portfolios, may facilitate becoming desirable partners in alliances and coalitions (Bekkers, Duysters, and Verspagen 2002). Similarly, larger firms have more resources and market power to become central players in cooperative networks. Additionally, the incentives driving the behavior of R&D or software providers may be different from those driving telecom operators or equipment manufacturers. The industry segment is thus a relevant control variable.

Three distinct measures of intellectual property rights (IPR) are considered. The number of patents registered in the United States Patent and Trademark Office (USPTO) can be seen to measure very general technological strength. The number of declarations of essential IPRs in ETSI, the European Telecommunications Standards Institute, may be somewhat biased toward European companies, but that number includes only patents concerning telecommunications technology. The third measure is the number of IPR declarations in ETSI that relate to the UMTS standard. This indicator measures strength in the relevant technological field. The measures are thus complementary, and none of them alone is perfect: ideally, we would like to measure firms' strength in intellectual property related to wireless telecommunications technologies, without focusing too narrowly on the European market. However, it is highly likely that firms contributing to technologies behind the UMTS standard declare them with ETSI, independently of their home region.

Table 16.5 presents descriptive statistics of the dependent and independent variables. Firms in the sample range from small and specialized software or R&D houses to global equipment suppliers and large telecom operators. Twenty-two percent of the firms are North American, and 15 percent are Asian. Most of the remaining firms are European, but the dataset also contains one Israeli firm and a handful of privately held companies whose origins are unknown. Firm size, measured by the number of employees, ranges between 190 and 440,000. Many of the equipment firms and operators have hundreds of thousands of employees.

To provide the first view of national differences in standardization strategies, the various strategies of different nationalities are displayed in Table 16.6. In absolute terms, Asian firms are clearly concentrated on alliances and SDO memberships, but their centrality within 3GPP is very low. American firms take the middle ground in all strategies, while European firms are by far the least central in alliances and the most central within the 3GPP coalitions.

Partial correlations, presented in Table 16.7, shed light on the interrelationships among cooperation strategies while controlling for other relevant

TABLE 16.5   Descriptive statistics

	Mean	Std. dev.	Minimum	Maximum	Cases
SDO membership	4.44	4.09	1	16	85
Alliance degree centrality	6.34	13.94	0	66	85
Alliance betweenness centrality	68.47	174.48	0	1025.48	85
3GPP coalition degree centrality	119.79	253.11	0	1204	85
3GPP coalition betweenness centrality	196.41	476.48	0	2404.61	85
IPR @ USPTO	178.22	432.33	0	1706	85
IPR @ ETSI	63.87	247.37	0	1584	83
UMTS IPR @ ETSI	8.07	35.28	0	212	85
Employees	73 109	88 936	190	430 200	57
R&D firm	10%				77
Equipment supplier	38%				77
Operator	31%				77
Software supplier	9%				77
North American	22%				85
Asian	15%				85
European	48%				85

TABLE 16.6   Means of standardization strategies by region of origin

Strategy	Asia	North America	Europe	All
Alliance network:				
Degree centrality	19.2	7.2	3.8	6.3
Betweenness centrality	81.9	121.7	59.6	68.5
SDO memberships	7.4	5.6	4.0	4.4
3GPP coalition network:				
Degree centrality	66.7	148.0	158.6	119.8
Betweenness centrality	94.5	197.1	285.9	196.4
Observations	13	19	41	85

factors. The first three columns report the correlations using degree centrality measures, and the last three report the correlations using betweenness centrality measures. The results suggest that large firms and firms with an extensive patent portfolio indeed tend to be more central in alliances and active in more standards coalitions and industry organizations. However, the impact of firm size is strongest with respect to alliance centrality, but variable with respect to SDO memberships and coalition centrality. Also, different types of IPR are relevant in different cooperative arrangements. Patents registered at

TABLE 16.7   Summary of partial correlation analysis

	Degree centrality			Betweenness centrality		
	Alliance network centrality	SDO membership activity	3GPP coalition network centrality	Alliance network centrality	SDO membership activity	3GPP coalition network centrality
Firm size (# employees)	+ +	+	o	+ +	o	+
IPR at the USPTO	+ +	+ +	− −	+ +	+ +	− −
IPR at ETSI	o	+ +	−	−	+	o
UMTS IPR at ETSI	o	o	+ +	o	o	+ +
Alliance network centrality	1	o	+ +	1	+	+ +
SDO memberships	o	1	+ +	+	1	+ +
3GPP coalition nw centrality	+ +	+ +	1	+ +	+ +	1
R&D firm	o	o	o	o	o	o
Equipment manufacturer	o	o	o	o	o	o
Telecom operator	−	− −	+ +	o	o	+
North American	o	+ +	− −	+ +	+ +	− −
Asian	o	+ +	− −	− −	+ +	− −

NB: + + implies a strong positive correlation ($\rho > 0.20$), + implies weaker positive correlation ($0.15 < \rho < 0.20$). Similarly, − − implies strong negative correlation, and − implies weaker negative correlation. o refers to no statistical relationship. The reference group for industry segments are software firms, and the reference group for geographic origins are European and other firms. The full results are available from the author on request.

the USPTO are associated with private alliances and SDO memberships, while declarations concerning UMTS-patents are relevant in 3GPP coalition networks. IPR declarations at ETSI more generally are also significantly associated with SDO memberships.

After controlling for size, patent portfolios, and industry segment, Asian firms are not any more likely to be central in the alliance network than European firms, and North American firms are more central only with respect to the betweenness measure. In fact, Asian firms are less central than European and other firms in terms of betweenness. However, American and Asian companies are significantly more actively engaged in SDOs and significantly less central in the 3GPP coalition networks than European firms, independent of the centrality measure. It thus seems that the European tradition of cooperation within the ETSI and, more recently, 3GPP is still strong. In contrast, American and Asian companies are both compensating by investing in other SDO memberships, and American firms also by private alliance activity.

The second research question concerns the relationships among alliance and coalition network positions and SDO memberships. The simple correlation analyses here suggest that these cooperative strategies are complements rather than substitutes. In particular, coalition network centrality has a strong positive correlation with both alliance network centrality and SDO membership activity. The correlations between alliance network centrality and SDO memberships are also positive but weaker. In other words, firms attempt to be central in 3GPP coalitions and simultaneously pursue alliance strategies or SDO strategies, but do not necessarily try to engage in both alliances and SDOs.

To summarize the empirical findings, IPR portfolio measures are positively associated with all cooperative arrangements, but firm size is an important correlate of alliance network positions only, not that of 3GPP coalition network positions or SDO memberships. Second, firms of different nationalities have followed different cooperative strategies. North American firms are more central in terms of betweenness in alliance networks, and both North American and Japanese firms are more likely than European firms to participate in a multitude of industry associations, while European firms put effort into activities within the 3GPP. This may partly derive from European companies' close relationships that date from the period when the predecessor of 3GPP was still under the umbrella of ETSI. However, during the time period of interest here, 3GPP had already become a globally operating and competing standards development organization. Nevertheless, it still appears more difficult (or less worthwhile) for non-European firms to become central actors in the coalition network. Finally, the evidence suggests that 3GPP coalition positions complement rather than substitute for SDO and alliance activities. However, firms may substitute SDO and alliance strategies for one another.

CONCLUSION

This chapter has discussed firms' standardization strategies using a new dataset of companies that are members in the Third Generation Partnership Project (3GPP). 3GPP is one of the main standardization organizations developing technical specifications for third-generation wireless telecommunication systems. The purpose of the study is to compare European, American, and Asian firms' strategies toward influencing standardization outcomes. It is argued that because of the different institutional histories on these continents, firms have different approaches to standardization. Domestic policies and institutions regarding standardization and industrial cooperation influence firms' accumulated capabilities and connections in this area. On one hand, if the national government has traditionally led standards development, companies may not have built the expertise and linkages necessary to be effective in the current standardization networks. On the other hand, in the absence of government participation in standardization, firms may have become highly

356   skilled in forming competing private alliance constellations but may not have
expertise in or understanding of formal and public standard-setting processes.
Additionally, firms representing different industrial segments may behave differently with respect to standardization. For example, many of the American
companies that currently are 3GPP members are at the core of their operations computer-oriented information technology firms (e.g., Cisco, IBM,
Apple, and Sun Microsystems). These firms have entered the wireless telecommunication scene because of the ongoing convergence of computing and communications. Thus, they are not telecommunication equipment suppliers in
the same sense as, for instance, Ericsson, Nokia, and Motorola are. Firms'
technological origins undoubtedly affect their standardization strategies. Indeed, in the network analyses the computer hardware suppliers stand out because of their dense alliance network.

A conceptual discussion in this chapter identifies the various strategies that
firms deploy in trying to influence and learn from standardization in wireless
telecommunications. The focus here is on cooperative arrangements to standardize technology through strategic alliances, industry associations, and
coalitions in a formal standards development organization. First, organization
theories do not explicate the observed multitude of organizational arrangements to achieve standardization outcomes. Previous research has not identified the roles played by alliances, industry associations, and coalitions within
SDOs in setting standards cooperatively. In particular, are these arrangements
substitutes or complements? This chapter argues that the various types of cooperative arrangements complement one another in the political game of
influencing other actors in standard setting.

In addition to their size (reflecting market share) and technological capabilities, firms' national origins affect their standardization approaches. Firms'
resources and strategies are influenced by the institutions and incentives
prevalent in the national origin. For example, European countries started
multinational cooperative standardization efforts in the wireless telecommunications area long before either North American or Asian countries did.
Partly for this reason, European firms are likely to be more central in the
standards development organization that develops specifications for the
UMTS system. Moreover, connections and cooperative arrangements forged
in the home base tend to persist even in the globalized telecommunications
industry. This is clearly the case regarding the alliances of Japanese firms in
the dataset. Therefore, North American wireless telecommunications companies were hypothesized to rely more strongly on private arrangements for
technology coordination, while European and Asian firms were expected to
emphasize public and formal cooperative arrangements.

An empirical analysis of firms active within the technical specification groups and working groups of 3GPP lends support for some of the ideas. North American firms are indeed more involved in strategic alliances than European and Asian firms, and they are more central (measured by betweenness centrality) in the resulting alliance network. On the other hand, European firms are the most central within the standards forum 3GPP, while Asian and North American firms operate more often than their European counterparts through a wide variety of industry associations and standards development organizations. Thus, it seems that even though the sample firms currently are global in their operations, institutional legacies from an earlier time continue to influence their behavior. Additionally, cultural factors may support certain cooperative arrangements and partnership choices better than others.

The results of the analysis also indicate that private alliance activity and industry association memberships facilitate becoming a central participant within 3GPP and vice versa. This supports the hypothesis that the forms of cooperation studied here are complements rather than substitutes. However, it is possible for firms to choose to pursue either SDO or alliance strategies, but not both. Correlation is weaker between these two types of cooperation. Nevertheless, the positive partial correlations among these three cooperative forms for standardization, controlling for a host of other firm- and industry-level factors, suggest that firms active in standard setting tend to employ multiple organizational approaches. They are highly likely to engage in private coalitions or semi-public industry associations and fora in addition to formal and public standards bodies such as 3GPP. Thus, it can be argued that the reason we see so many different organizations developing standards for the same third-generation wireless system is political rather than economic. The primary purpose of these cooperative technical organizations appears to be to rally competitors', complementors', and users' support for technical solutions early on, rather than simply to develop technical specifications most efficiently.

The empirical analysis is tentative and should be viewed as raising questions for further research rather than providing conclusive tests for the currently developed hypotheses. Future research efforts could examine the relationships among the different forms of cooperative standardization in more depth, preferably with longitudinal data. Also, an important policy question for future research is how firms' strategies to standardize technologies affect subsequent market evolution in terms of the speed of market emergence and the nature of competition. It is conceivable that when firms operate through open SDOs as opposed to private alliances, a more level playing field is created, facilitating rapid entry of small players. Nevertheless, from this study it is clear that large technological powerhouses are the most central in the stan-

358   dards negotiations and that it can be difficult, if not impossible, for smaller
technology specialist firms to have an effect on the outcome.

This study suggests that the organization of cooperative standard setting be
viewed as a game of political influence rather than a problem of maximizing
organizational efficiency. Moreover, there are significant differences in cooperation behavior among firms from different geographic regions. Controlling
for a number of other firm characteristics, Japanese and North American firms
are finding approaches to influence standardization outcomes that differ from
those of their European counterparts. Traditions, existing relationships, and
institutional legacies affect the behavior of global high-technology companies.

NOTE

1. British firms are included in the group of European firms, even though according to Tate (2001) they may traditionally behave more like North American firms
because of their liberal market economy institutions. However, in the case of wireless
telecommunications, British firms have a long history of participation in the
ETSI/SMG in very central roles, and therefore they are expected to behave more like
continental European firms than North American firms.

IV.   SOCIAL TRANSFORMATIONS

17

WEAVING THE AUTHORITARIAN WEB

*The Control of Internet Use
in Nondemocratic Regimes*

Taylor C. Boas

In the preparatory meetings leading up to the December 2003 World Summit on the Information Society in Geneva, the delegations of several authoritarian regimes reacted strongly to the hands-off approach to Internet regulation promoted by the United States and other advanced democracies. Saudi Arabia, for instance, proposed that the development of the information society "shall be done without any prejudice whatsoever to the moral, social, and religious values of all societies"—values to which the Saudi government has appealed when justifying its own regime for Internet censorship. The Chinese delegation campaigned strongly against a statement of support for the principles of free speech enshrined in the Universal Declaration of Human Rights. Ultimately, the summit's final declaration disregarded the objections that these and other authoritarian governments had voiced during negotiations, but their positions stand as a strong statement that not all countries accept a laissez-faire vision for the future of the Internet.

At first glance, the negotiating positions taken by China and Saudi Arabia might suggest that authoritarian leaders in the information age face a stark choice: promote the development of an Internet that remains free from extensive government control, or exert control over the technology by restricting its diffusion. Whether because of inherent technological characteristics that complicate efforts to censor the Internet or because countries are under pressure to align their policies with those preferred by the international community, many scholars have assumed that the only effective way to control the Internet is to limit its growth or even keep it out entirely. Milner (2003a, 2003b), for instance, hypothesizes that authoritarian leaders will be less likely than democratic ones to promote Internet development, and she uses indicators of diffusion (such as users or hosts per capita) as proxies for government policy toward the Internet. Franda (2002) interprets national policies to restrict the free flow of information as being deviant and "isolationist" with respect to the international regime for Internet governance, and he expresses skepticism that they will be sustainable. Kedzie (1997) argues that the technology poses a "dictator's dilemma" to autocrats, who must either connect to the Internet and democratize or shun the information revolution and accept economic decline.

Although its use can undoubtedly pose challenges to authoritarian rule, the Internet is an attractive technology to all governments, democratic and authoritarian alike, and hardly any dictator has been willing to ignore it entirely. Internet diffusion offers substantial economic benefits in terms of the potential development of an e-commerce sector, establishing conditions conducive to foreign investment, and stimulating existing domestic industries. Use of the Internet within government itself (e.g., procurement) can facilitate the development of a rational bureaucracy, reducing opportunities for corruption and graft. Implementing government services online (payment of income taxes and the like) can increase efficiency and boost public satisfaction with the regime. If it is possible for authoritarian rulers to have the best of both worlds — reaping the benefits of Internet diffusion while staving off any potentially destabilizing political effects — they will certainly want to do so.

In this chapter I argue that, contrary to the assumptions of many studies of the Internet in authoritarian regimes, governments can establish effective control over the Internet while simultaneously promoting its development. Indeed, China and Saudi Arabia are two of the most prominent examples of this phenomenon. Although they may have been less influential than they had hoped in negotiations over the global governance of the Internet, both have long sought to implement within their own borders the principles they recently espoused in Geneva. Far from trying to regulate the Internet by merely restricting its diffusion, these countries have employed both technological and institutional means to control use of the Internet while also encouraging its growth. In doing so, they stand as counterevidence to much of the optimistic thinking about the Internet and democratization that was voiced by pundits and politicians during the early days of the Internet and the technology boom of the late 1990s.

In the first section of this chapter I address the technological bases of Internet control in authoritarian regimes. Much of the early scholarship on the feasibility of government regulation of the Internet pointed out that the network was initially designed as a technology that would be difficult to control at a centralized level (D. Johnson and Post 1996; Froomkin 1997; for a review, see Boas 2004). I argue, however, that this control-frustrating characteristic of the early Internet is not necessarily locked into place as the technology diffuses around the globe. On the contrary, the logic of Internet diffusion means that the global network is quite flexible and capable of being modified in new environments, allowing authoritarian regimes to embed control-facilitating technological features into the portions of the global Internet that fall within their borders.

Although most authoritarian regimes have exploited the flexibility of Internet technology to implement technological measures of control, determined users have almost always found ways to circumvent these barriers. In the second section of the chapter, therefore, I distinguish between perfect and effective control — the former being what matters for tech-savvy individuals who want to gain unfettered access to the Internet, the latter being what authoritarian regimes actually pursue. It is in establishing effective control over Internet use that institutional constraints on behavior — law, social norms, and the market — come most clearly into play. By manipulating the architecture of a flexible technology and by leveraging influence over laws, social norms, and the market in ways that supplement these architectural constraints, the leaders of authoritarian regimes can exert control over the use of a supposedly control-frustrating technology.

Throughout the chapter I illustrate these conceptual and theoretical arguments about the Internet in authoritarian regimes with evidence from the cases of China and Saudi Arabia. As the two countries that have developed what are probably the world's most extensive technological mechanisms for Internet censorship, China and Saudi Arabia are not intended to be representative of authoritarian regimes as a whole. Rather than showing what is *typical* of nondemocratic governments, these extreme cases of Internet regulation illustrate what is *possible*. If each has largely succeeded in establishing control over the Internet, others may prove similarly capable in the future.[1]

INSTITUTIONAL AND TECHNOLOGICAL
CONSTRAINTS ON INTERNET USE

In evaluating the potential for establishing control of the Internet in authoritarian regimes, it is useful to consider the means by which authorities might seek to do so. In his study of Internet regulation in advanced democracies, Lessig (1999) has identified four specific mechanisms — law, social norms, the market, and architecture — that governments can employ to control Internet use. The first three can be loosely grouped together as *institutional* constraints — "the humanly devised constraints that shape human interaction" (North 1990, 3). The manner in which they influence behavior is fairly straightforward: laws threaten punishment for prohibited activities, violators of social norms may incur ostracism, and the market can encourage or discourage particular activities based on their cost. As societal constructs, each of these institutional constraints is capable of evolution and change over time. Laws are challenged and overturned; social norms evolve; markets fluctuate, and the degree to which any individual is constrained by them varies with wealth.

Architectural means of regulation occupy a somewhat different category than institutional constraints. In the case of the Internet, architectural constraints consist of the technological characteristics that make certain types of Internet use easier, more difficult, or impossible. In contrast to institutional constraints on Internet use, the technological architecture of the Internet is not as obviously capable of significant evolution. The Internet is a technology whose diffusion is characterized by increasing returns to scale; historically, many such technologies have been examples of path-dependent development and the lock-in of technological characteristics that remain static over time (David 1985; Arthur 1994). If we accept that the Internet's fundamental characteristics initially made it difficult to control, and if the diffusion of the Internet does indeed give rise to technological lock-in, then the lack of an effective architectural constraint on Internet use might actually be quite *incapable* of change over time.

If true, the potential persistence of a control-frustrating Internet architecture bears special significance for the regulation of Internet use in authoritarian regimes. When effectively implemented, architectural constraints are the only type of regulation that can exert immediate and absolute control over human behavior (Lessig 1999). Laws and social norms can be violated at will; sanctions for such violations are imposed by a government or community only after the fact. Market constraints can be violated in the form of theft; market actors must rely on both social norms and the legal system for effective enforcement. But a technological architecture that makes certain types of Internet use impossible cannot be circumvented even at the risk of future sanctions, and the effectiveness of this constraint does not depend on support from the community or the legal system.

Conversely, if the Internet's architecture is inherently unable to prevent certain types of online behavior, it is impossible for governments to place absolute constraints on Internet use. The combination of law, social norms, and market constraints can discourage the prohibited activity, but they can never render it impossible. Thus, the supposed rigidity of the Internet's technological architecture is a cornerstone of the argument that the medium inherently frustrates governments' efforts at control. To determine whether the development of the Internet in authoritarian regimes does in fact involve the replication of its initial control-frustrating characteristics, it is useful to see how well the dynamics of path dependence describe this technology's global diffusion.

Path Dependence and the Internet's Control-frustrating Characteristics

The concept of path dependence in technological development describes a pattern in which the particular configuration for a new technology becomes locked in over time as increasingly widespread use raises the cost of switching to another alternative. In particular, the diffusion of such technologies involves increasing returns to scale, which derive from at least one of several characteristics (Arthur 1994). The technology may have a large ratio of fixed to marginal cost, so that the production cost per unit declines as production increases. The technology's adoption may also be characterized by learning effects—the more it is used, the more its efficiency can be improved vis-à-vis other alternatives. Finally, path-dependent technologies often display network effects, in which the demand for the technology (and its value to each current user) increases with each additional unit sold.[2] The chosen technology constitutes a standard around which users coordinate, and though any one of them might *ceteris paribus* prefer a different technological configuration, the benefits of standardization outweigh the benefits of switching.[3]

The Internet shares each of these characteristics, making it a technology whose adoption generates increasing returns to scale. Establishment of the Internet's physical infrastructure and development of its core protocols involved significant fixed costs, which were underwritten by both the U.S. government and AT&T (which had already built many of the transmission lines upon which Internet traffic would flow). In contrast to these high fixed costs, the marginal cost of connecting additional users to the Internet is relatively low. Use of the Internet also involves learning effects, as with any complex technology. Most significantly, the development of the Internet generates especially strong network effects. Telecommunications technologies derive their entire value from the ability to interconnect with others; a single fax machine has no utility if there are no other fax machines to receive transmissions. Similarly, the value of the Internet is largely dependent upon the number of people and information resources that are connected to it.

Not only is the Internet a technology subject to increasing returns, but it was initially designed as a technology that would be resistant to centralized control. The original engineering decisions which gave rise to this characteristic were a product of the specific economic, political, and social environment in which the Internet was created. In part, the technological characteristics of the early

Internet derived from the norms of its designers and initial user community. The technology was originally the tool of a small group of engineers and academics who were wary of bureaucracy, trusted each other, and worked well through consensus. In light of this culture, they made specific choices about the design of the technology that rendered the network resistant to efforts at centralized control (Abbate 1999).

An even more important influence on the technological configuration of the early Internet was the military imperative for its development (Abbate 1999). The Internet has its origins in technology funded by and developed for the U.S. Department of Defense — packet-switching networks designed in the early 1960s and their first large-scale implementation in the ARPANET. The rationale for packet-switching technology was to design a communications network that could not be controlled from any single, centralized point, so that communications capacity could not be disabled by an enemy attack on a key portion of the network. With both the ARPANET and the later development of protocols for the Internet, survivability was the paramount goal, thus ensuring that these computer networks would not lend themselves to centralized control (Clark 1988).

The particular characteristics of the Internet that served to frustrate attempts at centralized control involve the end-to-end arguments in network design (Lessig 1999; Lemley and Lessig 2000; Blumenthal and Clark 2001). As guidelines for the design of computer networks, the end-to-end arguments state that complexity and control should be implemented at the "ends" of the network—the multiple computers and individual users that are interconnected (Saltzer, Reed, and Clark 1984). Meanwhile, the core of the network performs simple data transfer functions that do not require knowledge of how the ends are operating. In contrast to the telephone network, in which complex call routing is performed by a small number of centralized switching stations, the core infrastructural and computing elements of a "stupid network" such as the Internet simply move packets of information indiscriminately (Isenberg 1997). Because the Internet was built around an end-to-end design, one cannot control the entire network through control of a small number of centralized nodes. Control can be exerted at the ends of the network, but as these ends multiply, controlling the entire network by controlling the ends becomes less and less feasible.

A control-frustrating technological architecture suited the needs and preferences of the Internet's designers and initial user community, but the technology has since spread into a number of environments in which centralized control of information is a more desirable feature. One of the most important of these major shifts involves the global diffusion of the Internet. Today, the most rapid growth of the Internet is taking place in the developing world, including a number of authoritarian regimes where standards of information control are quite different from those in the United States. The leaders of these countries generally recognize the tangible benefits that the Internet has to offer, yet they worry that Internet use might pose a political threat, challenge state control of economic resources, or offend local cultural sensitivities. To reap the benefits of the technology while avoiding what they see as negative ramifications, their leaders would prefer to exert greater centralized control over Internet use.

If the dynamics of Internet development mean that its control-frustrating characteristics are locked into place as it diffuses around the world, the task of authoritarian leaders is a difficult one. Without recourse to an effective architectural constraint, authorities would have no means of exerting absolute control over use of the medium. Meanwhile, the economic logic of the technology's diffusion implies that there are few attractive alternatives to connecting to this control-frustrating Internet. The value of a single standardized network used by millions of people around the globe far exceeds the value of any alternative network that authoritarian governments might choose to construct within their own borders.

COMPOSITE STANDARDS, MACRO-LEVEL FLEXIBILITY, AND THE POSSIBILITIES FOR INTERNET CONTROL

When viewed through the lens of path-dependent technological development, the case for an inherently control-frustrating Internet may appear solid. This argument, however, rests upon the assumption that the architecture of the Internet is incapable of fundamental change. In this section, I delve deeper into the nature of the Internet's technological architecture, demonstrating that the composite nature of the "standard" which generates increasing returns to Internet diffusion actually gives the technology a great deal of flexibility at the macro-level. This capacity for evolution means that authoritarian leaders may be able to adapt this malleable technology for their purposes, embedding technological measures of control within the national computer networks that connect their citizens to the Internet.

To see how the architecture of the Internet might be characterized by flexibility rather than stasis, it is useful to consider the nature of the standard around which users of the Internet coordinate. In many traditional cases of pathdependent technology development, coordination around a single, simple standard (e.g., the QWERTY typewriter keyboard or the VHS format for videocassette tapes) is what generates network effects and contributes to lock-in

368 through increasing returns. The Internet, however, involves a whole series of separate standards at different layers of the network, working together in a complex fashion to facilitate communication. The value of connecting to the Internet is not simply derived from coordination around the core TCP/IP standard as a way of exchanging data traffic. Rather, network effects in the case of the Internet are derived from coordination around the entire package — standards for e-mail, Web browsing, streaming audio, encryption, and many more. At the macro-level, therefore, the Internet can be thought of as constituting a *composite* standard, with hundreds of simple standards as its constituent parts.

The composite nature of the standard involved in Internet diffusion lends great flexibility to this technology, allowing it to be adapted to meet the operating demands of new environments. At the micro-level, the individual standards for particular Internet services display a fair amount of inflexibility; once implemented and employed by millions of computers worldwide, these individuals protocols are very difficult to change.[4] At the macro-level, however, the combination of parts that make up the Internet's composite standard has changed significantly over time. The hypertext transfer protocol (HTTP) for the World Wide Web, for instance, was not a part of the Internet at its origins, but it is an essential component of the Internet's composite standard today. Indeed, both e-mail and the Web — two of the Internet's most popular applications — were not originally envisioned by the Internet's creators but rather resulted from processes of informal experimentation. The Internet's macro-level flexibility has allowed it to incorporate these and other new applications as its operating environment changes over time.

Like the characteristics that rendered the Internet challenging to centralized control, the Internet's flexibility is not inherent but was rather explicitly designed into the network. Many of the same characteristics that made the Internet hard to control make it a flexible technology as well. Unlike the telephone network, which was designed specifically for voice traffic, the core of the Internet was not optimized for any particular service. At the time of its creation, there was little sense of what services the Internet would need to support in the future, so the core of the network was built as a set of simple, flexible tools. Any service that conforms to the published protocols for addressing and transmitting information can be implemented at the ends of the network without altering the center. The Internet's central mechanisms simply move information indiscriminately; the core of the network does not need to know if it is transmitting packets from an e-mail, a Web site, streaming audio, or some as yet uninvented service. Thus, the characteristics of the Internet as a whole can be altered by adding new protocols that will help the technology meet the needs of operating in new environments.

Controlling the Ends of the Internet

As the Internet spreads to authoritarian regimes around the world, its macro-level flexibility suggests that their leaders may be able to adapt this malleable technology for their own purposes. To see exactly how this might occur, it is useful to reconsider the notion of the end-to-end arguments. As principles of network design, end-to-end arguments place users at the ends of the network. In reality, however, the Internet is much less a single network of individual users than it is a network connecting separate computer networks. Networks are interconnected through a gateway; behind the gateway, each individual network can be configured in any number of ways as long as it is compatible with the TCP/IP protocols. Conceptually, therefore, it may well make more sense to think of the Internet's component networks as its ends than to think of individual users as the outer edge of a single, seamlessly interconnected Internet.

When separate networks are conceived of as the ends of the Internet, new meaning is lent to the maxim that one can control the Internet only by controlling its ends. Exerting technological control of the Internet at the user level, in keeping with the end-to-end design principles, constitutes a quite daunting task; it would be akin to mandating that foolproof censorship software be installed on every user's computer. It is much more feasible, however, to exert control over individual networks connected to the Internet, especially where traffic passes through a single or small number of choke points.

Rather than controlling the *entire* Internet, governing authorities always attempt to control a relevant subset of Internet users. The administrators of corporate computer networks, for instance, often monitor employees' Internet usage and block certain types of non-work-related traffic. Users who have a choice of network will always be able to switch to a more liberal environment. For those with no realistic choice, however, the distinction between control of the Internet and control of a network attached to the Internet is largely irrelevant. For them, the choice is between access to a restricted Internet and access to nothing at all.

Such is the situation in many authoritarian regimes that are developing national computer networks with connections to the Internet. While in most democracies a number of individual Internet service providers (ISPs) maintain separate links to the global Internet, in authoritarian regimes all Internet users may effectively be members of a single national network. Even when there are multiple ISPs within a country, international connections to the global Internet are often channeled through a single government-controlled gateway.

Moreover, architectural constraints on the Internet at the national level can be supplemented by additional measures of technological control implemented by individual ISPs, Internet cafés, and online chat rooms. Each of these entities constitutes an additional "end" of the Internet at a level more diffuse than the national gateway but still closer to the Internet's core than the individual user. Although governments may have less direct control over the technological configuration of Internet access at these levels, they can leverage their control of law and their influence over markets and norms in ways that will encourage private entities to establish their own architectural constraints on Internet use.

Technological Control of Internet Use in Authoritarian Regimes: Saudi Arabia and China

Given the political, economic, and social conditions prevailing in many authoritarian-ruled countries, one should not be surprised to find that their governments have sought to establish technological measures of control over the portions of the Internet within their borders. The governments of authoritarian regimes are typically central players in the growth of their own information infrastructures, and one would expect them to build architectures of control into their "ends" of the Internet. In the section that follows, I show how the governments of Saudi Arabia and China have sought to develop national computer networks that facilitate rather than frustrate efforts at state control.

Saudi Arabia. Saudi Arabia's approach to the Internet has been strongly influenced by its conservative society, in which there is significant public concern over pornography and material offensive to Islam and strong support for censorship of this type of content on the Internet. In addition, Saudi Arabia is a monarchy whose royal family is quite sensitive to criticism and dissent; it is particularly cognizant of the threat posed by overseas opposition groups such as the Committee for the Defense of Legitimate Rights and the Movement for Islamic Reform in Arabia, which seek to turn public sentiment against the regime.

Saudi Arabia has therefore moved very slowly in its approach to the Internet. The country's first connection was established in 1994, but public access was delayed until 1999 while authorities perfected their technological mechanism for Internet control. Since then, public use of the Internet has grown steadily: from 690,000 users in April 2001 to 1.46 million (5.7 percent of the population) in September 2003.[5] Saudi Arabia has chosen to permit multiple, privately owned ISPs, but all international connections to the global Internet pass through a gateway maintained by the Internet Services Unit (ISU) of the King Abdulaziz City for Science and Technology, the Internet's governing authority in the country.

The concentrated national network structure has facilitated the technolog-ical control of Internet content, a goal about which Saudi authorities have been quite open.[6] Since the debut of public access in Saudi Arabia, all traffic to the global Internet has been filtered through a set of proxy servers managed by the ISU that are intended to block information authorities consider socially and politically inappropriate. Market conditions have facilitated the imposi-tion of censorship: since 1999, Saudi Arabia has outsourced the provision of censorship software to U.S.-based Secure Computing. Saudi authorities cur-rently rely on the preset list of sexually explicit sites in Secure Computing's SmartFilter software, which is customized with the addition of political and religious sites (Zittrain and Edelman 2002a). In addition, the ISU's Web site includes forms where the public can request that sites be blocked or un-blocked; officials report an average of five hundred block requests and one hundred unblock requests per day.

China. In its approach to the Internet, China has sought a strategy which will allow it to promote widespread, market-based diffusion of the technology while still retaining governmental control. Internet growth in China has con-tinued steadily since public access was first introduced in the mid-1990s; as of December 2004, the government estimated that there were 94 million users, representing 7.2 percent of the population.[7] Because filtering so much traffic through a single international gateway would be nearly impossible, Internet control in China is more diffuse than in Saudi Arabia. It is difficult to ascer-tain the specific technological details, as China has been much less open about the configuration and extent of its censorship regime. All evidence sug-gests, however, that China employs multiple, overlapping layers of Internet control which have been quite effective at limiting the access of the majority of users. Zittrain and Edelman (2002b) describe a number of ways in which the architecture of the Internet in China has been modified to implement technological control. Blocking specific Web pages on the basis of IP address has been the most common. In September 2002, however, authorities imple-mented a more sophisticated system capable of blocking pages dynami-cally, based on either keywords in the URL (prohibiting Google searches on specific terms, for instance) or in the actual Web page requested. These meth-ods of blocking are a step beyond earlier strategies and mechanisms em-ployed elsewhere as they do not rely on a preexisting blacklist of prohibited Web sites.

At the level of the international gateway, the cornerstone of China's Inter-net control has been its system of interconnecting networks. Though China has promoted rapid proliferation of the ISPs that provide Internet access to end users, actual connectivity to the global Internet has long been channeled through a small number of interconnecting networks with ties to government ministries or important state companies. Four interconnecting networks were initially established in 1996; the number has since grown to nine, though because the Ministry of Information Industries has licensed additional networks it has made certain that they are under effective state control (Harwit and Clark 2001). Moreover, the structure of this market is more concentrated than the number of interconnecting networks implies: the top two networks, ChinaNET and China169, jointly control 88 percent of international bandwidth.[8] This structure facilitates the implementation of censorship at the national level. Chase and Mulvenon (2002), for instance, report that most national-level Internet filtering is implemented by the International Connection Bureau, a set of computers belonging to ChinaNET's owner, China Telecom. Moreover, the major networks routinely exchange information about specific Web sites that they seek to block.

China has also augmented its control over Internet architecture by establishing control at the level of ISPs, Internet cafés, and chat rooms. Such points of access to the Internet number in the thousands, and most are thoroughly private entities without the same ties to the regime as the interconnecting networks. At this more diffuse level, authorities can implement an architecture of control indirectly, through their legal influence over intermediaries and the creation of a market environment in which cooperation with authorities is good business practice.

China's Internet regulations make ISPs, Internet cafés, and chat rooms responsible for online content, and the threat of sanctions (along with occasional large-scale crackdowns) has encouraged these entities to implement their own technological measures of control. It is likely that at least some of the filtering methods described by Zittrain and Edelman (2002b) are implemented by ISPs instead of (or in addition to) the interconnecting networks. For their part, many Internet cafés have chosen to install blocking software to limit what their patrons can view, and chat rooms use a technology that scans for potentially sensitive postings and sends them to a Webmaster for review (Chase and Mulvenon 2002). In addition to these filtering measures, ISPs and Internet cafés have been required to implement technological architectures that facilitate government surveillance. Regulations introduced in October 2000 require ISPs to keep logs of Internet traffic for sixty days and deliver the information to authorities on request (Harwit and Clark 2001). For their part, many Internet cafés have installed software that allows public security bureaus to track user records and monitor Internet traffic remotely (Kalathil and Boas 2003).

The cases of Saudi Arabia and China confirm the expectation that the architecture of the Internet is not inherently control frustrating, even if this characteristic was a feature of the early Internet in the United States. Rather, 373 the logic of end-to-end network design shows that authoritarian governments can construct national computer networks attached to the Internet in ways that facilitate technological control.

PERFECT VERSUS EFFECTIVE CONTROL: THE IMPORTANCE OF INSTITUTIONAL CONSTRAINTS

Though undoubtedly effective for the majority of users, the technological measures of control implemented by authoritarian regimes such as Saudi Arabia and China still fall short of an absolute constraint on Internet use. Internet controls are never 100 percent secure; they can almost always be circumvented by determined, tech-savvy users willing to run risks and pay the possible costs of alternative access channels. In this section, I address these inherent imperfections in technological measures of Internet control and examine the ways in which authoritarian governments have sought to supplement them by leveraging a combination of legal, normative, and market-based constraints. Although perfect technological control over the Internet may never be possible, these institutional constraints are essential for establishing effective control over Internet use — a level of control that is sufficient for the political, economic, and social goals that the authoritarian leaders seek to meet.

Those skeptical of arguments about Internet control routinely point to the myriad ways determined users can circumvent technological measures of control. Saudi authorities have acknowledged that many users are finding ways to access forbidden Web sites, often through the use of overseas proxy servers (Kalathil and Boas 2003). Wealthy Internet users who find this avenue blocked can always dial into unrestricted accounts in neighboring Bahrain — a common practice in the days before public access was permitted in Saudi Arabia. In the Chinese case, ongoing arrests of online dissidents confirm that people are successfully engaging in types of Internet use the government seeks to block. Zittrain and Edelman (2002b) and Chase and Mulvenon (2002) detail a number of ways Chinese Internet users can attempt to circumvent controls, from the use of peer-to-peer file-sharing systems to entering the URLs of blocked pages in ways that may fool censorship mechanisms.

In addressing the implications of these inevitable cracks in national firewall systems, it is important to distinguish between perfect control and effective control of the Internet. Libertarian perspectives on Internet control are essentially concerned with the individual: will the government be able to prevent *me* from doing what I want to do online? Only perfect architectural constraints will be able to control the online activity of the most determined and tech-savvy users. But the perspective of authoritarian governments, or of any authority seeking to exert control over the Internet, is different. For them, the goal is almost never perfect control, attempting to thwart the evasive maneuvers of every enterprising, tech-savvy individual. Rather, authoritarian leaders seek to exert control with an external referent — control that is "good enough" with respect to any number of important objectives, including regime stability and protection of local culture. Effective control of this sort may not be able to change the behavior of the last 0.1 percent of Internet users, but this small number is rarely enough to seriously challenge the goals that most authoritarian regimes are trying to pursue.

It is in establishing and enforcing effective control over the Internet that institutional constraints on Internet use come most clearly into play. To understand the interplay of technological and institutional constraints, an economic interpretation is useful, with unrestricted Internet access thought of as a good demanded by different numbers of users depending on the price. Although perfect architectural constraints, if they existed, could control the behavior of every user, institutional constraints are best seen as raising the cost of circumventing control. The cost may be literal in terms of market constraints — for instance, a satellite connection necessary to circumvent national restrictions on the Internet. In terms of law or social norms, users face the metaphorical (but still very real) costs of ostracism or punishment when they are caught.

In this economic model, most consumers are quite happy using the Internet for entertainment, online games, communication with friends, and access to officially sanctioned news sources; they place a low value on circumventing controls, especially with regard to political information. Similarly, some percentage of users will always demand unrestricted access to the Internet even at extremely high prices; they will spend money for technology to circumvent censorship, engage in illegal political communication at the risk of punishment, and ignore disapproval from members of society who frown on lawless activity. As these costs are raised, however, demand for unrestricted Internet access shrinks. The government's goal is not to set the cost so high that demand is completely eliminated; rather, authorities seek to reduce this demand to the point of political insignificance.

Leveraging law, social norms, and the market to raise the cost of unrestricted Internet use allows for a much more effective implementation of control than do architectural constraints alone. If firewalls can be circumvented with fancy technology or international phone calls, the high price of these activities helps to render this architectural constraint effective. If tech-savvy patrons of Internet cafés can configure their browsers to access pornographic or dissident Web sites, they will be stopped only by the ingrained knowledge that such behavior is socially unacceptable or that café managers may be observing their Internet use and could report their transgressions to authorities.

Establishing Effective Control in Saudi Arabia and China

The cases of Saudi Arabia and China illustrate how governments can leverage institutional constraints to establish effective control over Internet use. In Saudi Arabia, the government has found support for its censorship regime among conservative Islamist groups that are primarily concerned about pornography. Social norms against viewing material deemed offensive to Islam encourage self-censorship among users, as do legal prohibitions on accessing forbidden content and the possibility that surveillance mechanisms can identify violators. Attempts to view blocked sites are greeted with a message that all access attempts are logged; ISPs are required to keep records on the identity of users and provide such information to authorities if requested. In addition to these legal and normative sanctions, market conditions (such as the high price of dialing into an ISP outside of the country) have also discouraged those who would seek to obtain unrestricted access to the Internet in Saudi Arabia.

In China, the use of institutional constraints on Internet access has been even more extensive, likely owing to the greater challenge of exerting purely technological control over a broader and more diffuse Internet. One of the main ways China promotes self-censorship involves legal regulation of users. Authorities have engaged in high-profile crackdowns on various dissidents and individuals who run afoul of the regulations by engaging in politically sensitive communication. Chase and Mulvenon (2002) have offered numerous examples, from Huang Qi, who operated a Web site with news about the Tiananmen Square massacre, to members of the Falun Gong, who disseminate their materials online. Sentences of several years in prison are common for such offenses, undoubtedly deterring others who might have inclinations to engage in similar activity.

Similarly, periodic crackdowns on the Internet cafés and chat rooms that allow patrons to engage in prohibited activities have encouraged these intermediaries to police their own users. In addition to implementing the technological measures of censorship and surveillance detailed above, they have added elements of human control to comply with regulations. Internet café managers have tended to closely observe their users' surfing habits, especially after a series of crackdowns and closures of Internet cafés in 2001. Similarly, most chat rooms employ censors known as "big mamas" who screen postings and delete those that touch on prohibited topics. The operators of major Internet portals, who are forbidden to post information that "undermines social stability," have steered clear of anything potentially sensitive, offering primarily entertainment, sports information, and news from official sources. Even where regulations do not specifically require it, market conditions have encouraged the private sector to comply with the state's broad goals for the Internet. Doing business in China

376 means maintaining good relations with the government. In early 2000, for instance, more than a hundred of China's most important Internet entrepreneurs signed a pledge to promote self-discipline and encourage the "elimination of deleterious information [on] the Internet" (Kalathil and Boas 2003).

CONCLUSION

China and Saudi Arabia's experiences with the control of public Internet use offer a common lesson about the Internet in authoritarian regimes. Ultimately, the Internet is a tool, a medium of communication much like any other; it has no inherent political logic, no "built-in incompatibility [with] non-democratic rule" (Taubman 1998, 256). As a tool, its political impacts will depend largely on who controls the medium and in what manner they seek to use it. The Internet was initially considered an inherently control-frustrating form of communication because of features incorporated into the network by its designers. However, nothing in the technological architecture of the Internet ensured that it would remain difficult to control as it spread around the world. Rather, the architecture of the Internet is characterized by great flexibility at the macro-level, and the leaders of authoritarian regimes can take advantage of this flexibility to embed elements of control into their portions of the Internet. When leverage over Internet architecture is combined with legal, normative, and market constraints, authorities can exert effective control over the use of the Internet, preventing serious challenges to the economic and political goals that they pursue.

It is important to recognize that China and Saudi Arabia's efforts at controlling use of the Internet do not constitute mere restrictions on the diffusion of the technology within their borders. Although some authoritarian regimes such as Cuba and Burma have sought to control the Internet by regulating access, China and Saudi Arabia have been enthusiastic about promoting widespread access to their national networks (though the latter did so only after perfecting its mechanism for content censorship). Rather than clamping down on Internet growth in a reactive fashion, they have sought proactive measures of control over the technology that are consistent with its rapid growth. In doing so, they are able to gain many of the economic benefits that accompany greater Internet access, as well as the improved legitimacy that may come from establishing online government services and reducing corruption and graft. Although these two extreme cases of Internet control are not necessarily representative of a general trend among authoritarian regimes, they do illustrate a direction in which other countries may move in the future as they seek to emulate these successful examples of Internet control.

Indeed, there is evidence that increasing government control of the Internet is a trend not only in authoritarian regimes, but among advanced industrial democracies as well. In the international security environment that has followed the terrorist attacks of September 11, 2001, the United States has placed much less emphasis on the free flow of information abroad and at home and has sought greater control over the Internet within its own borders. The USA PATRIOT Act legalizes a certain degree of Internet surveillance without a warrant or the establishment of probable cause. Moral concerns have also encouraged greater control: federal E-Rate funding for Internet access in public libraries depends on the implementation of filtering schemes to limit access to pornography. Finally, influential corporate interests such as the Recording Industry Association of America (RIAA) have successfully lobbied for government crackdowns on file sharing and other technologies that could be used for copyright infringement.

In speculating about the more long-term prospects for control of the Internet, one should recall that accurately predicting the impact of a flexible technology is an inherently difficult enterprise. Thanks to the Internet's flexibility, its specific technological characteristics in any given environment will be largely contingent upon the political, economic, and social conditions that prevail. Moreover, the institutional constraints that influence Internet use — law, the market, and social norms — are similarly capable of change over time even when they exhibit a certain degree of stickiness. To say that China's laws and market environment or the social norms prevailing in Saudi Arabia currently support government control of Internet use does not mean that they will continue to do so fifty years hence. Although the Internet is not an automatically control-frustrating technology, a more liberal future for it is certainly possible. Such an outcome, however, will depend largely on the institutional variables shaping the evolution of Internet technology and the manner in which it is used — not on any inherent characteristic of the Internet itself.

NOTES

1. Indeed, the governments of many authoritarian regimes have sought to emulate the tactics of those most successful at controlling Internet use (Kalathil and Boas 2003, 138), so China and Saudi Arabia may well serve as practical examples for others.

2. Although often conflated with the effects of high fixed and low marginal costs, network effects are a separate mechanism in that they involve increasing demand for a more widely used technology rather than a lower cost to supply that technology in the marketplace (Lemley and McGowan 1998).

3. Arthur's fourth characteristic, adaptive expectations, is not considered here because it is not a characteristic of a technology per se but rather of its adoption

process. Moreover, adaptive expectations in this context are largely a result of network effects.

4. This does not mean, of course, that the Internet's micro-level standards are impossible to change. Indeed, there have been initiatives to alter some of the network's core protocols (Lemley and Lessig 2000; Blumenthal and Clark 2001). But this task is a more difficult one than altering the Internet by adding new functions and applications.

5. See http://www.isu.net.sa/surveys-&-statistics/num-users.htm (accessed November 9, 2005). More recent figures were not available.

6. See the description of Saudi Arabia's censorship regime on the ISU Web site, http://www.isu.net.sa/ (accessed November 9, 2005).

7. See http://www.cnnic.net.cn/en/index/index.htm (accessed December 14, 2005). An estimated 74 million Chinese citizens (5.7 percent of the population) were using the Internet as of September 2003. In relative terms, therefore, the numbers of users in China and in Saudi Arabia are quite comparable.

8. China Internet Network Information Center, "15th Statistical Survey Report on the Internet Development in China," January 2005, http://www.cnnic.net.cn/en/index/index.htm (accessed December 14, 2005).

18

Copyright's Digital Reformulation

Brodi Kemp

Digital technologies sent a shock wave through the political economy of content production and distribution, raising difficult questions and striking at the heart of intellectual property law. These technologies permit the wide distribution of perfect copies at virtually no marginal cost. This posed a problem for content providers: how could they make money if their product was freely available after its first sale?

Reframing the copyright laws became the answer. Notably, these revisions were an integrated international policy campaign, not exclusively national fights. The newly extended control, based on legally reinforced digital "containers" and trade law, arguably permits those who sell content effectively to "enclose" the public domain, to insulate their business models, and to define technological development.

As others have noted, the political and legal fights of the 1990s did not spawn revolutionary changes in intellectual property law (Burstein, DeVries, and Menell 2004). Instead, traditional legal concepts have been extended to this new medium. The resulting landscape feels familiar: copyright owners defend their property rights against infringing use in court.

Although these legal developments seem less than spectacular, a simple translation of old law to new technology, there are several pieces of the story that should be elaborated: first, the *way* in which the changes were secured, and second, the potent, if subtle, nature of these legal changes.

In this chapter, I will argue that content providers are "recreating the bottle" around their intellectual property, using digital technologies to reinforce their business models and supplant copyright. The content industries have successfully driven political fights, dramatically strengthening their control of content in the digital era. International treaties and agreements have been leveraged to strengthen and enforce intellectual property protection, forcing a globally "harmonized" reformulation of national laws. The resulting copyright policies have not been a simple translation of the old laws and enforcement mechanisms to a new technological era. In the revision of the intellectual property laws, the content industries claimed new power to control their intellectual property.

Secondly, I will show that the new policies adopted have undermined the traditional balance in intellectual property law between creator compensation and limits on the creator's exclusive rights. IP law was created to foster a vibrant public domain by encouraging the creation and exchange of knowledge. Recent developments have shifted that balance with a dramatic and one-sided strengthening of intellectual property rights. These policies empowered digital containers, or code, and trade law as the new enforcers of intellectual property rights but did not pay complementary attention to user rights and the public domain.

Finally, I will argue that this particular resolution of the copyright debate may have powerful implications beyond the content industries or the balance of intellectual property. It could influence the trajectory of technological innovation, indeed shaping the network's architecture itself and the business models that harness its capacities. Consider as only one example that many contend that network expansion is driven not by content distribution, but by the expansion of point-to-point communications. Yet, the intellectual property rules concocted for content will powerfully shape the architecture of the network.

Furthermore, it appears that the major firms in the content industries have the power to insulate themselves against competitive pressures that would force change in their strategies and business models. Rather than being forced to adapt and innovate, they have entrenched their position and set the stage for its reinforcement, the continuous expansion of intellectual property rights. At the moment it appears that the walls around the content industry incumbents are very powerful — are there holes through which newcomers can enter?[1] Would such entrants break the mold? For example, could peer-to-peer unravel the existing deals? Will affirmative policy action be required to assure ongoing innovation in business models and technology?

THE ARGUMENTS ELABORATED

Copyright enforcement, as well as the balance between content providers and the public, was predicated on a tangible balance of powers between creators and consumers, a metaphorical bottle in which a substance is at once contained and yet available to be circulated. But some fear that the bottle is vanishing, that the emergence of networked digital technologies has challenged and then changed copyright's original deal. Copyright is a delicate balance, addressing information's duality as both input and output of knowledge creation: copyright reserves rights for creators to incentivize *production* and limits those rights to facilitate the *exchange* of ideas.

Copyright was enforced and its delicate balance upheld largely by default. Large-scale copyright infringement was mainly precluded by the difficulty of replication and distribution, which left the market for authentic (creator-licensed) versions intact. Limitations on the exclusive rights of copyright were also realized by default: once a work was sold, its producers could no longer control the work's private use.

With little significant alteration, copyright has proven remarkably adaptable to technological change, and content producers have been forced to adjust to evolving technologies.[2] They could invent new business models to harness the capacities of the innovations, but could not use the law to insulate themselves against innovation (Lessig 2001).

Digital technologies have two components that undermine the enforcement mechanisms inherent in tangible media. First, information goods can now be perfectly replicated by users, or with such a marginal loss of quality as to render a near-perfect copy. In itself perfect replicability would generate a real challenge to those who hold rights to content. Amplifying this effect, however, is the capacity to distribute those ones and zeros across the network.

The formerly noncommercial act of infringement may pose a disproportionately large threat in the modern era: in the digital era, you don't have to own a factory to reproduce and distribute pirated music —you just need a computer and a phone cord (Samuelson and Davis 2000). Unlike previous challengers, such as the VCR and photocopier, networked digital technologies and peer-to-peer capacities exponentially increase the impact of a single violation.

Simultaneously, these technologies open new capacities for architectural control of the information flows they facilitate, empowering a new regulator: code (Lessig 2001; see also Samuelson 1996; Stefik 1998). Code is the stuff of which digital infrastructure and software applications are made. Digital products are constructed entirely of digitized elements from this programming; their encoded architectures have the power to set and enforce a particular set of terms and conditions. This regulatory mechanism differs sharply from conventional law in that it is perfectly self-enforcing (Lessig 2000). Thus, not only do encoded technical architectures set the norms and rules of access and usage, they enforce them independently.

PROPOSITION 1: RECREATING THE BOTTLE

Content owners responded quickly to what they called the "digital threat," arguing that these new capacities for individuals to privately reproduce and distribute copyrighted material would destroy the market for sales of their intellectual property. Furthermore, intellectual property owners and distributors are concerned that national variations in intellectual property rights (IPRs) and enforcement undermine the value of their property. Weaker standards for legal protection and enforcement permit unauthorized use and copying, or "piracy," which, they argue, translates into lost revenues. The content industries' concerns have taken on a new urgency with increased economic globalization.

Responding to this two-part threat, content owners have pushed new standards for IP protection. In both domestic and international fora, their successful lobby has produced a strikingly different approach to copyright, regulating technologies themselves and allowing copyright holders to insulate themselves against change.

Two major developments mark the content industry's victory and have permitted them to remake the bottle. The first set of policies reinforces the new digital capacities to control content, empowering privately constructed code. The same technologies that seemed to pose a digital threat were transformed into mechanisms of IP control. Content owners can now use *code* to control their intellectual property: new anticircumvention provisions prohibit technologies that could be used to circumvent measures used to protect copyrighted material. Unlike traditional containers, such as books and analog tapes, digital media are constructed from a highly structured architecture. Copyright owners may no longer need the formal law of copyright: the code will enforce itself, according to rules and standards set by the owners.

Second, the newly created WTO-TRIPS sets and enforces global standards for IP.[3] WTO-TRIPS legitimizes and institutionalizes the content industry's long-standing effort to strengthen and enforce IP protection globally, reframing IP as a trade issue. Some would argue that the WTO will take on its own independent institutional capacity to govern IP issues. At a minimum, as a treaty and a court, it will frame the debate and structure the fights.

REGULATING THROUGH CODE

After the content industry's first efforts to protect their works technologically failed,[4] a broad industry coalition (hereinafter "Content") pressured the Clinton administration and Congress for legislation that would make digital media safe for online distribution of their works. Content producers succeeded in capturing the administration's attention by characterizing new "user capacities" as a deadly threat.

Content's agenda quickly became the driver of U.S. intellectual property policy-making effort. Content targeted the Clinton administration's working group on intellectual property, whose 1995 White Paper articulated the U.S. digital agenda that has driven policy-making efforts, both domestically and internationally, to date (U.S. Information Infrastructure Task Force 1995). That agenda formed the basis of the outcomes in three critical arenas: the World Intellectual Property Organization (WIPO) integrated the agenda in its 1996 Copyright Treaty; the United States codified the agenda in its 1998 Digital Millennium Copyright Act (DMCA); and the European Union followed suit in 2001 with its Copyright Directive. The EU had been anxious to be the first to codify the implementation and to set the legislative precedent for copyright's adaptation. Despite the European Commission's rushed efforts to draft a copyright treaty, the United States set the precedent with its DMCA.

The ostensible goal of the agenda was to make the digital environment safe for the sale of copyrighted works (U.S. Information Infrastructure Task Force 1995). These policies went beyond a mere extension of traditional copyright to digital media (Samuelson 1997, 1999). Calling on the potential for new technical capacities to encode architectural protections, Content convinced policy makers that legal reinforcement for such technical protection systems was necessary to bridge the transition to the network era.

Voicing Content's concerns, these policies argue that the nature of network technologies demands technical *incapacitation* of possible violators, rather than reliance on the threat of liability as a deterrent. The anticircumvention measures are intended to reinforce technical protection for copyrighted works by making it illegal to circumvent such efforts (U.S. Information Infrastructure Task Force 1995). The anticircumvention provisions prohibit the manufacture or distribution of any device, technology, or service whose primary purpose or effect is to circumvent (without the authority of either the copyright owner or the law) any mechanism that protects an exclusive right of copyright (the National Infrastructure Information Copyright Protection Act of 1995). In its broad reinforcement of technical protection schemes, this policy approach gives copyright owners the right to define and enforce privately architected terms of access and usage, whose variable conditions may extend far beyond the exclusive rights of copyright. This approach gives copyright holders control over any digital transmission of their works, restricting intellectual property to an unprecedented degree (Samuelson 1999).

The process by which these policies were crafted is worth noting. Content providers arguably leveraged an international institution to reopen its domestic battle. The Clinton administration had planned to first seek domestic legislation of the agenda and then press the agenda abroad at the upcoming WIPO meeting.[5] Though the first effort to adopt these recommendations failed in both houses of Congress, the administration did not reformulate the agenda. Instead, the administration reversed its course, successfully pressing its agenda at the WIPO meeting. The agenda that so heavily favored the content coalition found new life at WIPO's meetings and became the basis for the WIPO Copyright Treaty (1996). The administration was then able to return to the United States for domestic implementation of the treaty rather than making policy from scratch, and at this stage, a version of the Content-proposed solution was probably unavoidable.

It is worth noting that an international organization was not merely the vehicle for reconciling competing international positions, but rather became another channel for a domestic fight. The WIPO Copyright Treaty functioned as both a vehicle for extending a national agenda abroad and the tool of a set of domestic interests to force a second round in a domestic fight.

Trade policy is another important instrument in the copyright wars: though content providers have consistently pushed to raise standards for intellectual property protection, their efforts have found new success recently. As IP has taken on increasing economic significance, national differences in IP protection have become a source of tension in international economic relations. WTO-TRIPS was created to address and remedy these differences. TRIPS establishes international rules to set and enforce minimum standards for IP protection, and acceptance of the agreement *in full* is compulsory upon joining the WTO.

TRIPS was not the first international attempt to harmonize standards for IP protection. The Berne and Paris Conventions, administered by WIPO, set forth minimum standards for IP protection. These conventions and their appendices are upheld by member states' voluntary acceptance—member states can choose the treaties with which they wish to comply—and cooperative reciprocity.

Building on the standards articulated in these conventions, TRIPS has been dubbed by some the "Berne and Paris-plus Agreement." WTO-TRIPS incorporates these conventions[6] and adds two significant elements to the package. First, TRIPS establishes new and strengthened IPRs, since those of the Paris and Berne Conventions were considered "inadequate." Second, and more important, administration under the WTO includes new mechanisms for formal oversight and dispute settlement.[7]

With the creation of WTO-TRIPS, we have shifted from WIPO's cooperative treaty system to a rule-based trade system, newly enforceable in the "Trade Supercourt." International rule of law for intellectual property now has bite.

PROPOSITION 2: UNDOING THE BALANCE

The tactics employed by content owners not only recreated the IP bottle but also dramatically shifted the balance of control between creators' protections and consumers' rights.

First, it is the privately architected nature of code that gives it such power: at present, there are no rules as to what code must allow, no body of rights and regulations to govern these digital walls, passages, and checkpoints. Content owners can set their own terms of access and use, terms that may effectively enclose the public domain within private holdings of the copyright owner. As such, copyright's crucial limitations and exceptions are facing a stealth attack embedded in the structure of the media themselves.

Second, TRIPS's trade-based approach to intellectual property undermines the complex balance of values IP was created to protect and uphold and marks a conscious and deliberate effort to reframe these issues according to a narrow set of economic preferences. Compensation for creation, designed to be merely a means to achieve an enriched public domain, has now become the focus and goal of intellectual property protection.

To maintain a vibrant public domain, do we need to translate copyright's limitations, such as the principle of fair use, from their traditional form to create equivalents for a digital era? Does this require affirmative policy action, or, as some argue, will the market achieve IP's underlying goals of diversity in information production and an enriched public domain?

The first policies and treaties to reformulate copyright for the digital era claimed to be mere translations and moderate adaptations of copyright's traditional balance, an update for new technologies. Changes in these two domains shift the balance between copyright holders and the public domain. First, although these policies reinforce Content's new capacities to digitally control content, copyright's crucial limitations are wilting without viable reinforcement. Second, the shift to a trade-based regime may provide content owners with a tool to consistently strengthen intellectual property rights.

HARNESSING CODE TO SUPPLANT COPYRIGHT'S LIMITATIONS

Copyright was designed to promote the exchange of ideas. To incentivize creation, copyright grants authors specific rights in their work, but these rights are bounded by key limitations that protect public access to and use of the intellectual property. First, "fair use" privileges exempt certain types of use from copyright infringement, without the prior permission of the copyright holder. These privileges serve to protect personal and educational uses whose social value outweighs the author's interests. Second, after copyright's expiration, public usage of the work is entirely unrestricted.

Thus, intellectual property was never "propertized" in a traditional sense. Rather, its balance was carefully crafted to create a public domain, a virtual space in which ideas, knowledge, and expression are free for public appropriation. The public domain underpins the cumulative creation of knowledge, building upon the body of knowledge and information that already exists.

Despite their crucial function, many of the former limits on copyright have been functionally ignored in debates about how to reformulate intellectual property for a digital era (Samuelson 1994). The recent anticircumvention policies reinforce the new capacities for increased control over content, giving copyright owners the right to use code to prohibit particular uses of their works or preclude access entirely. The reinforcement effectively strengthens copyright, but this strengthening has not been matched by comparable reinforcement of copyright's limitations (Burk and Cohen 2001).

Responding to concerns raised by Content's opposition, these laws did officially address copyright's balance and limitations, affirming the need to uphold a balance between content owners and users and extending the traditional limitations into the digital era.[8] The policies intentionally ignored the matter of real importance, however: privately constructed code changes the game, creating mechanisms for near-perfect control of information goods and services.

Copyright's limitations, recall, hinge on access to the content in question. Copyright's limitations were not affirmative rights, however, because they didn't need to be. In tangible media, users had the right and means by which to claim their privileged use. You could make use of copyrighted materials if you could get your hands on them. In some cases, this usage would be infringement, in others fair use, but the first decision was the users'. Users who made the wrong choice were liable for copyright infringement only after the fact. Thus, copyright's limitations functioned primarily as a guideline and a defense (Merges and Menell 2000).

Many argue that copyright's digital update has undermined this balance. Content owners can use digital technologies to build elaborate fences around content, defining and technically enforcing the terms of use and access. Unlike the traditional methods of copyright enforcement, encoded architectures do not have to comply with any law or standard, thus superseding copyright's limitations.

Because the architect sets the rules, these systems can be used to control content in radically new ways, including, for example, the enclosure of intellectual property in private holdings (Samuelson 1996; Lessig 1999). First, encoded architectures do not expire. Digital works elude any publication date from which expiration could be calculated. Second, fair use can be entirely precluded. A mere statement of rights as legal defense is powerless: if one can't access code-protected works, one can't claim a legitimate use (Nimmer 2000). Fair use, the longtime counterbalance to the exclusive rights, is now subject to the discriminate authorization of private actors.

The evident question is whether the concerns of the Content coalition could have been met without undermining either side of the present balance between protection of copyright holders and users. Did we have to make a stark practical choice between protection and fair use? An alternate strategy would require more than a reassertion of the rights of fair use and the importance of the public domain. The balance had to be reconstructed.

In addition to reinforcing Content's new capacities to protect their exclusive rights, policy makers needed to include an equally innovative mechanism by which to protect fair use, which now meant a means to generate and maintain the possibility of fair use. Nothing, however, was included to provide for the realization of the traditional exceptions, nor have the policies compensated for these losses.[9] The end result: public access is now the incidental by-product of the market for intellectual property sales, rather than its primary justification.[10]

USING TRADE AS AN INSTRUMENT TO STRENGTHEN INTELLECTUAL PROPERTY PROTECTION

Empowered code is only the first part of the story of strengthened IP protection and its shifting balance. The WTO-TRIPS enforcement mechanism may be used as a tool for content producers to systematically strengthen IPRs globally and may erode copyright's underlying balance.

A trade-based approach assumes and imposes a set of economic assumptions on IP. TRIPS reframes IP according to this narrow economic framework, legitimizing content producers' pressure on other nations to strengthen their IPRs. Indeed, TRIPS provides content owners the mechanism by which to drive this strengthening. As such, a trade-based regime may unravel the complex balance of values IP was created to protect and uphold. What is more, because the IP deals vary cross-nationally, this externally crafted compromise will have different consequences for each of the WTO's member states and their national polities.[11]

In TRIPS, content producers won endorsement for their reformulation of the commitment of the General Agreement on Tariffs and Trade (GATT) to reduce trade barriers. In theory, the reduction of trade barriers should increase global trade, benefit all participants, and facilitate the diffusion of wealth across borders.[12] Traditionally, this meant encouraging the equal treatment of goods, whether foreign or domestic in origin, and reducing tariffs. In the Uruguay round, IP owners argued that national differences in the level of protection for IP are a barrier to trade; content producers would be more willing to produce and distribute their products abroad if rules were uniform. Embedded in their argument, however, is the assumption that insufficient, rather than merely variable, IPRs were the barrier. Though this argument is controversial, TRIPS incorporates the notion that strengthened IPRs will encourage trade and economic development.

The content industries can now use the WTO to ratchet up IPRs by playing one jurisdiction off the other: TRIPS institutionalizes and legitimates the use of trade sanctions to strengthen and enforce IP protection. This trick is not a new one. IP owners have regularly pressured trade representatives to impose unilateral trade sanctions against countries with weaker IP protection. Their battles were fought on multiple fronts, however, and their outcomes were less significant; IP owners pressed for bilateral agreements, but their victories were narrow.

TRIPS consolidates these battles, channeling them into two institutions. First, TRIPS assigned WIPO and its conventions a new legislative significance: its treaties fall within the TRIPS standards.[13] Second, member states can now use the WTO's dispute resolution mechanism to "regulate" compliance with these standards.[14] Content's battles are now fought, and have been won, in the policy-making processes at the WTO and WIPO.[15]

While TRIPS resolves some trade tensions, it introduces new frictions by imposing a narrowly construed version of economic efficiency on a matter of cultural and social welfare. Like other policies for updating copyright to the digital era, TRIPS acknowledges the need to strike a balance between IP producers and users. What TRIPS does not acknowledge is that it shifts that balance in favor of IP owners. The economic preferences according to which these issues will be settled reframe the matter entirely. A trade regime tends to systematically neglect those issues it deems economically inconsequential or unquantifiable. As a result, a trade-based approach may undermine the purposes for which IP protection was crafted, diversity in information production and an enriched public domain.

Compensation was formerly a tool. Now it seems to be the end itself.

NOTES

This chapter is adapted from a student note that originally appeared in the *Yale Journal of Law and Technology* 5 (2002–2003): 141–153.

1. The most powerful of these incumbents are conglomerate producers and distributors of content, such as AOL Time Warner in print media, the Motion Picture Association of America, and the Recording Industry Association of America.

2. *Sony v. Universal City Studios*, 464 U.S. 417 (1984); *Galoob v. Nintendo*, 507 U.S. 985 (1993). See also Dratler 2000.

3. The initialism stands for World Trade Organization "Agreement on Trade-Related Aspects of Intellectual Property Rights," which was created out of the Uruguay round ending in 1994. Intellectual property is now considered one of the three pillars of the trade organization, joining goods and services. See http://www.wto.org/english/tratop_e/trips_e/trips_e.htm (accessed November 10, 2005).

4. Mark Solomons, "Hackers Crack Digital Music Codes," *Financial Times*, October 14, 2000.

5. The World Intellectual Property Organization is the administering body of the Berne Convention, an international treaty established to set minimum standards for intellectual property laws in all member nations.

6. Notably, TRIPS incorporates all aspects of the Paris and Berne Conventions except the sections relating to "moral rights" of authorship, a strong tradition in continental copyright law that has been rejected by common-law jurisdictions such as the United States.

7. The Council for Trade-Related Aspects of Intellectual Property administers TRIPS and monitors national implementation of and compliance with the agreement. National governments are required to notify the council of any change in their IP law, and the council serves as a forum for member review and consultation on TRIPS. All dispute resolution is conducted under the WTO formal mechanism.

8. The anticircumvention provision of the DMCA, for example, stipulated that "nothing in this section shall affect rights, remedies, limitations or defenses applicable to copyright infringement, including fair use."

9. Post-adoption joint study, anticircumvention hearings: comments from the Electronic Frontier Foundation and the D.C. Library Association.

10. With the DMCA, Congress affirmed and legitimized content owners' increased control and abandoned copyright's traditional technological neutrality. In addition, the DMCA lends the weight of the state to closed encoded architectures, which may have adverse effects on the architecture of the network.

11. Many argue that TRIPS will systematically transfer resources from developing (IP consumer) to industrialized (IP producer) countries.

12. The advantages of trade-based agreements — access to other countries' markets and equal treatment within those markets — supposedly outweigh the costs, which include ceding some control over the rules and dynamics of national economies.

13. These treaties were never crafted to come before a formal enforcement mechanism, however.

14. TRIPS is not a powerful institution in its own right: its treaties must be adopted by consensus, and its dispute resolution is member instigated. Thus, with TRIPS, the WTO becomes the mechanism by which countries adjudicate their differences over IP. Many argue that the WTO dispute resolution mechanism will become a tool by which the industrialized countries export stronger standards for IP protection. For example, the United States can threaten India with cross-product sanctions on textiles as punishment for lackadaisical IP enforcement. India holds no such trump card. Interview of Peter Holmes by the author (August 10, 2001).

15. Whether the United States continues to use its "special 301" process to achieve higher standards of protection or more favorable terms in negotiations with less powerful states is an interesting question. A special 301 requires the U.S. Trade Representative to scrutinize the intellectual property practices of other nations annually. A negative finding can result in trade sanctions. Though arguably its sanctions are no longer necessary, because IP can now be enforced through a legitimate WTO decision, the United States refused to remove the 301 sanctions from its laws, promising to use them only as authorized by the WTO.

19

Transforming Politics in the Digital Era

Abraham Newman and John Zysman

With the rise of digital technology, advanced democracies are in the process of transitioning from industrial to knowledge-based economies. The combination of binary knowledge expression, immense processing power, and digital networks has created the basis for a fundamental shift in economic and political organization domestically and internationally. This digital era holds tremendous opportunity for society and the global economy as new markets, business models, and means of organization emerge. At the same time, however, innovation has an intensely destabilizing effect for broad societal bargains, such as the notion of property, as well as real competitive position, as new entrants take advantage of technology to challenge incumbents. Our concern is how politics shapes this technologically instigated shift in society and the economy.

Borrowing from Karl Polanyi, we employ the metaphor of a second great transformation to drive a discussion about the political economy of the digital era.[1] By great transformation, we mean a fundamental and basic shift in the rules of society that alters the way economy and polity operate. There is one classic example: the great transformation that began in England in the sixteenth century. In this case, the commodification of land, labor, and money by the state defined the establishment of a market society. Before that transformation, markets were more of an adjunct to society. Those early markets, created by traders and burghers, were in a secondary position to landowners. For landlord, peasant, and burgher, position in a politically defined social community defined access to opportunities and to earning income. When the market system was endorsed by the state, it stood these relations on their head. Land and labor became commodities to be bought and sold in the market. Social position could move in relation to what was captured in the market.

392   Although some technological fanatics have naively downplayed the importance of politics, it is our contention that political actors — government, business interests, and public interest groups — serve an integral role in the evolution of the digital transformations transpiring across the globe. Counter to
claims that technology vitiates state power in the international economy, we
believe that the state has a critical role to play. Just as in the classic example,
governments are actively involved in creating a market for the fictitious commodity of information. The crux of this essay is to use the metaphor of the
great transformation to demonstrate how politics shapes and influences the
digital era. Governments have acted to mediate the transformation: promoting the technological infrastructure necessary for the digital era, establishing
the fictitious commodity of information through intellectual property, and
embedding digital markets in social norms. Far from being a neutral intervention, state actions influence the character of the transformation, including
its political effects. At the same time, governments have confronted well-
organized interest groups that have played an active role in shaping policy
outcomes.

This chapter analyzes the politics of the digital transformation in four parts.
The first section describes the scope and scale of the technological change in
more depth. This is followed by a brief presentation of the naive technologist
view that relegates politics to a peripheral role. The third section uses the lens
of the great transformation to explore in more detail the means by which the
state interacts with the digital world. The fourth section offers a set of policy
strategies that governments may employ to resolve digital challenges. The fifth
section highlights the dynamics of the political process that influence government policy and then presents some thoughts for future research.

THE RENEWABLE REVOLUTION

Just as the Industrial Revolution grew out of a revolution in tools and power,
the core of the information technology sector is the creation and production
of a new tool set, which Steve Cohen, Brad DeLong and John Zysman have
termed tools for thought (Cohen, DeLong, and Zysman 2000, 7–8). These
tools manipulate, organize, transmit, and store information in digital form,
thereby creating a set of information services and products that allow the application of information to industrial as well as machine processes (Weiner
1954). The digital transformation and the tools for thought can be broken
down into three fundamental elements:

- *The concept.* Information technology begins with the notion of information
  as something that can be expressed in binary form (Weiner 1954, 1965;

Shannon 1993). Data ranging from supermarket purchases to fingerprints can be represented in digital code.

- *The equipment.* Software consists of written programs, including procedures and rules, that guide how equipment processes information. The hardware, or equipment, that executes the processing instructions has evolved from vacuum tubes through individual silicon transistors to integrated circuits implemented on silicon wafers and may evolve into still other physical manifestations.
- *The networks.* Data networks interlink the processing nodes of individual computers, and the network of networks creates a digital community and society.

Tools for thought create the capabilities to process and distribute digital data, multiplying the scale and speed with which thought and information can be applied. Information Technology (IT) affects economic activity in which information sensing, organizing, processing, or communication is important — in short, every single economic activity.

What seems most significant is that information technology represents not one but a sequence of revolutions. It is a continued and enduring unfolding of digital innovation sustaining a long process of industrial adaptation and transition. The IT revolution is a recurrent one:

> In the 1960s Intel Corporation co-founder Gordon Moore projected that the density of transistors on a silicon chip — and thus the power of a chip — would double every eighteen months. Moore's law, as it came to be called, has held. Today's chips have 256 times the density of those manufactured in 1987— and 65,000 times the density of those of 1975. This continued and continuing every-eighteen-month doubling of semiconductor capability and productivity underpins the revolution in information technology. The increase in semiconductor density means that today's computers have 66,000 times the processing power, at the same cost, as the computers of 1975. In ten years computers will be more than 10 million times more powerful than those of 1975 — at the same cost. We now expect — routinely — that today's $1,000 personal computer ordered over the Internet will have the power of a $20,000 scientific workstation of five years ago. And what was once supercomputing is now run-of-the-mill. The past forty years have seen perhaps a billion-fold increase in the installed base of computing power. (Cohen, DeLong, and Zysman 2000, 13–14)

The conventional economic explanation of a leading sector is that the original innovation creates a set of opportunities, somewhat like distributing money on the ground. The original technological revolution loses force as the most valuable opportunities are picked up and implemented. The notion argued in this chapter is, of course, that the revolution is renewed — if not with each cycle of Moore's law, certainly with the radical increases in computing power generated every few years. An original transistor, a single bit, bears little relationship to a 16 kilobit integrated memory chip. A gigabyte chip with a billion transistors is another thing altogether, and it is seven Moore cycles, less than two decades, along the road from the 16k. And Moore's law has at least several more cycles to run. The technological revolution is renewed every decade. The currency is redistributed on the ground as the bills themselves get larger. The fundamental question, then, is how these resources will be distributed.

PUTTING THE STATE ON THE SIDELINES

With the creation of the Internet, the U.S. and European governments effectively laid the foundation for self-regulating groups. For those early "net" pioneers, it seemed as though the government was an interloper in a system run by technologists for technologists. These technological enthusiasts argued that the architecture of the Internet made it impossible to regulate. The famous claim made by the early Internet advocate John Gilmore that "the Net interprets censorship as damage and routes around it" epitomized the beliefs of many technologists.[2] They viewed cyberspace as beyond the reach of state controls. National governments would be forced to cede much of their regulatory authority to the cybercommunity, eroding traditional notions of sovereignty (Post and Johnson 1996).

The digital era posed a dual challenge to the state. First, IT reduced the costs associated with conducting international business, part of a phenomenon popularly labeled globalization (Weber 2001). Firms that took advantage of digital technology to expand their geographic reach in turn limited the efforts of public officials to manage their economies. Stringent domestic regulations, it was believed, encouraged footloose capital to relocate to more hospitable institutional environments, forcing governments around the globe to engage in a race to the bottom in economic intervention (Tonnelson 2000). State autonomy supposedly fell victim in the digital era to firm mobility (Ohmae 1993).

Second, the decentralized, non-hierarchical character of digital networks was viewed as incompatible with the rigid, inflexible governance tools available to the state. As Virginia Haufler has argued in the case of information privacy, "The decentralized, open, global character of one of the main transmission sources for personal information — the Internet — makes it difficult to design and implement effective regulations through top-down, government-by-government approaches" (2001, 82). The governance problems raised by digital technologies threatened to further erode state autonomy, as nonstate actors were empowered to resolve major societal disputes (Rosenau and Singh 2002).

Operating from this perspective, Debora Spar describes a cyberworld of state regulation displaced by industry self-regulation:

> Fundamentally, I argue, governments cannot set the rules of cyberspace. That is because cyberspace, unlike governments, slips seamlessly and nearly unavoidably across national boundaries. . . . With governments pushed effectively to the sidelines, firms will have to write and enforce their own rules, creating private networks to facilitate and protect electronic commerce. (1999, 82)

Politicians in the United States relied heavily on the private sector to navigate the first years of the digital era. A chorus of business lobbies argued that government regulation would crush this vital infant industry. The role of the government, if any, was to get out of the way of the sector's development. The Clinton administration did not stray far from these demands, insisting that the government should not interfere with the development of information technology. The administration believed that private sector self-regulatory mechanisms would guarantee the successful construction of information markets. In their 1997 Framework for Global Electronic Commerce, President Clinton and Vice President Gore asserted that "the Administration . . . will encourage the creation of private fora to take the lead in areas requiring self-regulation such as privacy, content ratings, and consumer protection and in areas such as standards development, commercial code, and fostering interoperability." Echoing the U.S. view, Europe's then Commissioner for Telecommunications, Martin Bangemann, suggested that business should take the lead in developing an "International Charter for Electronic Commerce" that would rely heavily on "market-led, industry-driven self-regulatory models" (Commission of the European Communities 1998).[3]

Of course, the early notions that the Internet should be free of government, like the mythical Wild West, ignored the fact that western settlements required a local sheriff; they required governments. When the Internet was transferred to the commercial world, those requirements for legal structure in the operation of the network became more evident, more urgent, and the rule making for the Internet became, at least in part, rule making for the economy. The issues were no longer simply technical ones of how to operate the network or communicate across this global network of networks. Suddenly, all the questions of an operating marketplace had to be addressed; appropriate domestic and international rules had to be defined for domains from privacy to taxation. The results of these decisions have real distributional and societal consequences. And states were not about to abdicate this responsibility to private actors. In fact, despite the decentralized, international character of digital networks, governments have played an instrumental role in shaping the character of forming digital societies.

396   The emergence and evolution of the digital era has not been the product of purely neoliberal strategies, on the one hand, nor of purely interventionist strategies, on the other hand. Governments acted simultaneously to subsidize infrastructure development, extract themselves from direct market control, and forge new rules to promote economic transactions. Framing the role of the state in ideological terms confounds the multitudinous and seemingly contradictory strategies governments undertook. Like all markets, cybermarkets require definitions of property, exchange, and competitive market structure. And all of this requires rules.

THE ROLE OF THE STATE IN THE DIGITAL TRANSFORMATION

The drama of the great transformation itself was the shift from a traditional society, in which markets fitted within social order and economic activity bowed to the confines of social rules, to a market society, in which land, labor, and capital became commodities moving in response to price signals from the market. That transition was the product of a series of political battles that redefined England, including the enclosures, the Poor Laws, and the repeal of the Corn Laws. Enclosures transformed community public lands into private farming lands, beginning the creation of a market for land. The series of Poor Laws, culminating in the elimination of the Speenhamland system in 1834, created a labor market. The 1834 Poor Law Reform broke the link for survival between individual and local community, making the individual worker's well-being dependent on wages obtained in the labor market. The repeal of the Corn Laws in 1846 opened British agricultural markets, limiting trade protection so that lower-cost grain could feed the emerging industrial workforce. That political decision marked a shift in power from the landed classes to the emerging industrial bourgeoisie (K. Polanyi 1944).

In the contemporary era, as during the Industrial Revolution, the state has played a vital role in the construction of the digital economy through policies of deregulation, market making, and reregulation.[4] This effort has focused around two central questions. First, what are the rules that should underpin new digital markets? As digital technologies diffuse, businesses in industries ranging from financial services to telecommunications search for market advantage. At the same time, these innovations have the potential to disrupt the current distribution of power within a sector and across polities. Incumbents simultaneously see lucrative market potential and economic disruption in digital advances. Market rules, drafted and enforced by the state, fundamentally shape the distribution of economic gains and modulate the extent of the transformation domestically and internationally.

The second question confronting state authorities concerns the implications of new market rules for society more generally. As the digital economy is constructed, decisions about market rules inherently structure information flows, influencing the character of the political community. The state must manage the social externalities that arise in parallel to the digital economy.

The role of the state in the digital debate has a peculiar form, in that the rules of digital information, hence of a digital polity, are embedded not only in convention or in the law, but in the computer code itself (Lessig 1999). Just as highway architecture dictates where you can get on and off the freeway, computer architecture and the code implementing applications dictate what is and is not possible in a digital era. In the early years of the Internet, the open, user-controlled architecture led to the sense of cyberspace as a domain outside the control of governments or physical communities. Hence Stewart Brand's infamous remark that "information wants to be free" reflected the particular architecture of the early Internet. But that early Internet was only one potential architecture; other, more controlled or more restricted networks are also possible. Digital information wants nothing at all; it flows where the network architecture permits. And network architecture is a product of the code writers. To say that we must regulate the code, hence the code writers, is not to say that there is a single technologically dictated outcome. Although politics is always about values and outcomes, about who gets what, for such choices to have meaning in a digital world they must inevitably be embedded in code and respect the technological logic of the tools for thought. Law and code then interact to establish the rules of the digital era.

At the dawn of the digital era, governments have played a critical part in the creation of the fictitious commodity of information. In this effort, they have used public policy to build the infrastructure for and remove barriers to the new market. Government legislation has shaped the way that information technology has interacted with production patterns, influencing the success of emerging business models and modes of industrial organization. At the same time, state initiatives have been instrumental in navigating the complex political fights that surround the digital transformation.[5] Government legislation is critical in order to embed the new markets in social norms and to limit the inevitable pushback by the losers of the new era. In short, the character of the digital era has been modulated by government interventions. The following section highlights several pathways by which the state has shaped the digital transformation. Chief among these are establishing the digital infrastructure and removing barriers to market evolution, constructing the commodity of information, and embedding new markets in social norms.

Building the Digital Infrastructure and Removing Barriers to New Market Evolution

In Polanyi's England, the government promoted enclosure at the same time it repealed protections that hindered the market's development. Similarly, in the digital era, governments have simultaneously developed the infrastructure for and removed barriers to commodified information markets. In the United States, the creation of the Internet was the product of both purposive intervention—government action by the Defense Department's Advanced Research Projects Agency (DARPA)—and aggressive deregulation and reregulation.

Seeking to promote communication among scientific researchers, DARPA funded the creation of the underlying conception and protocols of the Internet in what was called ARPANET (Hafner and Lyon 1998). This Internet prototype refined the technology necessary to transmit data through packets of information, providing the fundamental architecture of the current Internet. In contrast to telephone lines, which traditionally used switches to directly connect the receiver to the sender, packet switching decomposes information into its components and sends them through the network before recombining them at their destination. The government managed ARPANET through the National Science Foundation, then prepared it for transfer to commercial use.

The government laid the groundwork for the digital era through regulatory reform as well as infrastructure investment. The aggressive introduction of competition in the telecommunications sector, highlighted by but not confined to the break-up of AT&T, unleashed user-led and consumer-based innovation in data networks. Inexpensive local flat-rate fees, for example, gave consumers the ability to experiment with data networking at relatively minimal cost. That opened the way to user-generated networks and facilitated the radical and rapid spread of Internet technology (Hafner and Lyon 1998).

The European story likewise displays these twin roles of the state. Simplified, one part of the story is the deregulation of the telecommunications system led by the European Commission. The Commission created national coalitions for Europe-wide rules that would compel the transformation of state administrations responsible for post and telegraph into regulated companies in at least partly competitive markets.[6] The other side of the story is an array of directed state actions intended to develop and diffuse digital technology. The foundations of the World Wide Web were developed at the European Organization for Nuclear Research (CERN) in Geneva, Switzerland. A pan-

European nuclear physics lab, CERN faced the dilemma of bringing together a geographically very dispersed European nuclear physics community. An information systems scholar from CERN, in response to this organizational demand, developed the architecture of the World Wide Web based on a language for constructing Web pages, the protocol for transmitting these pages, and a browser system for reading the transmitted information. This last innovation resulted in the highly accessible browser system that has facilitated the rapid diffusion of the Web (Gribble 2004).

Government intervention has continued, but taken on a different flavor, with the state-sponsored transition to high-speed broadband connectivity. The reason the original consumer use of the Internet expanded so suddenly was that it could be deployed over the existing telephone infrastructure. However, new uses of the Internet, such as downloading music or playing videos, require a different infrastructure. That infrastructure is loosely called broadband, a term that typically refers to anything faster than telephone lines, whether a network of fiber or DSL technology. The fact that the next-generation consumer network requires an infrastructure other than the traditional copper-wire phone system has posed new policy problems.

Although there is international agreement on the need for rapid deployment of broadband, the policies to accomplish that rapid deployment vary radically by country. The question remains whether this build-out should be a purely private decision of local providers or should be encouraged and subsidized by the government. The answers around the world are quite varied. Korea, for example, is a story of stunning penetration of broadband services into the society. Of the 16 million Korean households, 78 percent have broadband access, compared to roughly 20 percent in the United States (Shameen 2004).[7] A conscious decision was made to subsidize the broadband build-out by redirecting funds from wireless spectrum auctions. In total, the Korean government has spent nearly $3 billion on broadband diffusion. This effort has been carried out through an aggressive campaign, including direct subsidies, loan programs, and research and development funding. The government has even adjusted its housing ratings system so that units with broadband systems may be priced at a higher rate.[8] In the United States, by contrast, we have left the effort to a competition amongst the cable TV companies, phone companies offering DSL services, and potentially even power companies offering access over electricity lines. The result is less overall coverage but more diversity in network forms. In both the United States and Korea, government policy has been instrumental in shaping the technological infrastructure for information markets, albeit in very different ways.

Commodifying Information

Cyber law has for the most part focused on creating a market world in cyberspace. The concept of intellectual property (IP), though not new, plays a particularly critical role in this effort and in the digital era more broadly. We know that property is always a legal fiction involving the specification of enforceable rules about what a person can have, hold, and dispose of. Hence, in a fundamental way, property and its rules of use are always a political creation.

We also know that physical property and intellectual property have different characteristics. In the case of tangible goods with a physical existence, the rules of property set the terms of use and disposition. Since physical property cannot be simultaneously shared, some rules of use and disposal are necessary, whether those rules constitute private property or not. With the great transformation in England, the enclosure movement closed off common public lands, converting them into private holdings. By contrast, intellectual property is a nonexcludable good — its use by one does not preclude use by another. Hence, intellectual property as economic property, that is, something one is willing to buy because one cannot have its use without payment, is an entirely political creation, a fictitious commodity. The very "good" is a product of a rule. Thus the rules of intellectual property are in a digital era absolutely central.

Digital technology radically changes the logic of control and distribution of intellectual property. Whatever the cost of developing intellectual property, be it a movie or a software product, the marginal cost of precise reproduction and distribution is almost zero. Since media products are so immediately affected, it is evident why media companies have driven the reformulation of intellectual property law to permit them to recreate control over the distribution of their products.

The most blatant example of the effort to recreate traditional notions of property in the digital era is the Digital Millennium Copyright Act (DMCA) of 1998 (Kemp, this volume). Content providers ranging from Hollywood to the publishing industry lobbied to rebuild walls around their intellectual property preserves. The DMCA contained two critical provisions. First, it created criminal penalties for the circumvention of encryption programs, which hide the underlying software code from the user, preventing the reproduction and distribution of the purchased product. Second, the DMCA outlaws the manufacturing or sale of code-breaking software. The digital nature of the media, however, does not just recreate past IP protection. Regulating code has broader implications for society more generally. The notion of "fair use," which allows the holder of intellectual property to make that information available in a noncommercial manner, has been severely curtailed. A digital recording or book, for example, may be encrypted so that it can never be duplicated, eliminating the consumer's ability to share a purchase with friends or colleagues, even though that practice is fully legal. Digital rights management software may stop the consumer from duplicating downloaded music, preventing the customer from listening to the recording on multiple personal entertainment devices — another legal practice. Implanting intellectual property protection into the product through encryption systems permits perfect control over use, replication, and distribution.[9]

The digital era allows new forms of intellectual property to be created. For example, many types of data can now be easily packaged and sold. Expressed in digital form, information becomes a commodity that can be transmitted, manipulated, stored, and sold as an object. Argued most generally, in a digital era commodified explicit knowledge becomes pervasive. As knowledge, including digital instructions for physical control, becomes explicit and expressible in useful ways, the possibility and importance of protecting that knowledge as property increases. As patent offices recognize the legitimacy of business model patents, processes increasingly expressed in digital form have become property. Patent disputes over the eBay auction or the Amazon checkout strategy provide troubling examples of how previously shared knowledge may become commodifiable through law. The process of establishing a checkout procedure on the Web, rather than the intellectual property behind the book being purchased, receives proprietary protection (Preston 2004).[10] The seemingly inexorable expansion of the protectable is the issue.

Intellectual property rules inevitably affect more than just the media industries or the business possibilities of sectors that use digitized information and programs. Intellectual property has always been about balancing the need to reward those who generate knowledge against the desire for widespread distribution and use. Digital technology makes more information more easily accessible; offsetting that, technology and law create new boxes to control that information. The texture of social and political debate is powerfully influenced by who owns and can use content generated by others. The political community is thus shaping and shaped by the rules of intellectual property.

Embedding Markets in Social Norms

The digital era has radically altered the types and amount of information in the economy as well as the ability of actors to transmit and use that information. With data cheaply passing over digital networks, long-nourished business dreams become reality. Yet the question arises: who will capture the benefits of these innovations, and what threats do they pose to society? Government

402   stands at the crossroads of the digital era, constructing the rules that underpin these emerging markets and mitigating negative social externalities. As in the case of the Industrial Revolution, the immensity of the technological change creates tremendous instability and displacement, which could derail the transformation. The state then steers Polanyi's "self-regulating market" so as to assure its viability. What is new about the digital state is less some technologically augmented or vitiated state authority than the ability of the state to influence the resolution of fundamental societal bargains that have been reopened by changes in information technology.

Two debates appear most critical. The first is privacy — that which permits us to remain in our personal domains, secluded from the view of others. The second debate concerns speech — that which we can say and debate in the public arena. These debates began in the marketplace over how to use information to economic advantage and spilled over into society, addressing how our communities and political processes will be organized. The state's failure to address these conflicts risks derailing the transformation and therefore demands government attention. Although the state has long been seen as the number 1 enemy of civil rights such as privacy, the digital transformation has ironically positioned the state as a critical defender of these very freedoms. As businesses augment their power to collect data and control the dissemination of information, new private sector threats emerge. Although fears concerning potential government abuse persist, the state also has the capacity to construct the rules to mitigate individual exploitation, formulating a consumer protection regime for the digital world.

The rules and norms associated with the collection, processing, and exchange of personal information, which come under the banner of privacy, are essential to the digital world. With the rise of digital technologies, both the quantity and the quality of personally identifiable information have shifted. As each credit card purchase, Web visit, and mobile phone log creates a new bit of data, behavior becomes easily tracked. New moments of personal life become monitorable. From the Webcam in the taxi to emerging genetic tests, these technologies erode the barriers between the knowable and the unknowable. They also permit the networking of previously discrete data. Information-intensive sectors, such as telecommunications, banking, and health care, are the first to rely on this wealth of personal information to customize products, rationalize costs, and minimize fraud. The supermarket club card typifies this line of innovation. With each swipe, the company is better able to target customers and lock in loyalty. The shift in a range of service industries, from marketing products to marketing customers, further demonstrates this trend. Where once an insurance firm marketed homeowner policies, it now attempts to understand individual customer needs across a wide array of company products. Reversing the logic of traditional credit cooperatives or risk-pooling efforts, complex individuation offers firms the ability to profit from extreme differentiation.

The opportunities inherent in personal information processing threaten to erode personal privacy, however. As digital technology expands the quantity and quality of personal information available, individuals lose the capacity to control information flows. The boundary between public knowledge and private secrets shifts, leaving less and less room for the private. What worries privacy advocates most is the networking of formally discrete personal information for third-party economic gain. Information privacy deals fundamentally with an individual's ability to control what is known about him or her, and not just what is published about him or her. And therefore, it addresses at root how individuals construct their identity. If credit data banks cement early risky consumer behavior into a widely distributed consumer report, it is difficult for individuals to be free of the negative data profile. In short, a major concern of the digital age is the inability to forget, a fundament of most healthy societies. Similarly, one could imagine car insurance firms using mobile phone logs to track commuting patterns and potentially changing rates of individuals traveling through high-risk areas. The flip side of customization and risk reduction is potential discrimination against those who are most vulnerable.[11]

As the amount of information held by the private sector rises, the possibility also exists that governments will look to private sector data warehouses to enhance public sector surveillance needs. A scandal involving JetBlue vividly illustrates the potential harm that lies in the linkage between private sector firms collecting information and government bureaucracies hoping to advance security interests. In this case, the airline transferred millions of personal customer files to a defense department contractor, who linked the airline data to commercial databanks in order to construct risk profiles.[12] Far from an isolated incident, governments across the globe are looking to private sector data files such as telephone or ISP records to monitor citizens' behavior. As the line between public and private enforcement breaks down, traditional checks against government abuse are neutralized. The traditional fear of a government-dominated Orwellian world is replaced by the specter of public-private partnerships of control.[13]

Such partnerships can operate in both directions. If the JetBlue scandal is a case of personal information gathered by a company being made available to the state, public policy has also compelled companies to make private information available to other business actors. For example, in attacking music downloading, the Recording Industry Association of America entered a series of lawsuits. To obtain the information on which the suits were based, the trade association required access to the records of the ISPs. The law as now written, the DMCA, compels the ISP to provide access to that information on the basis of suspicion of IP violations, without court authorization or review. This constitutes the creation of a private posse enforcing its will in civil courts.

Such threats have not gone unnoticed by state authorities. Since the 1970s, with the proliferation of computer technology, lawmakers have recognized the danger inherent in the collection and storage of personal information. In response, nations across the industrial world have adopted data privacy legislation. These rules have varied considerably, with the United States focusing on public sector data usage and Europe constructing comprehensive regulatory institutions for the public and private sectors (Bennett 1992; Regan 1995). As data processing has left the confines of a small number of government agencies in the mainframe era, the comprehensive structure has shown itself better suited at dealing with the explosion of data collection inherent in the digital era. In European countries, for example, it is difficult for private sector companies to routinely share personal information with government security agencies. Data protection rules often prevent the collection of detailed information by firms, limiting the amount of information available for sharing. Recent disputes between the United States and Europe over data privacy issues, ranging from telecommunications information to airline passenger flight records, demonstrate the importance of national government policy in the development of very distinct information societies (Newman 2005).

Like notions of privacy, questions concerning free speech have been reopened by the emergence of digital technologies. Though they receive fewer headlines than the economically more potent cases of property or privacy, speech issues lie at the cornerstone of modern political communities. By defining what can be said to whom, free speech rules shape an individual's ability to express himself or herself, maintain social networks, and organize politically. Free speech is invariably included in the catalog of basic democratic rights as the most critical right held by opponents of established power. Yet free speech is far from uncontroversial. By altering patterns of communication and the capacity to transmit content, digital technology has transformed global debates about free speech. With the rise of international Internet connectivity, a resident of the United States can as easily transmit information to a fellow netizen in Europe as to a local neighbor. As a result, differing cultural norms concerning criminal speech have come into conflict with one another. Most common among these are forms of obscenity, hate speech, and political protests. As digital connectivity permits citizens from one nation easy access to the media of another, jurisdictional conflicts emerge ("Recent Developments" 1999).

Technology has changed not only patterns of communication, but also the ability to control content. As previously described in the discussion of intellectual property, digital goods are naturally nonrival and replicable at no marginal cost. This has challenged traditional business models, spurring industry to use digital tools to increase the controllability of content. Through legislation and code, content providers have attempted to minimize the amount of IP available in the digital commons. This "second enclosure" restricts the fair use of information and in turn limits the free flow of ideas essential to free speech (Boyle 2003). It becomes much more difficult for activists to circulate news updates, for example, when they have to pay distribution fees to use digital information. The state is left in the delicate position of determining which types of content individuals should be allowed to share.

Potentially just as important for free speech have been policies concerned with harmful content. A distinct feature of modern society is the belief that certain information is dangerous and should be controlled through public policy. European governments, including Germany and France, have applied existing content laws to the digital era — for example, banning the sale of Nazi paraphernalia on the Internet. The firms selling such products are often located in countries with different laws. As a result, the application of content laws can take on an extraterritorial flair (Beesom and Hansen 1997). Far from a technologically driven race to the bottom in standards, firms playing in international markets have been confronted by the projection of national rules through digital networks.

At the same time that governments have moved to control content, they have also actively participated in the dissemination of information technology. Access is no doubt a precursor for communication and participation. In the United States, the E-Rate program was established, which subsidized broadband technology access in schools and rural communities (Newman 2003). Similarly, as discussed previously, the Korean government has been active in subsidizing the rollout of broadband technology. Though access is a critical component in overcoming the digital divide, the dissemination of digital technology should not be strictly equated with the promotion of free speech. Governments can promote technology diffusion while at the same time controlling (or supporting companies to control) the manner in which it is used (Boas, this volume).

GOVERNMENT EFFORTS TO MEDIATE THE TRANSFORMATION

Given the continued importance of the state in the political economy of the global digital era, it is important to consider how state intervention shapes markets and societies. We identify three basic strategies that states have

406   adopted in response to the challenges posed by information technology. First, governments may intervene to promote competition in the new marketplace, as technological change disrupts existing business strategies. States intervene to secure fair ground rules for the fights between dominant players and new entrants. These rules may emphasize equal market access, level regulatory playing fields, and transparency. The European Union convergence effort in the communications regime typifies this policy strategy. As media including telecommunications, radio, cable, and satellite compete head-to-head with one another for core digital products, market disruptions result from regulatory legacies. Telecommunications companies, for example, face very different regulatory burdens when entering new markets than cable companies: universal service requirements mandate that telephone companies guarantee access to underserved communities, a cost not faced by cable companies looking to compete in broadband markets. The convergence process attempts to smooth over these regulatory differences and create a comprehensive regime for the digital communications industry. This strategy of getting the market rules right privileges procedural neutrality and long-term market competition over attempts to shield specific national champions.

In the second policy strategy, governments intervene to reassert incumbent market power. Digital innovations have the potential to upset existing business dynamics in a sector, threatening powerful industry groups. Policies in this strain attempt to shore up the predigital distribution of resources and prevent political coalitions from shifting. The DMCA offers the prototypical example of this form of state intervention. It criminalized the development and use of devices that may be used to break encryption systems. Technological solutions to intellectual property rights questions received legal support, consolidating the entertainment industry's effort to reassert property rules in the digital environment. Despite intense lobbying efforts by new entrants from the information technology sector to curb the legislation, the government attempted to reassure the entertainment industry as a critical interest group. Similar international efforts have been carried out by advanced countries vis-à-vis the developing world. The WTO's Agreement on Trade Related Aspects of Intellectual Property Rights (TRIPS) and World Intellectual Property Organization (WIPO) agreements further insulate the intellectual property warehouses of incumbent players. The agreements provide firms with international commitments and credible enforcement mechanisms to protect their intellectual property globally. Potentially viewed as a reactionary strategy, the second approach steers the political character of the digital transformation in favor of existing power centers.

A third strategy likewise attempts to shape the substantive character of emerging digital markets. Instead of bolstering existing interest constellations, however, the state recasts the balance of power in favor of public interests. The citizen consumer is empowered in the new digital environment, receiving increased control over information resources. Most easily identified with the mission of consumer advocates, this third strategy attempts to promote the public interest more broadly and to prevent digital innovations from further concentrating power in the hands of economic and government elites. Often motivated by political fears that individuals will reject new technologies and thereby stall economic development, this approach emphasizes state safeguards that protect and assure citizens. The EU data privacy directive provides an important international example. With the explosion of personal information in the digital age, the directive provides individuals with a clear level of control over industry and government data processing. Before a company or a bureaucracy may transmit personal information to a third party, it must obtain consent from the individual in question. If an organization fails to obtain consent, it can be punished by a data protection agency. Though they do not eliminate the commodification of personal information, the European regulations reset the default position in favor of consumers. Governments, by promoting the third, potentially populist or progressive option, channel the transformation so as to rebalance societal relationships prioritizing citizen concerns.

THE POLITICS OF STATE REGULATION
OF DIGITAL TECHNOLOGIES

The digital era has reopened fundamental societal debates and in turn brought on a reexamination of the role of the state in the emerging political economy. As firms use digital technologies to create advantage or position in their markets, old political economy bargains are undermined. Often new entrants see opportunity in the technological disruption, incumbents struggle to hold on to old business models, and public interest groups fight to maintain or expand consumer rights. Amidst the commotion, governments begin to formulate policy strategies that inevitably impact the distribution of business opportunities.

The dynamics of these political debates are complex. In addition to the state, business lobbies and public interest groups struggle within a given political institutional environment to construct the emerging rules of the digital economy. In order to understand the variation in policy results across countries and internationally, it is vital to identify the roots of business sector and public interest preferences. In short, we contend that the organization of economic and public interest sectors influences their preference formation and their stake in digital debates.

408   Several caveats about business and public interests are important to keep in mind, as we examine the preferences of various political actors. Business interests may be driving the process of reformulating rules for a digital age, but there is no unified business position. There is certainly no "digital sectoral" interest, let alone a class interest. Firms have different preferences and positions on the same issues; competitors in networks seek to turn the rules to their advantage; and companies building and using different technologies, or at different positions in the market, have quite distinct needs.

But there is more to the story. The business interests of financial institutions depend not only on market problems alone, but also on the corporate organization of the firms themselves (Newman 2005). This organization is partly a business choice and partly a result of regulation. Integrated financial institutions, such as France has, do not depend on customer information markets to gather the information they need to market to their customers. French financial institutions rely instead on their internal warehouse of information to target customer needs. By contrast, the highly fragmented character of financial services in the United States reinforces demands for a market in personal information. So interests may be definable, but they cannot be read off a market map in any simple way.

Similarly, public interest groups have been at the forefront of many digital policy debates across the globe. But their level of engagement, their policy goals, and their lobbying strategies differ dramatically across countries. Compare the work of the most active public interest groups in the United States, such as the Electronic Privacy and Information Center and the Electronic Freedom Foundation, to that of their counterparts in Europe, such as data protection or consumer protection bureaus. Though the goals appear identical, guaranteeing a social agenda for an information society, the logic of their tactics (e.g., class action suits and media scandals versus negotiated technocratic bargains) vary and are in a very real sense shaped by their institutional settings.

Not only do their tactics differ, but the capacity of players to influence legislative debates varies across policy landscapes. In the United States, broader public interests are represented in only a limited way in the struggles over digital rules. Certainly, the narrow business story of the emergence of electronic commerce and the tools to conduct commerce using networks has become entangled with the broader political struggle over fundamental values, goals, and processes and jurisdiction. But at least in the United States — to oversimplify — it is a story of business seeking new rules to implement digital technologies, with public interest groups seeking to influence the character of those rules.[14] More often than not, groups defending general principles, such as privacy or consumer protection on the network, enter the fights in response to business-initiated or -proposed rule changes. None have mobilized effectively on a mass basis and, as a result, there is no digital equivalent of the environmentalist movement.

The U.S. debate is driven by markets and market actors and therefore has the flavor of business dominating the political debate. Elsewhere, public interest voices are fitted differently into the political system, either through a formal institutional position or through political parties. This positioning may force trade associations to respond to legislative agendas pushed by consumer interests. Two examples prove illustrative. The role of the Green Party in Germany has radically altered the place of consumer groups. This small party, a member of the governing coalition between 1998 and 2005, has successfully raised consumer protection to a cabinet responsibility. At the level of the European Union, consumer interests have been institutionalized in the consumer protection directorate, elevating public interest demands within European policy debates. As a result, industry finds itself in the position of responding to positions placed on the table by consumer advocates, who at the same time often have the sympathetic ears of the European Commission or national governments (Young and Wallace 2000).

These differences force us to at least open the basic question of how political groups form and how their interests are defined. Political strategies will now involve cross-national coalitions and deals in international institutions to settle what were once exclusively domestic decisions. Indeed, the creation of interests in the whole array of digital cases emphasizes that interest groups are never mechanical functions of markets or institutional structures, but rather the product of political struggles.

CONCLUSION

Information and how it is used provide the very substance of communities, polities, and markets. Communities can be conceived and indeed expressed as the character and flow of communications amongst members, polities as systems of decisions based on information, and markets as architectures for exchange based on information. Consequently, even technical rules about digital technology and the digital market are directly and simultaneously decisions about the very nature of the community and the polity.

It is clear that new deals are being struck, but the content of these deals is not compelled in any consistent way by the digital tools and networks themselves. Rather, the state finds itself struggling to manage digitally inspired conflicts fueled by business and public interest groups. As technology reopens de-

410   bates, governments have varying policy tools at their disposal and confront distinct policy legacies. One should, therefore, expect to see different government approaches to basic digital fights. Not only will proposed government solutions vary, but these proposals will be filtered by each country's unique political configuration. The cross-national diversity of policy debates will reflect not only market conditions and problems, but also, and more fundamentally, the distinct organization of the public and private sector lobbies involved. Owing to the transnational character of digital markets, these varying state positions will naturally shape international negotiations.

The state has played a fundamental role in the emergence and development of the digital era. As in the case of the great transformation, government policy has created the infrastructure for the fictitious commodity of information. Through deregulation, market making, and reregulation, public policy has constructed the rules for the new market and managed conflicts that threatened to derail the digital transformation. These efforts have had important political consequences for the character of the contemporary era. Given the differing ways governments have dealt with the various challenges posed by this digital transformation, several distinct information societies will no doubt emerge.

NOTES

1. All due deference to Karl Polanyi (1944).

2. See John Gilmore as quoted in Peter Lewis, "Limiting a Medium without Boundaries: How Do You Let the Good Fish through the Net while Blocking the Bad?" *New York Times*, January 15, 1996. See also Barlow 1996. Manuel Castells (1996) makes a more nuanced argument that the rise of complex networks has limited the power of the state.

3. For a more general discussion of the role of self-regulation in the early years of the Internet, see Marsden 2000.

4. For a detailed analysis of the various roles the state can play in modern political economy, see J. Levy 2006.

5. For the important role states have played in resolving international disputes posed by digital technologies, see Drezner 2004.

6. For a discussion of the political development of telecommunications liberalization, see Cowhey 1990a.

7. Assif Shameen, "Korea's Broadband Revolution," *Chief Executive*, April 2004, http://www.chiefexecutive.net (accessed December 5, 2005).

8. See Tim Richardson, "South Korea Broadband in League of Its Own," *The Register*, October 14, 2002, http://www.theregister.co.uk (accessed December 5, 2005).

9. For a discussion of the DMCA as well as its implications for fair use, see Samuelson 1999; Nimmer 2000.

10. Available at http://www.nwc.com/showArticle.jhtml?articleID=20300119 (accessed December 5, 2005).

11. See Ron Lieber, "Banks Now Get Daily Reports on Their Customers," *San Francisco Chronicle*, August 4, 2003.

12. *New York Times*, "Two US Agencies Investigate JetBlue over Privacy Issues," September 23, 2003.

13. For a review of privacy concerns in a digital age, see Newman and Bach 2004.

14. There are exceptions, of course. One is the present debate about the effort to restrict telemarketing calls in the United States, which has mobilized a broad constituency. But even this issue, which receives almost universal popular support, has faced a harrowing road to implementation, including multiple court injunctions that threaten to derail a consumer-friendly outcome.

REFERENCE MATTER

Bibliography

Abbate, Janet. 1999. *Inventing the Internet*. Cambridge, MA: MIT Press.

Adler, Paul, William Mark Fruin, and Jeffrey Liker, eds. 1999. *Remade in America*. New York: Oxford University Press.

Ali-Yrkkö, Jyrki, and Raine Hermans. 2002. "Nokia in the Finnish Innovation System." ETLA Discussion Paper No. 811, Research Institute of the Finnish Economy, Helsinki.

———. 2004. "Nokia: A Giant in the Finnish Innovation System." In Gerd Schienstock (ed.), *Embracing the Knowledge Economy: The Dynamic Transformation of the Finnish Innovation System*. Aldershot, Hants.: Edward Elgar. 106–127.

Ali-Yrkkö, Jyrki, and Pekka Ylä-Anttila. 2002. "Globalization of Business in a Small Country: Does Ownership Matter?" In Ari Hyytinen and Mika Pajarinen (eds.), *Financial Systems and Firm Performance: Theoretical and Empirical Perspectives*. Helsinki: ETLA/Research Institute of the Finnish Economy. 249–292.

Alter, Karen J. 1996. "The European Court's Political Power: The Emergence of the European Court as an Influential Actor in Europe." *West European Politics* 19: 458–487.

American Institute of Certified Public Accountants. "Accounting Salaries." http://www.aicpa.org/nolimits/job/salaries/index.htm (accessed November 12, 2005).

Anchordoguy, Marie. 2001. Nippon Telegraph and Telephone Company (NTT) and the Building of a Telecommunications Industry in Japan. *Business History Review* 75 (Autumn): 507–541.

Anderson, Philip, and Michael Tushman. 1990. "Technological Discontinuities and Dominant Designs: A Cyclical Model of Technological Change." *Administrative Science Quarterly* 35: 604–633.

Anderson, Philip, and Michael L. Tushman. 1997. *Managing Strategic Innovation and Change: A Collection of Readings*. New York, NY: Oxford University Press.

Antonelli, Cristiano, and Michel Quéré. 2003. "The Economics of Governance: Transactions, Resources and Knowledge." Paper presented at the DRUID [Danish

416   Research Unit for Industrial Dynamics] Summer Conference 2003 on Creating, Sharing and Transferring Knowledge, Copenhagen, June 12–14.

Aoki, Masahiko. 1988. *Information, Incentives, and Bargaining in the Japanese Economy.* New York: Cambridge University Press.

Aoki, Mashaiko, and Ronald Dore, eds. 1994. *The Japanese Firm: The Sources of Competitive Strength.* New York: Oxford University Press.

Aronson, Jonathan, and Peter Cowhey. 2004. "Wireless Standards and Applications: Industrial Strategies and Government Policies." Paper for the Annenberg Research Network, October.

Arora, Ashish, and Suma Athreye. 2002. "The Software Industry and India's Economic Development." *Information Economics and Policy* 14, no. 2: 253–273.

Arthur, W. Brian. 1989. "Competing Technologies, Increasing Returns, and Lock-in by Historical Small Events." *Economic Journal* 99: 116–131.

———. 1994. *Increasing Returns and Path Dependence in the Economy.* Ann Arbor: University of Michigan Press.

Asplund, Rita. 2003. "Flexibility and Competitiveness: Labour Market Flexibility, Innovation and Organisational Performance." ETLA Discussion Paper No. 875, Research Institute of the Finnish Economy, Helsinki.

Austin, Marc, and Helen Milner. 2001. "Product Standards and International and Regional Competition." *Journal of European Public Policy* 8, no. 3: 411–431.

Axelrod, Robert, Scott Bennett, Erhard Bruderer, Will Mitchell, and Robert Thomas. 1995. "Coalition Formation in Standard-setting Alliances." *Management Science* 41, no. 9: 1493–1508.

Baccaro, Lucio. 2003. "What Is Alive and What Is Dead in the Theory of Corporatism." *British Journal of Industrial Relations* 41, no. 4: 683–706.

Bach, David. 2000. "International Cooperation and the Logic of Networks: Europe and the Global System of Mobile Communications (GSM)." Berkeley Roundtable on the International Economy Working Paper 139, University of California, Berkeley.

Baldwin, Carliss Y., and Kim B. Clark. 2000. *Design Rules.* Vol. 1, *The Power of Modularity.* Cambridge, MA: MIT Press.

Banasek, Ronald M. 2002. "The Future's in Our Hands." *Capacity.* January–February.

Bar, François. 2001. "The Construction of Marketplace Architecture." In BRIE-IGCC Economy Project Task Force (ed.), *Tracking a Transformation: E-commerce and the Terms of Competition in Industries.* Washington, DC: Brookings Institution Press.

Bar, François, and Michael Borrus. 1997. "The Path Not Yet Taken: User-driven Innovation and U.S. Telecommunications Policy." Paper presented at the Fourth Annual CRTPS Conference, University of Michigan Business School, Ann Arbor, June 5–6, 1998.

Baran, Barbara. 1986. "The Technological Transformation of White Collar Work: A Case Study of the Insurance Industry." Ph.D. diss., University of California, Berkeley.

Bardhan, Deo, and Cynthia Kroll. 2003. "The New Wave of Outsourcing." Paper for the Fisher Center for Real Estate and Urban Economics, University of California, Berkeley.

Barlow, John Perry. 1996. *Declaration of the Independence of Cyberspace.* http://www.eff.org/~barlow/Declaration-Final.html (accessed November 12, 2005).

Barr, Avron, and Shirley Tessler. "The Software Shortage." 1997. SCIP Software   **417**
Research Brief no. 97-1 (March 3). http://www.stanford.edu/group/scip/avsgt/
swlabor397.pdf (accessed December 13, 2005).

Beesom, Ann, and Chris Hansen. 1997. "Fahrenheit 451.2: Is Cyberspace Burning?"
White paper for the American Civil Liberties Union, March 17.

Bekkers, Rudi, Geert Duysters, and Bart Verspagen. 2002. "Intellectual Property
Rights, Strategic Technology Agreements, and Market Structure: The Case of
GSM." *Research Policy* 31, no. 7: 1141–1161.

Bell, Daniel. 1973. *The Coming of Post-industrial Society: A Venture in Social Forecasting.* New York: Basic Books.

Bendix, Reinhard. 1964. *Nation Building and Citizenship.* Berkeley: University of
California Press.

Benner, Mats, and Torben Vad. 2000. "Sweden and Denmark: Defending the Welfare
State." In F. Scharpf and V. Schmidt (eds.), *Welfare and Work in the Open Economy*,
vol. 2, *Diverse Responses to Common Challenges.* Oxford: Oxford University Press.

Bennett, Colin J. 1992. *Regulating Privacy: Data Protection and Public Policy in Europe
and the United States.* Ithaca, NY: Cornell University Press.

Bennett, Jonathan. 1971. *Locke, Berkeley, and Hume: Central Themes.* Oxford: Clarendon Press.

Berger, Suzanne, and Ronald Dore. 1996. *National Diversity and Global Capitalism.*
Ithaca, NY: Cornell University Press.

Berger, Suzanne, and Michael J. Piore.1980. *Dualism and Discontinuity in Industrial
Societies.* New York: Cambridge University Press.

Berggren, Christian, and Staffan Laestadius. 2003. "Co-development and Composite
Clusters: The Secular Strength of Nordic Telecommunications." *Industrial and
Corporate Change* 12: 91–114.

Bernstein, Lisa. 2001. "Private Commercial Law in the Cotton Industry: Creating
Cooperation through Rules, Norms, and Institutions." Olin Working Paper No. 133,
University of Chicago Law and Economics.

Besen, Stanley, and Joseph Farrell. 1991. "The Role of the ITU in Standardization: Preeminence, Impotence, or Rubber Stamp?" *Telecommunications Policy* (August):
311–321.

———. 1994. "Choosing How to Compete: Strategies and Tactics in Standardization.
Journal of Economic Perspectives 8: 117–131.

Bhattacharjee, Ashok. 2003. "India's Call Centers Face Struggle to Keep Staff as Economy Revives." *Wall Street Journal*, October 29.

Birkinshaw, Julian, Sumantra Ghoshal, Constantinos C. Markides, John Stopford, and
George Yip, eds. 2003. *The Future of the Multinational Company.* West Sussex:
Wiley Press.

Blumenthal, Marjory S., and David D. Clark. 2001. "Rethinking the Design of the
Internet: The End to End Arguments vs. the Brave New World." *ACM Transactions
on Technology* 1, no. 1: 70–109.

Boas, Taylor C. 2004. "Technology, Freedom, and Democracy: An Evolving Debate."
APSA-CP: Newsletter of the Organized Section in Comparative Politics of the American Political Science Association 15, no. 1: 18–23.

418   Bohlin, Erik, Karolina Brodin, Anders Lundgren, and Bertil Thorngren, eds. 2000. *Convergence in Communications and Beyond*. Amsterdam: North-Holland/Elsevier.

Boltho, Andrea. 1982. "Growth." In *The European Economy: Growth and Crisis*. Oxford: Clarendon Press. 9–39.

Borrus, Michael. 1997. "Left for Dead: Asian Production Networks and the Revival of US Electronics." Berkeley Roundtable on the International Economy Working Paper 100, University of California, Berkeley.

Borrus, Michael, and François Bar. 1993. *The Future of Networking*. Berkeley: BRIE.

Borrus, Michael, and Stephen S. Cohen. 1998. "Building China's Information Technology Industry: Tariff Policy and China's Accession to the World Trade Organization." *Asian Survey* 38, no. 11: 1005–1017.

Borrus, Michael, Dieter Ernst, and Stephan Haggard, eds. 2000. *International Production Networks in Asia: Rivalry or Riches?* London: Routledge.

Borrus, Michael, Jim Millstein, and John Zysman. 1982. *US-Japanese Competition in the Semi-conductor Industry*. Berkeley: Institute of International Studies.

Borrus, Michael, and John Zysman. 1997a. "Globalization with Borders: The Rise of Wintelism as the Future of Industrial Competition." *Industry and Innovation* 4, no. 2: 141–166.

———. 1997b. "Wintelism and the Changing Terms of Global Competition: Prototype of the Future?" Berkeley Roundtable on the International Economy Working Paper 96B, University of California, Berkeley.

Boyer, Robert, and J. Rogers Hollingsworth. 1997. *Contemporary Capitalism: The Embeddedness of Institutions*. New York: Cambridge University Press.

Boyer, Robert, and Yves Saillard. 2002. *Regulation Theory: The State of the Art*. New York: Routledge.

Boyle, James. 2003. "The Second Enclosure Movement and the Construction of the Public Domain." *Law and Contemporary Problems* 66: 33–74.

Braga, Carlos A. Primo, John A. Daly, and Bimal Sareen. 2003. "The Future of Information and Communication Technologies for Development." ICT Development Forum Working Paper. http://old.developmentgateway.org/download/221030/Future_ICT.pdf (accessed December 13, 2005).

Bresnahan, Timothy, and M. Trajtenberg. 1995. "General Purpose Technologies: Engines of Growth?" *Journal of Econometrics* 65, no. 1: 83–108.

Breuss, Fritz. 2003. "Austria, Finland and Sweden in the European Union: Economic Effects." *Austrian Economic Quarterly* 4: 131–158.

Bridges.org. 2001. "Spanning the Digital Divide." http://www.bridges.org/spanning/ (accessed December 13, 2005).

Brown, John Seely, and John Hagel III. 2005. "Innovation Blowback: Disruptive Management Practices from Asia." *McKinsey Quarterly* 1: 34–45.

Brown, John Seely, and Paul Duguid. 2000. *The Social Life of Information*. Boston: Harvard Business School Press.

Bruce, Robert, and Jeffrey Cunard. 1994. "Restructuring the Telecommunications Sector in Asia: An Overview of Approaches and Options." In Bjorn Wellenius and Peter Stern (eds.), *Implementing Reforms in the Telecommunications Sector: Lessons from Experience.* World Bank Regional and Sectoral Study Series. Washington, DC: World Bank. 197–231.

Brynjolfsson, Erik, and Lorin M. Hitt. 2004. "The Catalytic Computer: Information Technology, Enterprise Transformation, and Business Performance." In William Dutton, Brian Kahin, Ramon O'Callaghan, and Andrew Wyckoff (eds.), *Transforming Enterprise.* Cambridge: MIT Press.

Bunge, Mario. 1959. *Causality and Modern Science.* New York: Dover.

———. 2001. *Scientific Realism.* New York: Prometheus Books.

Burgelman, Robert, and Andrew Grove. 1996. "Strategic Dissonance." *California Management Review* 38: 8–28.

Burk, Dan L., and Julie E. Cohen. 2001. "Fair Use Infrastructure for Rights Management Systems." *Harvard Journal of Law and Technology* 15: 41–83.

Burstein, Aaron, Will Thomas DeVries, and Peter S. Menell. 2004. "FOREWARD: The Rise of Internet Interest Group Politics." *Berkeley Technology Law Journal* 19: 1–2.

Büthe, Tim, and Walter Mattli. 2003. "Setting International Standards: Technological Rationality or Primacy of Power?" *World Politics* 56 (October): 1–42.

Cahuc, Pierre, and Francis Kamartz. 2004. *De la précarité à la mobilité: Vers une sécurité sociale professionnelle.* Paris: Rapport au Ministre d'Etat, Ministre de l'Economie, des Finances et de l'Industrie et au Ministre de l'Emploi, du Travail et de la Cohésion Sociale.

Callaghan, G., P. Thompson, and C. Warhurst. 2001. "Ignorant Theory and Knowledgeable Workers: Interrogating the Connections between Knowledge, Skills, and Services." *Journal of Management Studies* 38, no. 7: 923–942.

Calmfors, Lars, and John Driffill. 1988. "Bargaining Structure, Corporatism, and Macroeconomic Performance." *Economic Policy* 6: 15–61.

Caloghirou, Yannis, Stavros Ioannides, and Nicholas Vonortas. 2003. "Research Joint Ventures." *Journal of Economic Surveys* 17, no. 4: 541–570.

Cameron, David. 1984. "Social Democracy, Corporatism, Labor Quiescence, and the Representation of Economic Interest in Advanced Capitalist Society." In John Goldethorpe (ed.), *Order and Conflict in Contemporary Capitalism.* Oxford: Clarendon Press. 143–178.

Campbell, David, and Andrew Faulkner, eds. 2003. "Strategic Innovation." In *Oxford Handbook of Strategy*, vol. 2. Oxford: Oxford University Press.

Carbaugh, John Jr. 2002. "Telecom Deregulation and Telecom Competition in World Markets." *Daily Report.* March 3, 1–4.

Castells, Manuel. 1996. *The Rise of the Network Society.* Cambridge: Blackwell.

Castells, Manuel, and Pekka Himanen. 2002. *The Information Society and the Welfare State: The Finnish Model.* Oxford: Oxford University Press.

Chaikin, S., and J. Lave, eds. 1993. *Understanding Practice.* New York: Cambridge University Press.

Chase, Michael S., and James C. Mulvenon. 2002. *You've Got Dissent! Chinese Dissident Use of the Internet and Beijing's Counter-strategies.* Santa Monica, CA: RAND.

Chesbrough, Henry. 2003. *Open Innovation.* Cambridge, MA: Harvard Business School Press.

420

Christensen, Clayton. 1997. *The Innovator's Dilemma: When New Technology Causes Great Firms to Fail*. Boston: Harvard Business School Press.

Clark, David D. 1988. "The Design Philosophy of the DARPA Internet Protocols." *Computer Communication Review* 18, no. 4: 106–114.

Coase, Ronald. 1937. "The Nature of the Firm." *Economica* 4: 386–405.

———. 1959. "The Federal Communications Commission." *Journal of Law and Economics* 2, no. 1: 1–40.

———. 1960. "The Problem of Social Cost." *Journal of Law and Economics* 3: 1–44.

Coffman, K. G., and A. M. Odlyzko, "Internet Growth: Is There a "Moore's Law" for Data Traffic?" June 4, 2001. http://www.dtc.umn.edu/~odlyzko/doc/complete.html (accessed December 14, 2005).

Cohen, Stephen S., J. Bradford DeLong, Steven Weber, and John Zysman. 2001. "Tools: The Drivers of E-Commerce." In BRIE-IGCC Economy Project Task Force (ed.), *Tracking a Transformation: E-commerce and the Terms of Competition in Industries*. Washington, DC: Brookings Institution Press.

Cohen, Stephen S., J. Bradford DeLong, and John Zysman. 2000. "Tools for Thought: What Is New and Important about the 'E-conomy'?" Berkeley Roundtable on the International Economy Working Paper 138, University of California, Berkeley.

Cohen, Stephen S., and John Zysman. 1987. *Manufacturing Matters: The Myth of the Post-industrial Economy*. New York: Basic Books.

Cole, Robert E. 1989. *Strategies for Learning: Small Group Activities in American, Japanese, and Swedish Industry*. Berkeley: University of California Press.

———. 1994. "Reengineering the Corporation: A Review Essay." *Quality Management Journal* 1, no. 4: 77–85.

Commission of the European Communities. 1996. *Illegal and Harmful Content on the Internet*. Brussels: European Communities.

———. 1998. *Globalization and the Information Society: The Need for Strengthened International Co-ordination*. Brussels: European Communities.

Cook, Karen S. 1977. "Exchange and Power in Networks of Inter-organizational Relations." *Sociological Quarterly* 18: 62–82.

Coriat, Benjamin. 1990. "The Revitalization of Mass Production in the Computer Age." Paper presented at the UCLA Lake Arrowhead Conference Center, Los Angeles, CA, March 14–18.

Cornford, F. M. 1935. *Plato's Theory of Knowledge*. London: Routledge and Kegan Paul.

Correa, Carlos M. 1995. "Innovation and Technology Transfer in Latin America: A Review of Recent Trends and Policies." *International Journal of Technology Management* 10, no. 7–8: 815–846.

Cowhey, Peter. 1990a. "The International Telecommunications Regime: The Political Roots of High Technology Regimes." *International Organization* 44: 169–199.

———. 1990b. "States and Politics in American Foreign Economic Policy." In John Odell and Thomas Willett (eds.), *International Trade Policies*. Ann Arbor: University of Michigan Press. 225–252.

———. 1990c. "Telecommunications." In Gary Hufbauer (ed.), *Europe 1992: An American Perspective*. Washington, DC: Brookings Institution.

————. 1995. "The Politics of U.S. and Japanese Security Commitments." In Peter Cowhey and Mathew McCubbins (eds.), *Structure and Policy in Japan and the United States*. Cambridge: Cambridge University Press.

————. 1998. "FCC Benchmarks and the Reform of the International Telecommunications Market." *Telecommunications Policy* 22, no. 11: 899–911.

Cowhey, Peter, and Mikhail M. Klimenko. 2001. "The WTO Agreement and Telecommunications Policy Reform." Policy Research Working Paper 2601, World Bank, Washington, DC.

Cowhey, Peter, and Mathew McCubbins (eds.). 1995. *Structure and Policy in Japan and the United States*. New York: Columbia University Press.

Cox, Gary, and Mathew D. McCubbins. 1993. *Legislative Leviathan*. Berkeley: University of California Press.

Damasio, Antonio. 2003. *Looking for Spinoza: Joy, Sorrow, and the Feeling Brain*. Orlando, FL: Harcourt.

David, Paul A. 1985. "Clio and the Economics of QWERTY." *American Economic Review* 75, no. 2: 332–337.

D'Costa, Anthony P. 2003. "Uneven and Combined Development: Understanding India's Software Exports." *World Development* 31, no. 1: 211–226.

Dedrick, Jason, and Kenneth L. Kraemer. 1993. "Information Technology in India: The Quest for Self-reliance." *Asian Survey* 33, no. 5: 463–492.

de la Mothe, John, and Albert Link, eds. 2002. *Networks, Alliances, and Partnerships in the Innovation Process*. Boston: Kluwer Academic.

Delaney, Kevin. 1999. "Telecom-equipment Concerns Focus on Software." *Wall Street Journal*, October 18.

————. 2002. "France Telecom Approves Plan to Reduce Debt." *Wall Street Journal*, December 5.

"Delegates Cast Vote for Paging." 2000. *Wireless Week* 6, no. 34 (August 21): 42–43.

DeLong, J. Bradford. 2004. "The Economics of the New Economy." Lecture to the Governance of the Economy, Berkeley, CA, Spring.

DeLong, J. Bradford, and Konstantin Magin. 2005. "The Last Bubble Was Brief, but It Was Still Irrational." *Financial Times*, April 19.

DeLong, J. Bradford, and Lawrence H. Summers. 2002. "Is the 'New Economy' a Fad?" Project Syndicate, April. http://www.j-bradford-delong.net/movable_type/archives/000773.html (accessed December 13, 2005).

Digital Opportunities Task Force. 2001. *Digital Opportunities for All: Meeting the Challenge*. Report of the Digital Opportunity Task Force. http://lacnet.unicttaskforce.org/Docs/Dot%20Force/Digital%20Opportunities%20for%20All.pdf (accessed December 13, 2005).

Dokko, Gina, and Lori Rosenkopf. 2003. "Social Capital for Hire? The Mobility of Technical Professionals and Firm Centrality in Wireless Standards Committees." Unpublished manuscript. University of Pennsylvania.

————. 2004. "Social Capital Formation in Standards Setting Committees." Paper presented at the Academy of Management Conference, New Orleans, LA, August 8–11.

Dore, Ronald. 1987. *Taking Japan Seriously*. Stanford, CA: Stanford University Press.

422   Dorgan, Stephen J., and John J. Dowdy. 2004. "When IT Lifts Productivity." *McKinsey Quarterly* 4. http://www.adamsmithesq.com/blog/pdf/McKinsey.When.IT.Lifts .Productivity.pdf (accessed December 13, 2005).

Dosi, Giovanni. 1982. "Technological Paradigms ands Technological Trajectories." *Research Policy* 11: 147–162.

Dossani, Rafiq. 2002. *Telecommunications Reform in India*. Westport, CT: Greenwood Press.

Dossani, Rafiq, and Martin Kenney. 2004. "Went for Cost, Stayed for Quality? Moving the Back Office to India." Berkeley Roundtable on the International Economy Working Paper 156, University of California, Berkeley.

Drake, William J., and Kalypso Nicolaides. 1992. "Ideas, Interests, and Institutionalization: 'Trade in Services.'" *International Organization* 46: 37–100.

Dranove, David, and Neil Gandal. 2003. "The DVD vs. DIVX Standard War: Empirical Evidence of Network Effects and Preannouncement Effects." *Journal of Economics and Management Strategy* 12: 363–386.

Dratler, Jay Jr. 2000. *Cyberlaw: Intellectual Property in the Digital Millennium*. New York: Law Journal Press.

Drezner, Daniel. 2004. "The Global Governance of the Internet: Bringing the State Back In." *Political Science Quarterly* 119: 477–498.

Dunning, John, and Gavin Boyd, eds. 2003. *Alliance Capitalism and Corporate Management*. Cheltenham, England: Edward Elgar.

Dunning, Thad. 2003. "The Political Economy of the International 'Digital Divide': A Tentative Outline of Some Issues and Concepts." Berkeley Roundtable on the International Economy Working Paper, University of California, Berkeley.

Dutta, Soumitra, Bruno Lanvin, and Fiona Paua, eds. 2003. *The Global Information Technology Report, 2002–2003: Readiness for the Networked World*. New York: Oxford University Press.

Duysters, Geert, and John Hagedoorn. 1996. "Internationalization of Corporate Technology through Strategic Partnering: An Empirical Investigation." *Research Policy* 25: 1–12.

———. 1998. "Technological Convergence in the IT Industry: The Role of Strategic Technology Alliances and Technological Competencies." *International Journal of the Economics of Business* 5, no. 3: 335–368.

Duysters, Geert, Charmianne Lemmens, and John Hagedoorn. 2003. "The Effect of Strategic Block Membership on Innovative Performance: A Dynamic Perspective." *Revue d'economie industriell* 103, no. 2–3: 59–70.

Ebbinghaus, Bernhard, and Anke Hassel. 1999. "Striking Deals: Concertation in the Reform of the Continental European Welfare States." Max Planck Institute for the Study of Societies Discussion Paper 99/3, Cologne.

Ebbinghaus, Bernhard, and Philip Manow, eds. 2001. *Comparing Welfare Capitalism: Social Policy and Political Economy in Europe, Japan, and the USA*. London: Routledge.

Economist. 2004a. "Profits and Poverty." August 19.

———. 2004b. "A World of Work: A Survey of Outsourcing." November 13.

Economist Intelligence Unit. 1995. *Country Briefing: Finland, 1995*.

————. 2000. *Country Briefing: Finland, 2000.*

Edvinsson, Leif, and Michael Malone. 1997. *Intellectual Capital.* New York: Harper-Collins.

"Electronics Giants to Develop High-Speed Internet Routers." 2003. *Nikkei Weekly,* December 22.

Esping-Andersen, Gøsta. 1990. *The Three Worlds of Welfare Capitalism.* Princeton, NJ: Princeton University Press.

Estevez-Abe, Margarita, Torben Iversen, and David Soskice. 2001. "Social Protection and the Formation of Skills: A Reinterpretation of the Welfare State." In Peter A. Hall and David Soskice (eds.), *Varieties of Capitalism: The Institutional Foundations of Comparative Advantage.* Oxford: Oxford University Press. 145–183.

European Commission. 1999. *The Economic and Financial Situation in Finland.* Luxembourg: Office for Official Publications of the European Communities.

————. 2003a. *A Pocketbook of Enterprise Policy Indicators.* Luxembourg: Office for Official Publications of the European Communities.

————. 2003b. *Jobs, Jobs, Jobs: Creating More Employment in Europe.* Luxembourg: Office for Official Publications of the European Communities.

————. 2004. *Joint Employment Report, 2003–2004.* Luxembourg: Office for Official Publications of the European Communities.

European Information Technology Observatory. 2004. *European Information Technology Observatory.* Frankfurt: Yearbook/EITO.

European Parliament. 1999. Decision No. 276/1999/EC of the European Parliament and of the Council of 25 January 1999 Adopting a Multi-annual Community Action Plan on Promoting Safer Use of the Internet by Combating Illegal and Harmful Content on Global Networks. January 25.

European Union. 1998. *The Radio Paging Market in Europe.* A Study by ECSC-EC-EAEC.

————. 2000. "Towards a European Research Area." COM (2000) 6, Commission of the European Communities, Brussels.

Evans, Peter B. 1995. *Embedded Autonomy: States and Industrial Transformation.* Princeton, NJ: Princeton University Press.

Evans, Peter B., Dietrich Rueschemeyer, and Theda Skocpol. 1985. "On the Road toward a More Adequate Understanding of the State." In Peter D. Evans, Dietrich Rueschemeyer, and Theda Skocpol (eds.), *Bringing the State Back In.* Cambridge: University of Cambridge Press.

Farrell, Henry. 2003. "Constructing the International Foundations of E-commerce: The EU-US Safe Harbor Arrangement." *International Organization* 57: 277–306.

Farrell, Joseph, and Garth Saloner. 1985. "Standardization, Compatibility, and Innovation." *RAND Journal of Economics* 16: 70–83.

————. 1988. "Coordination through Committees and Markets." *RAND Journal of Economics* 19: 235–252.

Federal Communications Commission. 2000. *Privacy Online: Fair Information Practices in the Electronic Marketplace: A Federal Trade Commission Report to Congress.* Washington, DC: Federal Trade Commission.

————. 2001a. *Annual Report and Analysis of Competitive Market Conditions with Respect to Commercial Mobile Services.* Sixth report. FCC 01-192.

424 ———. 2001b. Year 2000 Biennial Regulatory Review: Amendment of Part 22 of the Commission's Rules to Modify or Eliminate Outdated Rules Affecting the Cellular Radiotelephone Service and Other Commercial Mobile Radio Services. FCC 01–153.

Federal Trade Commission. 1999. *Self-regulation and Privacy Online: A Report to Congress*. Washington, DC: Federal Trade Commission.

Feyerabend, Paul K. 1975. *Against Method*. London: Verso.

Fields, Gary. 2003. *Territories of Profit: Communications, Capitalist Development, and the Innovative Enterprises of G. F. Swift and Dell Computer*. Innovation and Technology in the World Economy. Stanford, CA: Stanford University Press.

Finland in the Global Economy Steering Committee. 2004. *Finland's Competence, Openness, and Renewability: The Final Report of the Finland in the Global Economy Project*. Helsinki: Prime Minister's Office.

Fligstein, Neil. 1996. "Markets as Politics: A Political-cultural Approach to Market Institutions." *American Sociological Review* 61: 656–673.

Florida, Richard. 2002. *The Rise of the Creative Class and How It's Transforming Work, Leisure, Community and Everyday Life*. New York: Basic Books.

Florida, Richard, and Martin Kenney. 1991. "Transplanted Organizations: The Transfer of Japanese Industrial Organization to the U.S." *American Sociological Review* 56: 381–398.

Foss, Nicolai, ed. 1997. *Resources, Firms, and Strategies: A Reader in the Resource-based Perspective*. Oxford: Oxford University Press.

Foucault, Michel. 1994. *The Order of Things*. New York: Vintage Books.

———. 2001. *Talens forfatning: Nietzsche, Genealogien, Historien*. Copenhagen: Hans Reitzels Forlag.

Franda, Marcus. 2002. *Launching into Cyberspace: Internet Development and Politics in Five World Regions*. Boulder, CO: Lynne Rienner.

Fransman, Martin. 1995. *Japan's Computer and Communications Industry: The Evolution of Industrial Giants and Global Competitiveness*. New York: Oxford University Press.

Freeman, Christopher. 1991. "Networks of Innovators: A Synthesis of Research Issues." *Research Policy* 20: 499–514.

Froomkin, A. Michael. 1997. "The Internet as a Source of Regulatory Arbitrage." In Brian Kahin and Charles Nesson (eds.), *Borders in Cyberspace: Information Policy and the Global Information Infrastructure*. Cambridge, MA: MIT Press.

———. 2000. "Wrong Turn in Cyberspace: Using ICANN to Route around the APA and the Constitution." *Duke Law Journal* 50: 171–184.

Fujii, Koichiro. 2003. *NTT: Koroshita no wa Dare Da? [Who Killed NTT?]*. Tokyo: Kobunsha Paperbacks.

"Fujitsu, routa gaibu choutatsu [Fujitsu Will Procure Routers from Outside]." 2004. *Nihon keizai shimbun*, February 20.

Fuke, Hidenori. 2000. *Joho Tsushin Sangyo no Kozo to Kisei Kanwa: Nchibeiei Hikaku Kenkyuu [Structural Change and Deregulation in the Telecommunications Industry]*. Tokyo: NTT Shuppan.

———. 2003. "The Spectacular Growth of DSL in Japan and Its Implications." *Communications and Strategies* 52: 175–191.

Funk, Jeffrey L. 1998. "Competition between Regional Standards and the Success and Failure of Firms in the World-wide Mobile Communication Market." *Telecommunications Policy* 2, no. 4: 419–441.

———. 2002. *Global Competition between and within Standards: The Case of Mobile Phones.* Houndsmills, England, and New York: Palgrave.

Funk, Jeffrey L., and David T. Methe. 2001. "Market- and Committee-based Mechanisms in the Creation and Diffusion of Global Industry Standards: The Case of Mobile Communication." *Research Policy* 20: 589–610.

Gandal, Neil, David Salant, and Leonard Waverman. 2003. "Standards in Wireless Telephone Networks." *Telecommunications Policy* 27: 325–332.

Gao, Maija, Ari Hyytinen, and Otto Toivanen. 2005. "Demand for Mobile Internet: Evidence from a Real-world Pricing Experiment." ETLA Discussion Paper No. 964. Helsinki: Research Institute of the Finnish Economy.

Garner, C. Alan. 2004. "Offshoring in the Service Sector: Economic Impact and Policy Issues." *Federal Reserve Bank of Kansas City Economic Review* (third quarter): 5–37.

Gentle, Chris. 2003. "On the Cusp of a Revolution: How Offshoring Will Transform the Financial Services Industry." *Deloitte Research* March.

Georghiou, Luke, Keith Smith, Otto Toivanen, and Pekka Ylä-Anttila. 2003. *Evaluation of the Finnish Innovation Support System.* Publications 5/2003, Finland Ministry of Trade and Industry.

Gereffi, Gary. 1994. "The Organization of Buyer-driven Global Commodity Chains: How U.S. Retailers Shape Overseas Production Networks." In Gary Gereffi and Miguel Korzeniewicz (eds.), *Commodity Chains and Global Capitalism.* Westport, CT: Greenwood Press. 95–122.

Gereffi, Gary, and Miguel Korzeniewicz, eds. 1994. *Commodity Chains and Global Capitalism.* Westport, CT: Greenwood Press.

Gereffi, Gary, John Humphrey, and Timothy Sturgeon. 2005. "The Governance of Global Value Chains." *Review of International Political Economy* 12, no. 1: 78–104.

Gerlach, Michael L. 1992. *Alliance Capitalism: The Social Organization of Japanese Business.* Berkeley: University of California Press.

Glimstedt, Henrik. 2001. "Competitive Dynamics of Technological Standardization: The Case of Third Generation Cellular Communications." *Industry and Innovation* 8: 49–78.

Global Value Chain Initiative. http://www.globalvaluechains.org/ (accessed November 12, 2005).

Goodin, Robert, Bruce Heady, Ruud Muffels, and Henk-Jan Dirven. 1999. *The Real Worlds of Welfare Capitalism.* Cambridge: Cambridge University Press.

Goodman, Bill, and Reid Steadman. 2002. "Services: Business Demand Rivals Consumer Demand in Driving Job Growth." *Monthly Labor Review* 125, no. 4: 3–16.

Granovetter, Mark. 1985. "Economic Action and Social Structure: The Problem of Embeddedness." *American Journal of Sociology* 91: 481–510.

Green-Pedersen, Christoffer. 2001. "Welfare-state Retrenchment in Denmark and the Netherlands, 1982–1998." *Comparative Political Studies* 34: 963–985.

426   Greenstein, Shane. 1994. "Invisible Hand versus Invisible Advisors: Coordination Mechanisms in Economic Networks." Columbia Institute of Tele-information Working Paper, Columbia University.

Gribble, Cheryl. 2004. "History of the Web Beginning at CERN," July 13. http://www .hitmill.com/internet/web_history.html (accessed December 13, 2005).

Grimes, William. 2001. Unmaking the Japanese Miracle: Macroeconomic Politics. 1985–2000. Ithaca, NY: Cornell University Press.

Gruber, Harold, and Frank Verboven. 2001. "The Evolution of Markets under Entry and Standards Regulation: The Case of Global Mobile Telecommunication." International Journal of Industrial Organization 19: 1189–1212.

Gulati, Ranjay. 1995. "Social Structure and Alliance Formation Patterns: A Longitudinal Analysis." Administrative Science Quarterly 40: 619–652.

Gunningham, Neil, and Joseph Rees. 1997. "Industry Self-regulation: An Institutionalist Perspective." Law and Policy 19: 363–414.

Habermas, Jürgen. 1961. Strukturwandel der Öffentlichkeit. Frankfurt: Luchterhand Verlag.

Hafner, Katie, and Matthew Lyon. 1998. Where Wizards Stay Up Late: The Origins of the Internet. New York: Touchstone.

Hagedoorn, John. 2002. "Inter-firm Partnerships: An Overview of Major Trends and Patterns since 1960." Research Policy 31: 477–492.

Hagedoorn, John, Elias Carayannis, and Jeffrey Alexander. 2001. "Strange Bedfellows in the Personal Computer Industry: Technology Alliances between IBM and Apple." Research Policy 30: 837–849.

Hagedoorn, John, Albert Link, and Nicholas Vonortas. 2000. "Research Partnerships." Research Policy 29: 567–586.

Hagedoorn, John, and Jos Schakenraad. 1992. "Leading Companies and Networks of Strategic Alliances in Information Technologies." Research Policy 21: 163–190.

———. 1993. "A Comparison of Private and Subsidized R&D Partnerships in the European Information Technology Industry." Journal of Common Market Studies 31: 373–390.

Häikiö, Martti. 2001a. Nokia Oyj:n historia, vol. 1, Fuusio. Helsinki: Edita.

———. 2001b. Nokia Oyj:n historia, vol. 3, Globalisaatio. Helsinki: Edita.

———. 2001c. Nokia Oyj:n historia, vol. 2, Sturm und Drang. Helsinki: Edita.

———. 2002. Nokia: The Inside Story. London: Prentice Hall.

Hall, Peter. 1986. Governing the Economy. Oxford: Oxford University Press.

Hall, Peter, and Daniel W. Gingerich. 2001. "Varieties of Capitalism and Institutional Complementarities in the Macroeconomy: An Empirical Analysis." Paper presented at the annual meeting of the American Political Science Association, San Francisco, CA, August 30, 2001.

Hall, Peter A., and David Soskice. 2001. Varieties of Capitalism: The Institutional Foundations of Comparative Advantage. New York: Oxford University Press.

Hamm, Steve. 2004. "Tech's Future." Business Week, September 27, 82–89.

Hammer, Michael, and James Champy. 1993. Reengineering the Corporation: A Manifesto for Business Revolution. New York: HarperCollins.

Hannan, Michael T., and John Freeman. 1989. Organizational Ecology. Cambridge, MA: Harvard University Press.

Hardiman, Niamh. 2002. "From Conflict to Coordination: Economic Governance **427** and Political Innovation in Ireland." *West European Politics* 25, no. 3: 1–24.

Harwit, Eric, and Duncan Clark. 2001. "Shaping the Internet in China: Evolution of Political Control over Network Infrastructure and Content." *Asian Survey* 41: 377–408.

Haufler, Virginia. 2001. *Public Role for the Private Sector: Industry Self-regulation in a Global Economy.* Washington, DC: Carnegie Endowment for International Peace.

Hazlett, Thomas. 2001. "The Wireless Craze, the Unlimited Bandwidth Myth, the Spectrum Auction Faux Pas, and the Punchline to Ronald Coase's 'Big Joke': An Essay on Airwave Allocation Policy." Working Paper 01-01, AEI-Brookings Joint Center for Regulatory Studies.

Heeks, Richard. 1996. *India's Software Industry: State Policy, Liberalisation, and Industrial Development.* New Delhi: Sage.

Helpman, Elhanan, ed. 1998. *General Purpose Technologies and Economic Growth.* Cambridge: MIT Press.

Hernesniemi, Hannu, Markku Lammi, and Pekka Ylä-Anttila. 1996. *Advantage Finland: The Future of Finnish Industries.* Helsinki: Taloustieto.

Herrigel, Gary, Horst Kern, and Charles F. Sabel. 1989. *Collaborative Manufacturing: New Supplier Relations in the Automobile Industry and the Redefinition of the Industrial Corporation.* Cambridge, MA: International Motor Vehicle Program, Massachusetts Institute of Technology.

Hira, Ron. 2004. "U.S. Immigration Regulations and India's Information Technology Industry." *Technological Forecasting and Social Change* 71: 837–854.

Hira, R., and A. Hira. 2005. *Outsourcing America: What's Behind Our National Crisis And How We Can Reclaim American Jobs.* New York, NY: Amacom Books.

Hirst, Paul, and Jonathan Zeitlin. 1997. "Flexible Specialization: Theory and Evidence in the Analysis of Industrial Change." In J. Rogers Hollingsworth and Robert Boyer (eds.), *Contemporary Capitalism: The Embeddedness of Institutions.* Cambridge: Cambridge University Press. 220–239.

Hjelm, Bjorn. 2000. "Standards and Intellectual Property Rights in the Age of Global Communication: A Review of the International Standardization of Third Generation Mobile System." Paper presented at the Fifth IEEE Symposium on Computers and Communications, Antibus-Juan Les Pin, France, July 3–6.

Hjerppe, Reino, Riitta Hjerppe, Kauko Mannermaa, Olavi E. Niitamo, and Kaarlo Siltari. 1976. *Suomen teollisuus ja teollinen käsityö, 1900–1965.* Helsinki: Bank of Finland.

Hobday, Michael. 1995. "East Asian Latecomer Firms: Learning the Technology of Electronics." *World Development* 23: 1171–1193.

———. 2000. "East versus Southeast Asian Innovation Systems: Comparing OME- and TNC-led Growth in Electronics." In Linsu Kim and Richard R. Nelson (eds.), *Technology, Learning, and Innovation: Experiences of Newly Industrializing Economies.* Cambridge: Cambridge University Press.

Hollingsworth, J. R. 1997. *Contemporary Capitalism: The Embeddedness of Institutions.* Cambridge: Cambridge University Press.

Honkapohja, Seppo, and Erkki Koskela. 1999. "The Economic Crisis of the 1990s in Finland." *Economic Policy* 14: 399–436.

428   Hooghe, Liesbet, and Gary Marks. 2000. *Multi-level Governance and European Integration.* Lanham: Rowman and Littlefield.

Hudson, Heather. 1997. *Global Connections.* New York: Van Nostrand.

Hufbauer, Gary, ed. 1990. *Europe 1992: An American Perspective.* Washington, DC: Brookings Institution.

Hyytinen, Ari, Iikka Kuosa, and Tuomas Takalo. 2003. "Law or Finance: Evidence from Finland." *European Journal of Law and Economics* 16: 59–89.

Hyytinen, Ari, and Mika Pajarinen. 2002. "Financial Systems and Venture Capital in Nordic Countries: A Comparative Study." In Ari Hyytinen and Mika Pajarinen (eds.), *Financial Systems and Firm Performance: Theoretical and Empirical Perspectives.* Helsinki: ETLA/Research Institute of the Finnish Economy. 19–64.

———. 2005. "Financing of Technology-intensive Small Businesses: Some Evidence of the Uniqueness of the ICT Industry." *Information Economics and Policy* 17: 115–132.

Ikeda, Nobuo. 2003. "The Unbundling of Network Elements: Japan's Experiences." In *RIETI Discussion Paper Series 03–E-023.* Tokyo: Ministry of Economy, Trade and Industry.

Innovation Council. 2004. *Innovation Monitor: An Assessment of Denmark's Innovation Capacity.* Copenhagen: Ministry of Economic and Business Affairs.

International Data Corporation. 2000. "Worldwide Computer Networking Equipment Supplier Shipment Share." IDC company report.

International Telecommunications Union. 2001. "ITU Secretary General Utsumi Presses US on 3G Allocations." *Telecommunications Reports International,* June 1.

———. 2003. "Promoting Broadband: The Case of Japan." Report, International Telecommunications Union, Geneva, April.

Internet Engineering Task Force. 1996. "The Internet Standards Process: Revision 3." BCP: 9. http://www.ietf.org/rfc/rfc2026.txt (accessed November 12, 2005).

———. 2001. "The Tao of IETF." http://www.ietf.org/tao.html (accessed November 13, 2005).

"IP denwa demo bangou ga kawaranai riyuu: mittsu no kufuu de soumushou no jouken wo kuria [The Reason Why Your Phone Number Will Not Change Even If You Switch to an IP Telephone: Three Modifications Will Allow (IP Telephony) to Pass Soumusho's Conditions]." 2003. *Nikkei Communications,* November 24: 66–68.

"IP denwa 'Dounyu' 9 wari [Ninety Percent (of Large Firms) Have Installed IP Telephones]." 2005. *Nihon keizai shimbun,* March 3.

"IP denwa: Katei nimo shintou [IP Telephony Penetration Extends to Households]." 2005. *Nihon keizai shimbun,* March 29.

Isenberg, David. 1997. "Rise of the Stupid Network." *Computer Telephony* (August): 16–26.

Iversen, Torben. 1996. "Power, Flexibility, and the Breakdown of Centralized Wage Bargaining in Denmark and Sweden." *Comparative Politics* 28: 399–436.

Iversen, Torben, Jonas Pontusson, and David W. Soskice, eds. 2000. *Unions, Employers, and Central Banks: Macroeconomic Coordination and Institutional Change in Social Market Economies.* Cambridge: Cambridge University Press.

Jagannathan, Shankar, Stanislav Kura, and Michael J. Wilshire. 2003. "A Help Line for   **429**
European Telcos." *McKinsey Quarterly* 1: 86–97.

Jaikumar, Ramchandran. 1988. "From Filing and Fitting to Flexible Manufacturing:
A Study in the Evolution of Process Control." Working Paper 88–045, Division
of Research, Graduate School of Business Administration, Harvard University,
Boston.

Jochem, Sven. 2003. "Nordic Corporatism and Welfare State Reforms: Denmark
and Sweden Compared." In Frans van Waarden and Gerhard Lehmbruch (eds.),
*Renegotiating the Welfare State: Flexible Adjustment through Corporatist Concerta-
tion.* London: Routledge. 114–141.

John, Richard R. 1998. *Spreading the News: The American Postal System from Franklin
to Morse.* Cambridge, MA: Harvard University Press.

Johnson, Björn, Bengt-Åke Lundvall, Edward Lorenz, and Morten Berg Jensen. 2004.
"Codification and Modes of Innovation." Paper presented at the DRUID [Danish
Research Unit for Industrial Dynamics] Summer Conference 2004 on: Industrial
Dynamics, Innovation and Development. Elsinore, Denmark, June 14–16.
http://www.druid.dk/ocs/index.php?cf=1 (accessed December 13, 2005).

Johnson, Chalmers. 1989. "MITI, MPT, and the Telecom Wars: How Japan Makes Pol-
icy in High Technology." In Chalmers Johnson, Laura Tyson, John Zysman (eds.),
Politics and Productivity: How Japan's Developmental Strategy Works. New York:
Harper Business. 177–240.

Johnson, Chalmers, Laura Tyson, and John Zysman, eds. 1989. *Politics and Productivity: How Japan's Developmental Strategy Works.* New York: Harper Business.

Johnson, David R., and David Post. 1996. "Law and Borders: The Rise of Law in Cyberspace." *First Monday* 1, no. 1 (May). http://www.firstmonday.org/issues/issue1/
law/index.html (accessed December 13, 2005).

Jones, T. Daniel, Daniel Roos, and James P. Womack. 1991. *The Machine That
Changed the World.* New York: Harper Perennial.

"Juten go bunya no hyoka. IT senryaku honbu [Evaluation of Five Strategic Areas]."
2002. *Nikkei Communications,* July 1. 106–116.

Kagan, Robert A. 2000. "How Much Do National Styles of Law Matter?" In Robert A.
Kagan and Lee Axelrad (eds.), *Regulatory Encounters: Multinational Corporations
and American Adversarial Legalism.* University of California Series in Law, Politics,
and Society 1. Berkeley: University of California Press.

———. 2001. *Adversarial Legalism: The American Way of Law.* Cambridge, MA: Harvard University Press.

Kahler, Miles, and David Lake, eds. 2003. "Public and Private Governance in Setting
International Standards." In *Governance in a Global Economy: Political Authority
in Transition.* Princeton, NJ: Princeton University Press. 199–225.

Kalathil, Shanthi, and Taylor C. Boas. 2003. *Open Network, Closed Regimes: The Impact of the Internet on Authoritarian Rule.* Washington, DC: Carnegie Endowment
for International Peace.

Kant, Immanuel. 1956. *Kritik der reinen Vernunft.* Hamburg: Verlag von Felix Meiner.

Kato, Takao. 1993. "Internal Labor Markets for Managers and the Speed of Promotion
in the U.S. and Japan." In Japan Institute of Labour (ed.), *An International Com-*

430

parison of *Professionals and Managers: Their Job Careers and Quality of Working Life.* JIL Report Series 2. Tokyo: Japan Institute of Labour. 109–125.

Katz, Michael L., and Carl Shapiro. 1985. "Network Externalities, Competition, and Compatibility." *American Economic Review* 75: 425–440.

Katz, Richard. 2002. *Japanese Phoenix: The Long Road to the Economic Revival.* Armonk, NY: M. E. Sharpe.

Katzenstein, Peter. 1984. *Corporatism and Change: Austria, Switzerland, and the Politics of Industry.* Ithaca, NY: Cornell University Press.

———. 2003. "*Small States* and Small States Revisited." *New Political Economy* 8, no. 1 (March): 9–30.

Kauppinen, Timo. 2001. "The Transformation of Corporatist Industrial Relations in Finland." In György Szell (ed.), *European Labor Relations*, vol. 2, *Selected Country Studies.* Aldershot, Hants.: Ashgate. 46–62.

Kealey, Terence. 1996. *The Economic Laws of Scientific Research.* London: Macmillan.

Kearns, David T., and David A. Nadler. 1992. *Prophets in the Dark: How Xerox Reinvented Itself and Beat Back the Japanese* New York: Harper Business.

Kedzie, Christopher R. 1997. *Communication and Democracy: Coincident Revolutions and the Emergent Dictator's Dilemma.* Santa Monica, CA: RAND.

Kenney, Martin. 1997. "Value Creation in the Late 20th Century: The Rise of the Knowledge Worker." In John Davis, Tom Hirshl, and Michael Stack (eds.) *Cutting Edge: Technology, Information, Capitalism and Social Revolution.* London: Verso: 87-102.

———. 2004. Introduction to Martin Kenney and Richard Florida (eds.), *Locating Global Advantage.* Stanford, CA: Stanford University Press.

Kenney, Martin, and Rafiq Dossani. 2004. "The Impacts of Service Offshoring on Nonmetro America: Thinking about the Future." Paper presented at the Frederick H. Buttel Symposium, 2004 annual meeting of the Rural Sociological Society, Sacramento, CA, August 12–15.

Kenney, Martin, and Richard Florida. 1993. *Beyond Mass Production: The Japanese System and Its Transfer to the U.S.* New York: Oxford University Press.

———, eds. 2004. *Locating Global Advantage.* Stanford, CA: Stanford University Press.

Kenney, Martin, and David Mayer. 2002. "Economic Action Does Not Take Place in a Vacuum: Understanding Cisco's Acquisition and Development Strategy." Berkeley Roundtable on the International Economy Working Paper 148, University of California, Berkeley.

Kenney, Martin, and Urs von Burg. 2000. "Institutions and Economies: Creating Silicon Valley." In Martin Kenney (ed.), *Understanding Silicon Valley: Anatomy of an Entrepreneurial Region.* Stanford, CA: Stanford University Press. 218–240.

Kiander, Jaakko. 2002. Introduction to Jaakko Kiander and Sari Virtanen (eds.), *The Research Program on the Economic Crisis of the 1990s in Finland: Final Report.* Helsinki: VATT/Government Institute for Economic Research. 1–7.

Kiander, Jaakko, and Pentti Vartia. 1996. "The Great Depression of the 1990s in Finland." *Finnish Economic Papers* 9: 72–88.

Kim, Linsu, and Richard R. Nelson. 2000. Introduction to Linsu Kim and Richard R. Nelson (eds.), *Technology, Learning, and Innovation: Experiences of Newly Industrializing Economies.* Cambridge: Cambridge University Press. 1–9.

Kitschelt, Herbert. 1991. "Industrial Governance Structures, Innovation Strategies, and the Case of Japan: Sectoral or Cross-national Comparative Analysis?" *International Organization* 45: 453–493.

Klemperer, Paul. 2002. "How (Not) to Run Auctions: The European 3G Telecom Auctions." *European Economic Review* 46: 829–845.

Klepper, Steven. 1996. "Entry, Exit, Growth, and Innovation over the Product Life Cycle." *American Economic Review* 86: 562–583.

Kogut, Bruce. 2004. "From Regions and Firms to Multinational Highways: Knowledge and Its Diffusion as a Factor in the Globalization of Industries." In Martin Kenney and Richard Florida (eds.), *Locating Global Advantage*. Stanford, CA: Stanford University Press. 261–282.

Kogut, Bruce, and Udo Zander. 1992. "Knowledge of the Firm, Combinative Capabilities, and the Replication of Technology." *Organization Science* 3: 383–397.

Koski, Heli, and Tobias Kretschmer. 2005. "Entry, Standards, and Competition: Firm Strategies and the Diffusion of Mobile Telephony." *Review of Industrial Organization* 26: 89–113.

Koski, Heli, Petri Rouvinen, and Pekka Ylä-Anttila. 2002. "ICT Clusters in Europe: The Great Central Banana and the Small Nordic Potato." *Information Economics and Policy* 14: 145–165.

Koyre, Alexandre. 1945. *Discovering Plato*. New York: Columbia University Press.

Kraemer, Kenneth L., and Jason Dedrick. 1994. "Payoffs from Investment in Information Technology: Lessons from the Asia-Pacific Region." *World Development* 22: 1921–1931.

Krasner, Stephen. 1991. "Global Communications and National Power: Life on the Pareto Frontier." *World Politics* 43: 336–366. http://links.jstor.org/sici?sici= 00438871%28199104%2943%3A3%3C336%3AGCANPL%3E2.0.CO%3B2-P (accessed December 13, 2005).

Kripalani, Manjeet, and Pete Engardio. 2003. "The Rise of India." *Business Week*, December 8, 66–76.

Kuhn, Thomas S. 1962. *The Structure of Scientific Revolutions*. Chicago, IL: University of Chicago Press.

Kushida, Kenji. 2002. "The Japanese Wireless Telecommunications Industry: Innovation, Organizational Structures, and Government Policy." *Stanford Journal of East Asian Affairs* 2: 55–70.

———. 2004. "Globalization, Global Markets, and Exogenous Price Shocks: Japan's Telecommunications Industry and the State's Industrial Policies." Master's thesis, University of California, Berkeley.

———. 2006. "The Politics of Restructuring NTT: Historically Rooted Trajectories from the Creation Actors, Institutions, and Interests." Stanford Japan Center Discussion Paper, DP-2006-001-E.

Lakatos, Imre, and Alan Musgrave, eds. 1970. *Criticism and the Growth of Knowledge*. Cambridge: Cambridge University Press.

Lakoff, George, and Mark Johnson. 1999. *Philosophy in the Flesh*. New York: Basic Books.

Lall, Sanjaya. 1993. "Promoting Technology Development: The Role of Technology Transfer and Indigenous Effort." *Third World Quarterly* 14: 95–108.

432 ———. 1999. "Promoting Industrial Competitiveness in Developing Countries: Lessons from Asia." Commonwealth Economic Paper Series No. 39, Commonwealth Secretariat, London.

———. 2000. "Technological Change and Industrialization in the Asian Newly Industrializing Economies: Achievements and Challenges." In Linsu Kim and Richard R. Nelson (eds.), *Technology, Learning, and Innovation: Experiences of Newly Industrializing Economies.* Cambridge: Cambridge University Press. 13–68.

———. 2003. "Indicators of the Relative Importance of IPRs in Developing Countries." *Research Policy* 32: 1657–1680.

Lall, Sanjaya, and Carlo Pietrobelli. 2002. *Failing to Compete: Technology Development and Technology Systems in Africa.* Cheltenham, England: Edward Elgar.

Lane, Charles. 2002. "Justices Partially Back Cyber Pornography Law." *Washington Post,* May 14.

Lave, Jean, and Etienne Wenger. 1991. *Situated Learning: Legitimate Peripheral Participation.* Cambridge: Cambridge University Press.

Leachman, Robert C., and Chien H. Leachman. 2001. "E-commerce and the Changing Terms of Competition in the Semiconductor Industry." In BRIE-IGCC Economy Project Task Force (ed.), *Tracking a Transformation: E-commerce and the Terms of Competition in Industries.* Washington, D.C.: Brookings Institution Press.

Leander, Tom. 2004. "Does Microsoft Need China?" *CFO.com.* http://www.cfoasia.com/archives/200407-01.htm (accessed December 13, 2005).

Leib, Volker. 2002. "ICANN—EU Can't: Internet Governance and Europe's Role in the Formation of the Internet Corporation for Assigned Names and Numbers." *Telematics and Informatics* 19, no. 2: 159–171.

Lehmbruch, Gerhard. 1979. "Liberal Corporatism and Party Government." In Philippe Schmitter and Gerhard Lehmbruch (eds.), *Trends toward Corporatist Intermediation.* London: Sage. 147–183.

Lehr, William, and Lee W. McKnight. 2003. "Wireless Internet Access: 3G vs. WiFi?" *Telecommunications Policy* 27: 351–370.

Lehtoranta, Olavi. 2000. "Technology-based Firms and Public R&D Support: A Descriptive Introduction." In Rita Asplund (ed.), *Public R&D Funding, Technological Competitiveness, Productivity, and Job Creation.* Helsinki: ETLA/Research Institute of the Finnish Economy. 15–46.

Lembke, Johan. 2001. "Harmonization and Globalization: UMTS and the Single Market." *INFO: The Journal of Policy, Regulation, and Strategy for Telecommunications, Information, and Media* 3: 15–26.

———. 2002a. "The EU Regulatory Strategy for Mobile Internet." *Journal of European Public Policy* 9 (April): 273–291.

———. 2002b. "Global Competition and Strategies in the Information and Communications Technology Industry: A Liberal Strategic Approach." *Business and Politics* 4 (April): 41–69.

Lemley, Mark A. 2002. "Intellectual Property Rights and Standard-setting Organizations." *California Law Review* 90 (December). http://www.law.berkeley.edu/journals/clr/library/lemley03.html (accessed December 13, 2005).

Lemley, Mark A., and Lawrence Lessig. 2000. "The End of End-to-end: Preserving the

Architecture of the Internet in the Broadband Era." *Social Science Research Network Electronic Paper Collection.* http://papers.ssrn.com/sol3/papers.cfm?abstract _id=247737 (accessed November 13, 2005).

Lemley, Mark A., and David McGowan. 1998. "Legal Implications of Network Economic Effects." *California Law Review* 86: 481–611.

Lemola, Tarmo. 2002. Convergence of National Science and Technology Policies: The Case of Finland. *Research Policy* 31: 1481–1490.

Lessig, Lawrence. 1998. "What Things Regulate Speech: CDA 2.0 vs. Filtering." *Jurimetrics Journal* 38: 629–670.

———. 1999. "Internet Regulation through Architectural Modification." *Harvard Law Review* 112 (May): 1634–1657.

———. 2000. *Code and Other Laws of Cyberspace.* New York: Basic Books.

———. 2001. Expert Report in A&M Records v. Napster, 3–4. http://www.lessig .org/content/testimony/nap/napd3.pdf (accessed November 13, 2005)

Lester, Richard, and Michael Piore. 2004. *Innovation: The Missing Dimension.* Cambridge, MA: Harvard University Press.

Levin, Richard K., N. R. Alvin, and W. Sidney. 1987. "Appropriating the Returns from Industrial Research and Development." *Brookings Papers on Economic Activity* 3: 783–820.

Levy, Adam. 2001. "Inside Enron." *Bloomberg Markets,* May, p. 31.

Levy, Frank, and Richard J. Murnane. 2004. *The New Division of Labor: How Computers Are Creating the Next Job Market.* Princeton, NJ: Princeton University Press.

Levy, Jonah. 1999. *Tocqueville's Revenge: State, Society, and Economy in Contemporary France.* Cambridge, MA: Harvard University Press.

———. 2006. *The State after Statism: New State Activities among the Affluent Democracies.* Cambridge, MA: Harvard University Press.

Levy, Jonah, Robert Kagan, and John Zysman. 1997. "The Twin Restorations: The Political Economy of the Reagan and Thatcher 'Revolutions.'" In *Ten Paradigms of Market Economies.* Kyounggi-Do, South Korea: Korea Research Institute for Human Settlements. 3–57.

Levy, Jonah, Mari Miura, and Gene Park. 2006. "Exiting Etatisme? New Directions in State Policy in France and Japan." In Jonah Levy (ed.), *The State after Statism: New State Activities among the Affluent Democracies.* Cambridge, MA: Harvard University Press.

Li, Feng, and Jason Whalley. 2002. "Deconstruction of the Telecommunications Industry: From Value Chains to Value Networks." *Telecommunications Policy* 26: 451–472.

Lightman, Alex. 2002. *Brave New Unwired World: The Digital Big Bang and the Infinite Internet.* New York: John Wiley.

Liikanen, Jukka, Paul Stoneman, and Otto Toivanen. 2002. *Intergenerational Effects in the Diffusion of New Technology: Case of Mobile Phones.* ETLA Discussion Paper No. 809, Research Institute of the Finnish Economy, Helsinki.

Lohr, Steve. 2004. "Sale of IBM PC Unit is a Bridge between Cultures." *New York Times,* December 8.

Lorenz, Edward, and Antoine Valeyre. 2004. "Organisational Change in Europe: Models or the Diffusion of a New 'One Best Way'?" Paper presented at the DRUID [Danish Research Unit for Industrial Dynamics] Summer Conference

434   2004 on Industrial Dynamics, Innovation and Development. Elsinore, Denmark, June 14–16. http://www.druid.dk/ocs/viewpaper.php?id=12&cf=1 (accessed December 13, 2005).

Lukács, Georg. 1968. *Geschichte und Klassenbewusstsein*. Berlin: Luchterhand Verlag.

Lundvall, Bengt-Åke. 2002. *Innovation, Growth, and Social Cohesion: The Danish Model*. Cheltenham, England: Edward Elgar.

———, ed. 1992. *National Systems of Innovation: Toward a Theory of Innovation and Interactive Learning*. London: Pinter.

MacAfee, Preston E., and John McMillan. 1987. "Auctions and Bidding." *Journal of Economic Literature* 25: 699–738.

Mahmood, Ishtiaq P., and Jasjit Singh. 2003. "Technological Dynamism in Asia." *Research Policy* 32: 1031–1054.

Majone, Giandomenico. 1996. *Regulating Europe*. New York: Routledge.

Maliranta, Mika. 2000. "Privately and Publicly Funded R&D as Determinants of Productivity: Evidence from Finnish Enterprises." In Rita Asplund (ed.), *Public R&D Funding, Technological Competitiveness, Productivity, and Job Creation*. Helsinki: ETLA/Research Institute of the Finnish Economy. 47–86.

Maliranta, Mika, and Petri Rouvinen. 2003. "Productivity Effects of ICT in Finnish Business." ETLA Discussion Papers No. 852. Helsinki: Research Institute of the Finnish Economy.

Margolis, Michael, and David Resnick. 2000. *Politics as Usual: The Cyberspace "Revolution."* Thousand Oaks, CA: Sage.

Marsden, Christopher. 2000. *Regulating the Global Information Society*. New York: Routledge.

Maskus, Keith E., and Jerome H. Reichman. 2004. "The Globalization of Private Knowledge Goods and the Privatization of Global Public Goods." *Journal of International Economic Law* 7: 279–320.

Mastanduno, Michael. 1990. "Do Relative Gains Matter? America's Response to Japanese Industry Policy." *International Security* 16: 73–113.

Matloff, Norman. 1998. "Debunking the Myth of a Desperate Software Labor Shortage." Testimony before the U.S. House Judiciary Committee, Subcommittee on Immigration.

"Matsushita ga hokubei de keitai daikousei [Matsushita Aiming Major Offensive in North American Cellular Handset Markets]" 2004. *Nikkei Business*, April 26

Mattli, Walter, and Tim Büthe. 2003. "Setting International Standards: Technological Rationality or Primacy of Power?" *World Politics* 56: 1–42.

Mayer, David, and Martin Kenney. 2004. "Ecosystems and Acquisition Management: Understanding Cisco's Strategy." *Industry and Innovation* 8: 67–103.

Mayntz, Renate, and Fritz W. Scharpf, eds. 1995. *Gesellschaftliche Selbstregelung und politische Steuerung*. Frankfurt am Main: Campus.

McGillivray, Fiona. 2001. "Government Hand-outs, Political Institutions, Stock Price Dispersion." Unpublished manuscript, Yale University.

McNollgast [Mathew D. McCubbins, Roger Noll, and Barry Weingast]. 1999. "The Political Origins of the Administrative Procedure Act." *Journal of Law, Economics, and Organization* 15: 180–217.

Melody, William H. 2000. "Telecom Myths: The International Revenue Settlements Subsidy." *Telecommunications Policy* 24, no. 1: 1–3.

Merges, Robert P., and Peter S. Menell. 2000. *Intellectual Property in the New Technological Age.* 2nd ed. New York: Aspen Law and Business.

Merleau-Ponty, Maurice. 1962. *Phenomenology of Perception.* London: Routledge Classics.

———. 1964. *The Primacy of Perception.* Evanston, IL: Northwestern University Press.

Messerschmitt, David. 2000. *Understanding Networked Applications.* San Francisco: Morgan Kaufmann.

Migdal, Joel S. 1988. *Strong Societies and Weak States: State-society Relations and State Capabilities in the Third World.* Princeton, NJ: Princeton University Press.

Milner, Helen V. 2003a. "The Digital Divide: The Role of Political Institutions in Technology Diffusion." Paper presented at the annual meeting of the American Political Science Association, Philadelphia, PA, August 28–31.

———. 2003b. "The Global Spread of the Internet: The Role of International Diffusion Pressures in Technology Adoption." Working Paper, August 25. http://www.wws.princeton.edu/hmilner/working%20papers/internet_diffusion8-03.pdf (accessed December 13, 2005).

Mindel, Joshua L., and Marvin A. Sirbu. 2001. "Taxonomy of Traded Bandwidth." Work in progress, May 1. http://www.itc.mit.edu/itel/wp/btm_mindel_sirbu.pdf (accessed December 13, 2005).

Mjoset, Lars. 1987. "Nordic Economic Policies in a Comparative Perspective." *International Organization* 41: 403–456.

Morgan Stanley Research Department. 2005. Morgan Stanley to clients, memorandum. "Telecom Services and Equipment: Cross-industry Insights." February 1.

Mosher, Patrick, and Eric P. Gist. 2002. "The Customer Care Workforce: Driving More Profitable Customer Interactions." CRM Project, vol. 3, October 30. http://www.crmproject.com/documents.asp?grID=293&d_ID=1578 (accessed November 13, 2005).

Mowery, David. 1995. "The Practice of Technology Policy." In Paul Stoneman (ed.), *Handbook of the Economics of Innovation and Technological Change.* Oxford: Basil Blackwell. 513–557.

Mowery, David, Joanne Oxley, and Brian Silverman. 1998. "Technological Overlap and Inter-firm Cooperation: Implications for the Resource-based View of the Firm." *Research Policy* 27: 507–523.

Mowery, David C., and Joanne E. Oxley. 1995. "Inward Technology Transfer and Competitiveness: The Role of National Innovation Systems." *Cambridge Journal of Economics* 19: 67–93.

Mueller, Milton L. 2002. *Ruling the Root: Internet Governance and the Taming of Cyberspace.* Cambridge: MIT Press.

Murakami, Yasuske, and Kimiaki Yamamura. 1982. "A Technical Note." In Kozo Yamamura (ed.), *Policy and Trade Issues of the Japanese Economy.* Seattle: University of Washington Press.

Nachtigal, Jeff. 2003. "Microsoft Plans Largest Lay-off of Full-time Employees in Company History." *WashTech News.* http://www.techsunite.org/news/techind/030701_msjobsabroad.cfm (accessed November 12, 2005).

436   Nakamoto, Michiyo. 2001. "Japan Telecom Succumbs to a Gentle Touch: Michiyo Nakamoto on the Success of Vodafone's Diplomatic Approach to the Takeover." *Financial Times*, September 21, p. 35.

NASSCOM. 2005. "Indian IT-ITES: FY05 Results and FY06 Forecast." June 2. http://www.nasscom.org/download/Indian_IT_ITES_%20FY05_Results_FY06_Forecast.pdf (accessed December 13, 2005).

NASSCOM-McKinsey Report 2002. 2002. New Delhi: NASSCOM.

National Infrastructure Information Copyright Protection Act of 1995. http://fairuse.stanford.edu/primary_materials/legislation/niilegis/ (accessed November 12, 2005)

National Research Council, Charles W. Wessner (eds.). 2003. *Securing the Future: Regional and National Programs to Support the Semiconductor Industry.* Washington, DC: National Academies Press.

Natsuno, Takeshi. 2001. *I-mode Sutorateji: Sekai wa naze oitaukenai ka* [I-Mode Strategy: Why the World Cannot Catch Up]. Tokyo: Nikkei BP.

Naughton, John. 1999. *A Brief History of the Future: From Radio Days to Internet Years in a Lifetime.* New York, NY. Overlook Press.

Newman, Abraham. 2003. "When Opportunity Knocks: Economic Liberalization and Stealth Welfare in the United States." *Journal of Social Policy* 32: 179–197.

———. 2005. "Creating Privacy: The International Politics of Personal Information." Ph.D. diss., University of California, Berkeley.

Newman, Abraham L., and David Bach. 2004. "Privacy and Regulation in a Digital Age." In Brigitte Preissl, Harry Bouwman, and Charles Steinfield (eds.), *E-life after the Dot Com Bust.* Heidelberg: Physica Verlag.

Nezu, Risaburo. 2002a. *IT sengoku jidai* [IT's Warring Period]. Tokyo: Chuo Koron Shinsha.

———. 2002b. "Perspective and Strategies for Japanese Industry." Paper presented at the conference Prospects for Core Industries in Japan and Germany, Japanese-German Center, Fujitsu Research Institute, and German Institute for Economic Research, Berlin, November 28–29.

———. 2004. "Building an Information Infrastructure for Knowledge Based Economy." Draft paper presented at the Workshop on Japan as a Knowledge Economy — Assessment and Lessons, co-organized by Hitotsubashi University and World Bank Institute, Tokyo, November 13–14.

Nida, Eugene. 1969. *Towards a Science of Translating.* Leiden: E. J. Brill.

Nielsen, Klaus, and Stefan Kesting. 2003. "Small Is Resilient: The Impact of Globalization on Denmark." *Review of Social Economy* 61: 365–387.

Nielsen, Niels Christian, and Maj Cecilie Nielsen. 2004. "Spoken-about Knowledge: Why It Takes Much More than 'Knowledge Management' to Manage Knowledge." Berkeley Roundtable on the International Economy Working Paper 158, University of California, Berkeley.

Niles, Raymond C., and Dale F. Meyerhoeffer. 2000. "Bandwidth Trading: Enron." Schroder & Co. Working Paper, January 12, p. 13.

Nimmer, David. 2000. "A Riff on Fair Use in the Digital Millennium Copyright Act." *University of Pennsylvania Law Review* 148: 673–742.

Noam, Eli. 1993. *Telecommunications in Europe.* New York: Oxford University Press.

Noll, Roger, and Frances Rosenbluth. 1995. "Telecommunications Policy: Structure, **437** Policy, Outcomes." In Peter Cowhey and Mathew McCubbins (eds.), *Structure and Policy in Japan and the United States*. New York: Columbia University Press. 119–176.

Nonaka, Ikujiro, and Hirotaka Takeuchi. 1995. *The Knowledge-creating Company*. New York: Oxford University Press.

Nonaka, Ikujiro, Georg von Krogh, and Kazuo Ichijo. 2000. *Enabling Knowledge Creation*. New York: Oxford University Press.

Nørretranders, Tor. 1991. *Mærk Verden*. Denmark: Gyldendal.

North, Douglass C. 1990. *Institutions, Institutional Change, and Economic Performance*. New York: Cambridge University Press.

Oatley, Thomas, and Thomas Nabors. 1998. "Market Failure, Wealth Transfers, and the Basle Accord." *International Organization* 52: 35–54.

O'Donnell, Rory. 2001. "The Future of Social Partnership in Ireland." National Competitiveness Council Working Paper, Dublin.

Ohmae, Kenichi. 1993. "The Rise of the Region State," *Foreign Affairs* 72:78–87

Oie, Yuji, Goto Shigeki, Konishi Kazunori, and Nishio Shoujirou. 2001. *Intanettou Dai 1 Sho: Intanettou Nyumon* [The Internet: Volume 1, An Introduction to the Internet]. Tokyo: Iwanami Shoten.

Ojainmaa, Kaisa. 1994. *International Competitive Advantage of the Finnish Chemical Forest Industry*. ETLA C 66. Helsinki: Taloustieto.

Okimoto, Daniel I. 1989. *Between MITI and the Market: Japanese Industrial Policy for High Technology*. Stanford, CA: Stanford University Press.

Olson, Mancur. 1965. *The Logic of Collective Action: Public Goods and the Theory of Groups*. Cambridge, MA: Harvard University Press.

Organisation for Economic Co-operation and Development. 1980. *Guidelines on the Protection of Privacy and Transborder Flows of Personal Data*. Paris: The Organization.

———. 1999a. "Building Infrastructure Capacity for Electronic Commerce: Leased Line Developments and Pricing." Report to the Working Party on Telecommunications and Information Services Policies. Directorate for Science, Technology, and Industry. Paris: OECD.

———. 1999b. *OECD Economic Surveys: Finland*. Paris: OECD.

———. 1999c. *Science, Technology, and Industry Scoreboard: Benchmarking Knowledge-based Economies*. Paris: OECD.

———. 2000a. *Mobile Phones: Pricing Structures and Trends*.Paris: OECD.

———. 2000b. *A New Economy? The Changing Role of Innovation and Information Technology in Growth*. Paris: OECD.

———. 2001. *OECD Science, Technology, and Industry Scoreboard*. Paris: OECD.

———. 2002a. *Measuring the Information Economy*. Paris: OECD.

———. 2002b. *OECD Information Technology Outlook*. Paris: OECD.

———. 2003. *OECD Communications Outlook*. Paris: OECD.

———. 2004. *The Economic Impacts of ICT: Measurement, Evidence, and Implications*. Paris: OECD.

O'Riain, Sean. 2000. "The Flexible Development State: Globalization, Information Technology, and the Celtic Tiger." *Politics and Society* 28: 157–194.

438   Ornston, Darius. 2004. "The Politics of Perpetual Renegotiation: State Intervention and Social Concertation in Ireland." Paper presented at the 100th annual meeting of the American Political Science Association, Chicago, September 2–5.

Ostrom, Elinor. 1990. *Governing the Commons: The Evolution of Institutions for Collective Action*. Cambridge: Cambridge University Press.

Owen, Bruce, and Gregory Rosston. 2001. "Spectrum Allocation and the Internet." SIEPER Discussion Paper 01–09.

Paija, Laura. 2001. "The ICT Cluster in Finland: Can We Explain It?" In Laura Paija (ed.), *Finnish ICT Cluster in the Digital Economy*. Helsinki: Taloustieto.

Paija, Laura, and Petri Rouvinen. 2003. "Evolution of the Finnish ICT Cluster." In Gerd Schienstock (ed.), *Embracing the Knowledge Economy: The Dynamic Transformation of the Finnish Innovation System*. Aldershot, Hants.: Edward Elgar.

Palmberg, Christopher. 2002. "Technological Systems and Competent Procurers: The Transformation of Nokia and the Finnish Telecom Industry Revised." *Telecommunications Policy* 26: 129–148.

Palmberg, Christopher, and Olli Martikainen. 2003a. "The Economics of Strategic R&D alliances: A Review with Focus on the ICT Sector." ETLA Discussion Paper No. 881. Research Institute of the Finnish Economy, Helsinki.

———. 2003b. "Overcoming a Technological Discontinuity: The Case of the Finnish Telecom Industry and the GSM." ETLA Discussion Paper No. 855. Research Institute of the Finnish Economy, Helsinki.

———. 2005. "Nokia as an Incubating Entrant: The Case of Nokia's Entry to the GSM." *Innovation: Management, Policy, and Practice* 7, no. 1: 61–78.

Paoli, Pascal, and Damien Merllié. 2001. *Third European Survey on Working Conditions, 2000*. Luxembourg: Office for Official Publications of the European Communities.

Paré, Daniel J. 2002. *Internet Governance in Transition: Just Who Is the Master of This Domain?* Boulder, CO: Rowman and Littlefield.

Patterson, Mark R. 2002. "Invention, Industry Standards, and IP." *Berkeley Technology Law Journal* 17: 1043–83.

Pehkonen, Jaakko, and Kangasharju, Aki. 2001. "Employment and Output Growth in the 1990s." In Jorma Kalela, Jaakko Kiander, Ullamaija Kivikuru, Heikki A. Loikkanen, and Jussi Simpura (eds.), *Down from the Heavens, Up from the Ashes: The Finnish Economic Crisis of the 1990s in Light of Economic and Social Research*. Helsinki: Government Institute for Economic Research. 217–228.

Pelkmans, Jacques. 2001. "The GSM Standard: Explaining a Success Story." *European Journal of Public Policy* 8: 432–453.

Peltzman, Sam. 1976. "Towards a More General Theory of Regulation." *Journal of Law and Economics* 19: 211–240.

Pempel, T. J. 1998. *Regime Shift: Comparative Dynamics of the Japanese Political Economy*. Ithaca, NY: Cornell University Press.

Perez, Carlotta. 2003. *Technological Revolutions and Financial Capital: The Dynamics of Bubbles and Golden Ages*. New York: Edward Elgar.

Peteraf, Margaret. 1993. "The Cornerstones of Competitive Advantage: A Resource-based View." *Strategic Management Journal* 14: 179–188.

Piore, Michael, and Charles F. Sabel. 1984. *The Second Industrial Divide: Possibilities for Prosperity*. New York: Basic Books.

Polanyi, Karl. 1944. *Great Transformation: The Political and Economic Origins of Our Time*. Boston, MA: Beacon Press.

Polanyi, Michael. 1958. *Personal Knowledge*. Chicago: University of Chicago Press.

Pontusson, Jonas, and Peter Swenson. 1996. "Labor Markets, Production Strategies and Wage Bargaining Institutions: The Swedish Employer Offensive in Comparative Perspective." *Comparative Political Studies* 29: 223–250.

Posen, Adam. 1998. *Restoring Japan's Economic Growth*. Washington, DC: Institute for International Economics.

Post, David, and David Johnson. 1996. "Law and Borders: The Rise of Law in Cyberspace." *Stanford Law Review* 48: 1367. http://www.firstmonday.org/issues/issue1/law/index.html (accessed December 13, 2005).

Posthuma, A. 1987. "The Internationalization of Clerical Work: A Study of Offshore Work Services in the Caribbean." Occasional Paper, Science Policy Research Unit, University of Sussex, Brighton, UK.

Prahalad, C. K. 2005. *The Fortune at the Bottom of the Pyramid: Eradicating Poverty through Profits*. Upper Saddle River, NJ: Wharton School Publishing.

Prahalad, C. K., and Stuart L. Hart. 2002. "The Fortune at the Bottom of the Pyramid." *Strategy and Business* 26: 1–14.

Preston, Ron. 2004. "Software Patents Abused." *Network Computing*, May 13.

Prime Minister's Office. 2001. "IT Kakumei no Suishin ni Mukete: 'e-Japan senryaku' kettei" [Toward Facilitating an IT Revolution: "E-Japan Strategy" Decided]. *Tokino Ugoki*.

Quine, Willard Van Orman. 1953. "Two Dogmas of Empiricism." In *From a Logical Point of View*. Harvard University Press.

Randall, Marc. 2004. Lehman Brothers Semiconductor and Hardware Private Company Conference in San Francisco. Force10 Networks. May 11.

Raumolin, Jussi. 1992. "The Diffusion of Technology in the Forest and Mining Sector in Finland." In Synnove Vuori and Pekka Ylä-Anttila (eds.), *Mastering Technology Diffusion: The Finnish Experience*. ETLA B 82. Helsinki: Taloustieto.

"Recent Developments in the Law: The Law of Cyberspace." 1999. *Harvard Law Review* 112: 1574–1704.

Rée, Jonathan. 1974. *Descartes*. London: Allen Lane.

Regan, Priscilla. 1995. *Legislating Privacy*. Raleigh: University of North Carolina Press.

Regini, Marino. 2000. "The Dilemmas of Labour Market Regulation." In Gøsta Esping-Andersen and Marino Regini (eds.), *Why Deregulate Labour Markets?* Oxford: Oxford University Press.

Rehn, Olli. 1996. "Corporatism and Industrial Competitiveness in Small European States: Austria, Finland, and Sweden, 1945–1995." Ph.D. diss., Oxford University.

———. 2003. "Transforming Enterprise in Europe: A Comparison with the United States." Paper presented at the Transforming Enterprise Conference, Washington, DC, January 27–28.

Reich, Robert. 1991. *The Work of Nations*. New York: Knopf.

440   Rhodes, Martin. 2001. "The Political Economy of Social Pacts: 'Competitive Corporatism' and European Welfare Reform." In Paul Pierson (ed.), *The New Politics of the Welfare State*. Oxford: Oxford University Press. 195–196.

Ribeiro, John. 2003. "AOL Expands Indian Call Centre Staff." *ComputerWeekly.com*. http://www.computerweekly.com/Article123470.htm (accessed November 12, 2005).

Richards, John. 1999. "Toward a Positive Theory of International Institutions: Regulating International Aviation Markets." *International Organization* 53: 1–37.

———. 2004. "Clusters, Competition, and Global Players in ICT Markets: The Case of Scandinavia." In Timothy Bresnahan and Alfonso Gambardella (eds.), *Building High-tech Clusters: Silicon Valley and Beyond*. Cambridge: Cambridge University Press.

Richardson, Tim. 2002. "South Korea Broadband in League of Its Own." *Register*, October 14, 2002. http://www.theregister.co.uk (accessed November 11, 2005).

Rieke, Michael. Dow Jones Newswires, November 15, 2000.

Rosenau, James. 2002. "Information Technology and the Skills, Networks, and Structures that Sustain World Affairs." In James Rosenau and J. P. Singh (eds.), *Information Technologies and Global Politics: The Changing Scope of Power and Governance*. Albany: State University of New York Press.

Rosenau, James, and J. P. Singh. 2002. *Information Technologies and Global Politics: The Changing Scope of Power and Governance*. Albany: State University of New York Press.

Rosenkopf, Lori, and Michael L. Tushman. 1998. "The Coevolution of Community Networks and Technology: Lessons from the Flight Simulation Industry." *Industrial and Corporate Change* 7: 311–346.

Rostow, Walt W. 1962. *The Stages of Economic Growth*. New York: Cambridge University Press.

Rouvinen, Petri, and Pekka Ylä-Anttila. 1999. "Finnish Clusters and New Industrial Policy Making." In *OECD Proceedings: Boosting Innovation; The Cluster Approach*. Paris: Organisation for Economic Co-operation and Development.

———. 2003. "Little Finland's Transformation to a Wireless Giant." In S. Dutta, B. Lanvin, and F. Paua (eds.), *The Global Information Technology Report*, 2003–2004. New York: Oxford University Press.

Ruggles, Rudy L., ed. 1996. *Knowledge Management Tools*. Boston: Butterworth Heinemann.

Saari, Juho. 2001. "Bridging the Gap: Financing of Social Policy in Finland, 1990–1998." In Jorma Kalela, Jaakko Kiander, Ullamaija Kivikuru, Heikki A. Loikkanen, and Jussi Simpura (eds.), *Down from the Heavens, Up from the Ashes: The Finnish Economic Crisis of the 1990s in Light of Economic and Social Research*. Helsinki: Government Institute for Economic Research. 189–214.

Sabel, Charles. 1982. *Work and Politics*. Cambridge: Cambridge University Press.

———. 1994. "Flexible Specialization and the Re-emergence of Regional Economies." In Ash Amin (ed.), *Post-Fordism: A Reader*. Oxford: Blackwell.

Saltzer, J. H., D. P. Reed, and D. D. Clark. 1984. "End-to-end Arguments in System Design." *ACM Transactions in Computer Systems* 2: 277–288.

Samuelson, Pamela. 1994. "Legally Speaking: the NII Intellectual Property Report."

Communications of the ACM 47 (December). http://www.ifla.org.sg/documents/ **441**
infopol/copyright/samp1.htm (accessed November 12, 2005).

———. 1996. "Regulation of Technologies to Protect Copyrighted Works." *Communications of the ACM* 39 (July): 17–24.

———. 1997. "The U.S. Digital Agenda at WIPO." *Virginia Journal of International Law* 37 (Winter): 369–439.

———. 1999. "Intellectual Property and the Digital Economy: Why the Anti-circumvention Regulations Need to be Revisited." *Berkeley Technology Law Journal* 14: 519–566.

Samuelson, Pamela, and Randall Davis. 2000. "The Digital Dilemma: Intellectual Property in the Information Age." Written for presentation at the Telecommunications Policy Research Conference. http://www.sims.berkeley.edu/~pam/papers/digdilsyn.pdf (accessed November 12, 2005).

Sandholtz, Wayne. 1992. *High-tech Europe: The Politics of International Cooperation.* Berkeley: University of California Press.

Sandholtz, Wayne, Michael Borrus, John Zysman, Ken Conca, Jay Stowsky, Steven Vogel, and Steve Weber, eds. 1992. *The Highest Stakes: The Economic Foundations of the Next Security System.* New York: Oxford University Press.

Sandholtz, Wayne, and John Zysman. 1989. "Recasting the European Bargain?" *World Politics* 42: 1–30.

Saxenian, Annalee. 2003. "Taiwan's Hsinchu District: Imitator and Partner for Silicon Valley." Paper presented at the Conference on Institutional Change in East Asian Economies, Harvard University, Cambridge, MA.

Scharpf, Fritz W. 1997. *Games Real Actors Play: Actor-centered Institutionalism in Policy Research.* Boulder, CO: Westview Press.

Schienstock, Gerd, and Hämäläinen, Timo. 2001. *Transformation of the Finnish Innovation System: A Network Approach.* Helsinki: Sitra.

Schmidt, Susanne K., and Raymund Werle. 1998. *Coordinating Technology: Studies in the International Standardization of Telecommunications.* Cambridge, MA: MIT Press.

Schmitter, Philippe. 1979. "Still the Century of Corporatism?" In Philippe Schmitter and Gerhard Lehmbruch (eds.), *Trends toward Corporatist Intermediation.* London: Sage. 7–49.

Schoppa, Leonard. 1997. *Bargaining with Japan: What American Pressure Can and Cannot Do.* New York: Columbia University Press.

Schware, Robert. 1987. "Software Industry Development in the Third World." *World Development* 15: 1249–1267.

Schwartz, Andrew. Forthcoming. *The Politics of Greed: Privatization, Neo-liberalism, and Plutocratic Capitalism in Central and Eastern Europe.* Lanham: Rowman & Littlefield.

Scott, Allen. J. 2002. "A New Map of Hollywood and the World." *Regional Studies.* http://www.ersa.org/ersaconfs/ersa02/cd-rom/papers/521.pdf (accessed December 13, 2005).

Semilof, Margie. 2000. "Cisco's Challengers," *Computer Reseller News,* October 16: 1–5.

Shameen, Assif. 2004. "Korea's Broadband Revolution." *Chief Executive,* April. http://www.chiefexecutive.net (accessed November 11, 2005).

442   Shannon, Claude Elmwood. 1993. "A Mathematical Theory of Communication." In
       N. J. A. Sloane and Aaron D. Wyner (eds.), *Claude Elmwood Shannon: Collected
       Papers.* New York: IEEE Press.
       Shapiro, Carl. 2003. "Navigating the Patent Thicket: Cross Licenses, Patent Pools, and
       Standard-setting." In Adam Jaffe, Josh Lerner, and Scott Stern (eds.), *Innovation
       Policy and the Economy,* vol. 1. Boston: MIT Press.
       Shapiro, Carl, and Hal R. Varian. 1999. *Information Rules: A Strategic Guide to the
       Network Economy.* Boston: Harvard Business School Press.
       Sherman, Laura. 1998. "Wildly Enthusiastic about the First Multilateral Agreement on
       Trade in Telecommunications Services." *Federal Communications Law Journal* 51:
       61–110.
       Shy, Oz. 2001. *The Economics of Network Industries.* New York: Cambridge University
       Press.
       Simitis, Spiros. 1995. "From the Market to the Polis: The EU Directive on the Protec-
       tion of Personal Data." *Iowa Law Review* 80: 445–451.
       Simons, Kenneth. 2003. "Product Market Characteristics and the Industry Life Cycle."
       Unpublished manuscript. Rensselaer Polytechnic Institute.
       Singh, Nirvikar. 2002. "India's Information Technology Sector: What Contribution to
       Broader Economic Development?" Unpublished manuscript. University of Cali-
       fornia, Santa Cruz.
       Skilling, Jeffrey. Comments at Senate Bankruptcy Committee Hearing, February 26,
       2002.
       Skocpol, Theda. 1979. *States and Social Revolutions.* Cambridge: Cambridge Univer-
       sity Press
       Solomons, Mark. 2000. "Hackers Crack Digital Music Codes." *Financial Times,* Oc-
       tober 14.
       Soskice, David. 1999. "Divergent Production Regimes: Coordinated and Uncoordi-
       nated Market Economies in the 1980s and the 1990s." In Herbert Kitschelt, Peter
       Lange, Gary Marks and John D. Stephens (eds.), *Continuity and Change in Con-
       temporary Capitalism.* Cambridge: Cambridge University Press. 101–134.
       Soumusho. 2003. "White Paper: Information and Communications in Japan: Building
       a New, Japan-inspired IT Society." White Paper, Ministry of Public Manage-
       ment, Home Affairs, Posts and Telecommunications. http://www.johotsusintokei.
       soumu.go.jp/whitepaper/eng/WP2003/2003-index.html (accessed December 13,
       2005).
       ———. 2004. "Information and Communications in Japan: Building a Ubiquitous
       Network Society That Spreads throughout the World." White Paper, Ministry of
       Public Management, Home Affairs, Posts and Communications. http://www
       .johotsusintokei.soumu.go.jp/whitepaper/eng/WP2004/2004-index.html (accessed
       December 13, 2005).
       Spar, Debora. 1999. "Lost in (Cyber)space: The Private Rules of Online Commerce."
       In A. Claire Cutler, Virginia Haufler, and Tony Porter (eds.), *Private Authority in In-
       ternational Affairs.* Albany: SUNY Press.
       Stankiewicz, Rikard. 2003. *Digitalization-Induced Evolution of Innovation Systems:
       The Case of Biotechnology.* Part of the FP5 EU-project on Tracking the New Econ-
       omy Transformation, Florence Workshop, October 17–18.

Starbuck, William, and Frances J. Milliken. 1988. "Executives Perceptual Filers: What They Notice and How They Make Sense." In D. C. Hambrick (ed.), *The Executive Effect: Concepts and Methods for Studying Top Managers*. Greenwich, Conn.: JAI Press. 35–36.

Stefik, Mark. 1998. "Round One: Opening Remarks." *Atlantic Online*. http://www .theatlantic.com/issues/98sep/index.htm (accessed December 13, 2005).

Steinbock, Dan. 1998. *The Competitive Advantage of Finland: From Cartels to Competition?* Helsinki: ETLA/Research Institute of the Finnish Economy.

———. 2003. *Wireless Horizon: Strategy and Competition in the Worldwide Mobile Marketplace*. New York: Amacom.

Stewart, Thomas. 1997. *Intellectual Capital: The New Wealth of Organizations*. New York: Doub.

Stigler, George. 1971. "The Theory of Economic Regulation." *Bell Journal of Economics and Management Science* 2: 3–21.

Stowsky, Jay. 2003. "Secrets to Shield or Share? New Dilemmas for Dual Use Technology Development and the Quest for Military and Commercial Advantage in the Digital Age." Berkeley Roundtable on the International Economy Working Paper 151, University of California, Berkeley.

Streeck, Wolfgang. 1991. "On the Institutional Conditions of Diversified Quality Production." In Egon Matzner and Wolfgang Streeck (eds.), *Beyond Keynesianism*. Aldershot: Elgar. 21–61.

Streeck, Wolfgang, and Philippe C. Schmitter, eds. 1985. *Private Interest Government: Beyond Market and State*. Beverly Hills, CA: Sage.

Streeck, Wolfgang, and Kathleen Thelen, eds. 2005. *Beyond Continuity: Institutional Change in Advanced Political Economies*. New York: Oxford University Press.

Streeck, Wolfgang, and Kozo Yamamura. 2003. "Introduction: Convergence or Diversity? Stability and Change in German and Japanese Capitalism." In Wolfgang Streeck and Kozo Yamamura (eds.), *The End of Diversity? Prospects for German and Japanese Capitalism*. Ithaca, NY: Cornell University Press. 1–51.

Suchman, Lucy. 1987. *Plans and Situated Actions*. New York: Cambridge University Press.

Swire, Peter P., and Robert E. Litan. 1998. *None of Your Business: World Data Flows, Electronic Commerce, and the European Privacy Directive*. Washington, DC: Brookings Institution Press.

Tagliabue, John. 2002. "France Struggles to Ease Debt of Phone Company." *New York Times*, September 14.

Takano, Yoshihiro. 1992. *Nippon Telegraph and Telephone Privatization Study: Experience of Japan and Lessons for Developing Countries*. World Bank Discussion Paper 179.

Takemoto, Yoshifumi. 2004. "Fujitsu Expects 16% Rise in U.S. Sales." *International Herald Tribune*, May 13.

Tan, Xixiang. 2001. "Comparison of Wireless Standards-setting: United States versus Europe." Draft manuscript, September.

Tate, Jay. 2001. "National Varieties of Standardization." In Peter A. Hall and David Soskice (eds.), *Varieties of Capitalism: The Institutional Foundations of Comparative Advantage*. Oxford: Oxford University Press.

444

Tate, John Jay. 1995. *Driving Production Innovation Home: Guardian State Capitalism and the Competitiveness of the Japanese Automotive Industry.* Berkeley: BRIE.

Taubman, Geoffry. 1998. "A Not-so World Wide Web: The Internet, China, and the Challenges to Nondemocratic Rule." *Political Communication* 15: 255–272.

Taylor, Phil, and Peter Bain. 1999. "An Assembly Line in the Head: Work and Employee Relations in the Call Centre." *Industrial Relations Journal* 30, no. 2: 101–117.

Teece, David. 1984. "Economic Analysis and Strategic Management." *California Management Review* 26, no. 3: 87–110.

———. 1986. "Profiting from Technological Innovation: Implications for Integration, Collaboration, Licensing and Public Policy." *Research Policy* 15: 285–305.

———1992. "Competition, Cooperation, and Innovation: Organizational Arrangements for Regimes of Rapid Technological Progress." *Journal of Economic Behavior and Organization* 18: 1–25.

Thatcher, Mark, and Alec Stone Sweet. 2002. "Theory and Practice of Delegation to Non-majoritarian Institutions." *West European Politics* 25: 1–22.

Thelen, Kathleen. 1991. *Union of Parts: Labor Politics in Postwar Germany.* Ithaca, NY: Cornell University Press.

———. 2001. "Varieties of Labor Politics in the Developed Democracies." In Peter A. Hall and David Soskice (eds.), *Varieties of Capitalism: The Institutional Foundations of Comparative Advantage.* Oxford: Oxford University Press.

———. 2003. "How Institutions Evolve: Insights from Comparative Historical Analysis." In James Mahoney and Dietrich Rueschemeyer (eds.), *Comparative Historical Analysis in the Social Sciences.* Cambridge: Cambridge University Press.

———. 2004. *How Institutions Evolve: The Political Economy of Skills in Germany, Britain, the United States, and Japan.* Cambridge Studies in Comparative Politics. Cambridge: Cambridge University Press.

Thelen, Kathleen, and Sven Steinmo. 1992. "Historical Institutionalism in Comparative Politics." In Sven Steinmo, Kathleen Thelen, and Frank Longstreth (eds.), *Structuring Politics: Historical Institutionalism in Comparative Analysis.* Cambridge: Cambridge University Press.

Tilton, Mark. 2004. "Neoliberal Capitalism in the Information Age: Japan and the Politics of Telecommunications Reform." JPRI Working Paper 98, Japan Policy Research Institute. 1–11.

Todes, Samuel. 2001. *Body and World.* Cambridge, MA: The MIT Press.

Tonnelson, Alan. 2000. *The Race to the Bottom.* Boulder, CO: Westview Press.

Tsebelis, George. 1995. "Decision Making in Political Systems: Veto Players in Presidentialism, Parliamentarism, Multicameralism, and Multipartyism." *British Journal of Political Science* 25: 291–325.

———. 2002. *Veto Players: How Political Institutions Work.* Princeton, NJ: Princeton University Press.

"Tsushin Seigyo Souchi: NEC-Hitachi Kyodo Kaihatsu [Telecommunications Controlling Equipment: NEC and Hitachi to Develop Jointly]." 2003. *Nihon keizai shimbun,* December 17.

Tyson, Laura, and John Zysman. 1989. "The Politics of Productivity: Developmental Strategy and Production Innovation in Japan." In Chalmers Johnson, Laura Tyson,

and John Zysman (eds.), *Politics and Productivity: The Real Story of How Japan* **445**
Works. New York: Ballinger, 1989.

Uhmavaara, Heikki, Martti Kairinen, and Jukka Niemelä. 2000. *Paikallinen Sopiminen Työelämässä* [Local Bargaining in Working Life]. Turku: University of Turku Faculty of Law.

United Nations Conference on Trade and Development. 2002. *E-commerce and Development Report 2002.* New York: United Nations.

United Nations Development Programme. 2000. *Driving Information and Communications Technology for Development: A UNDP Agenda for Action, 2000–2001.* New York: United Nations.

United States. Bureau of Labor Statistics. Department of Commerce. 2004. http://www.bls.gov/jlt/home.htm (accessed November 12, 2005).

United States. Congress. Office of Technology Assessment. 1993. *The 1992 World Administrative Radio Conference: Technology and Policy Implications.* Washington, DC: GPO.

United States. Federal Government. 1998. *Notice of Proposed Rule Making: Improvement of Technical Management of Internet Names and Addresses.* Washington, DC: United States Department of Commerce.

United States. General Accountability Office. 2004. *International Trade: Current Government Data Provide Limited Insight into Offshoring of Services.* GAO-04–932 (September). http://www.gao.gov/new.items/d04932.pdf (accessed December 13, 2005).

United States. Information Infrastructure Task Force. 1995. *Intellectual Property and the National Information Infrastructure: Report of the Working Group on Intellectual Property Rights.* Washington, DC: U.S. Patent and Trademark Office.

Ure, John. 2003. "Deconstructing 3G and Reconstructing Telecoms." *Telecommunications Policy* 27: 187–206.

van Kersbergen, Kees. 1995. *Social Capitalism: A Study of Christian Democracy and the Welfare State.* New York: Routledge.

Vartiainen, Juhana. 1998. *The Labour Market in Finland: Institutions and Outcomes.* Helsinki: Prime Minister's Office.

———. 1999. "The Economics of Successful State Intervention in Industrial Transformation." In Meredith Woo-Cumings (ed.), *The Developmental State.* Ithaca, NY: Cornell University Press. 200–234.

Verma, Prachi. 2003. "EXL Service to Double Staff Count." *Financial Express.*

Virtanen, Aki. 2002. "Uudella koulutuksella uudelle vuosituhannelle." Ph.D. diss, University of Tampere, Helsinki.

Vogel, David. 1995. *Trading Up: Consumer and Environmental Regulation in a Global Economy.* Cambridge, MA: Harvard University Press.

Vogel, Steven K. 1996. *Freer Markets, More Rules: Regulatory Reform in Advanced Industrial Countries.* Ithaca, NY: Cornell University Press.

———. 1997. "Telecommunication Reform in Japan." In Japan Information Access Project (ed.), *Japanese Deregulation: What You Should Know.* Organized for the Japan–United States Friendship Commission. 143–153. http://www.jiaponline.org/about/06_programs.html (accessed December 13, 2005).

———. 2000. "Creating Competition in Japan's Telecommunications Market." Japan

446   Information Access Project Working Paper. http://www.jiaponline.org/whatsnew/events/2000/april10/VogelFinalPrintPaper060500.pdf (accessed December 13, 2005).

Vogel, Steven K., and John Zysman. 2002. "Technology." In Steven K. Vogel (ed.), *U.S.–Japan Relations in a Changing World.* Washington, DC: Brookings Institution Press.

von Hippel, Eric. 1988. *The Sources of Innovation.* New York: Oxford University Press.

Vonortas, Nicholas. 1997. *Cooperation in Research and Development.* Boston: Kluwer Academic.

Vuori, Synnöve, and Vuorinen, Pentti. 1994. "Outlines of the Finnish Innovation System: The Institutional Setup and Performance." In Synnöve Vuori and Pentti Vuorinen (eds.), *Explaining Technical Change in a Small Country: The Finnish National Innovation System.* Heidelberg: Physica-Verlag. 1–42.

Wallerstein, Michael, and Golden, Miriam. 1997. "The Fragmentation of the Bargaining Society." *Comparative Political Studies* 30: 699–731.

Wasserman, Stanley, and Katherine Faust. 1994. *Social Network Analysis: Methods and Applications.* Cambridge: Cambridge University Press.

Weber, Steven. 2001. Introduction to *Globalization and the European Political Economy.* New York: Columbia University Press.

———. 2004. *The Success of Open Source.* Boston: Harvard University Press.

Weber, Steven, and Naazneen Barma. 2003. "Open Source and Free Software: Development and Policy Implications." Draft chapter. In United Nations Conference on Trade and Development, *E-commerce and Development Report, 2003.* New York: United Nations.

Weber, Steven, and John Zysman. 2002. "The New Economy and Economic Growth in Developing Countries: Speculation on the Meaning of Information Technology for Emerging Markets." Unpublished manuscript. Berkeley Roundtable on the International Economy, University of California, Berkeley.

Weiner, Norbert. 1954. *The Human Use of Human Beings: Cybernetics and Society.* Boston: Da Capo Press.

———. 1965. *Cybernetics, or Control and Communication in the Animal and the Machine.* Cambridge: MIT Press.

Wenger, Etienne. 1998. *Communities of Practice: Learning, Meaning, and Identity.* Cambridge: Cambridge University Press.

Whorf, Benjamin Lee. 1956. *Language, Thought, and Reality.* Cambridge, MA: MIT Press.

Wilensky, Harold. 2002. *Rich Democracies: Political Economy, Public Policy, and Performance.* Berkeley: University of California Press.

Williamson, Oliver. 1975. *Markets and Hierarchies: Analysis and Antitrust Implications.* New York: Free Press.

———. 1985. *The Economic Institutions of Capitalism.* New York: Free Press.

———. 1991. "Comparative Economic Organization: The Analysis of Discrete Structural Alternatives." *Administrative Science Quarterly* 36: 269–296.

———. 1999. "Strategy Research: Governance and Competence Perspectives." *Strategic Management Journal* 20: 1087–1108.

Womack, James P., Daniel T. Jones, and Daniel Roos. 1991. *The Machine That Changed the World.* New York: Harper Perennial.

World Bank. 2001. "Intellectual Property: Balancing Incentives with Competitive Access." In *Global Economic Prospects*. Washington, DC: The Bank. 129–150.

———. 2002. *Information and Communication Technologies: A World Bank Group Strategy*. Washington, DC: The Bank.

World Economic Forum. 2004. *Global Competitiveness Report, 2003–2004*. New York: Oxford University Press.

Yamamura, Kimiaki. 1982. "Success That Soured: Administrative Guidance and Cartels in Japan." In Kozo Yamamura (ed.), *Policy and Trade Issues of the Japanese Economy*. Seattle: University of Washington Press.

Yamamura, Kozo. 2003. "Germany and Japan in a New Phase of Capitalism: Confronting the Past and Future." In Wolfgang Streeck and Kozo Yamamura (eds.), *The End of Diversity? Prospects for German and Japanese Capitalism*. Ithaca, NY: Cornell University Press. 115–146.

Yamashita, Masayoshi. 2004. "Towards the Emerging Ubiquitous Networks Society." *Japan Economic Currents* 43 (April). http://www.kkc.or.jp/english/activities/publications/economic-currents43.pdf (accessed December 13, 2005).

Yamazaki, Ryouhei. 2003. "Kesu Sutadei Shisuko Shisutemuzu: Gyakukyo o bane ni kousei [Case Study, Cisco Systems: Aggressiveness after the Crisis]." *Nikkei Business*, May 26, 49–50.

Ylä-Anttila, Pekka. 2000. "Globalization of Business in a Small Country: Implications for Corporate Governance and the National Innovation System." *Ekonomiska Samfundets Tidskrift* 1: 5–20.

Young, Alisdair R., and Helen Wallace. 2000. *Regulatory Politics in the Enlarging European Union: Weighing Civic and Producer Interests*. Manchester: Manchester University Press.

Yue, Chia Siow, and Jamus Jerome Lin, eds. 2002. *Information Technology in Asia: New Development Paradigms*. Singapore: Institute of Southeast Asian Studies.

Ziegler, J. Nicholas. 1997. *Governing Ideas: Strategies for Innovation in France and Germany*. Ithaca, NY: Cornell University Press.

Zittrain, Jonathan, and Benjamin Edelman. 2002a. "Documentation of Internet Filtering in Saudi Arabia." Berkman Center for Internet and Society, Harvard Law School. http://cyber.law.harvard.edu/filtering/saudiarabia/ (accessed November 13, 2005).

———. 2002b. "Empirical Analysis of Internet Filtering in China." Berkman Center for Internet and Society, Harvard Law School. http://cyber.law.harvard.edu/filtering/china/ (accessed November 13, 2005).

Zysman, John. 1977. *Political Strategies for Industrial Order*. Berkeley: University of California Press.

———. 1994. "How Institutions Create Historically Rooted Trajectories of Growth." *Industrial and Corporate Change* 3: 243–283.

———. 2002. "Production in a Digital Era: Commodity or Strategic Weapon?" Berkeley Roundtable on the International Economy Working Paper 147, University of California, Berkeley.

———. 2004a. "Manufacturing in a Digital Era: Strategic Asset or Vulnerable Commodity?" In *New Directions in Manufacturing: Report of a Workshop*. Washington, DC: National Academies Press.

448 ————. 2004b. "Transforming Production in a Digital Era." In William Dutton, Brian Kahin, Ramon O'Callaghan, and Andrew Wyckoff (eds.), *Transforming Enterprise*. Cambridge: MIT Press.

Zysman, John, and Eileen Doherty. 1996. "Leader or Strategic Follower: What Role for the Japanese State?" *Journal of Japanese Studies* 22, no. 1: 234–245.

INDEX

Bold page numbers indicate material in tables or figures.